Synthese Library

Studies in Epistemology, Logic, Methodology, and Philosophy of Science

Volume 429

The aim of *Synthese Library* is to provide a forum for the best current work in the methodology and philosophy of science and in epistemology. A wide variety of different approaches have traditionally been represented in the Library, and every effort is made to maintain this variety, not for its own sake, but because we believe that there are many fruitful and illuminating approaches to the philosophy of science and related disciplines.

Special attention is paid to methodological studies which illustrate the interplay of empirical and philosophical viewpoints and to contributions to the formal (logical, set-theoretical, mathematical, information-theoretical, decision-theoretical, etc.) methodology of empirical sciences. Likewise, the applications of logical methods to epistemology as well as philosophically and methodologically relevant studies in logic are strongly encouraged. The emphasis on logic will be tempered by interest in the psychological, historical, and sociological aspects of science.

Besides monographs *Synthese Library* publishes thematically unified anthologies and edited volumes with a well-defined topical focus inside the aim and scope of the book series. The contributions in the volumes are expected to be focused and structurally organized in accordance with the central theme(s), and should be tied together by an extensive editorial introduction or set of introductions if the volume is divided into parts. An extensive bibliography and index are mandatory.

More information about this series at http://www.springer.com/series/6607

Harald A. Wiltsche • Philipp Berghofer
Editors

Phenomenological Approaches to Physics

Springer

Editors
Harald A. Wiltsche
Department of Culture and Society (IKOS)
Linköping University
Linköping, Sweden

Philipp Berghofer
Department of Philosophy
University of Graz
Graz, Austria

Synthese Library
ISBN 978-3-030-46972-6 ISBN 978-3-030-46973-3 (eBook)
https://doi.org/10.1007/978-3-030-46973-3

This Springer imprint is published by the registered company Springer Nature Switzerland AG.
The registered company address is: Gewerbestrasse 11, 6330 Cham, Switzerland

Contents

Chapter 1
Phenomenological Approaches to Physics: Mapping the Field

Philipp Berghofer and Harald A. Wiltsche

Abstract Much ink has been spilled over the interrelations between philosophy and physics in the late nineteenth and early twentieth century as well as over the emergence of philosophy of science as an autonomous philosophical sub-discipline. Although our understanding of these issues is certainly more nuanced today than it was only a couple of years ago, more work needs to be done in order to arrive at an adequate picture of the intricate relations between philosophy and physics on the one hand and of how philosophical reflections on the physical sciences evolved during the last century on the other. This volume addresses one of the remaining blind spots, namely the role of phenomenology in the development of twentieth century (philosophy of) physics. In this introductory chapter, we shed light on the characteristics and historical development of phenomenological approaches to physics, indicate how current debates in philosophy of physics could benefit from phenomenological approaches, and provide summaries of the individual chapters.

1.1 Introduction

One of the more curious aspects of the development of twentieth century philosophy is the infamous continental/analytic-divide. Even though there are growing doubts about its philosophical significance, the continental/analytic-split continues to shape the face of professional philosophy. In many areas the reality is still that philosophers who feel at home in one tradition tend to ignore the other. This state of mutual ignorance is particularly noticeable in philosophy of science, where references to thinkers from outside the well-established canon of analytic philosophy are even scarcer than in other fields such as ethics, philosophy of mind, or aesthetics.

P. Berghofer
Department of Philosophy, University of Graz, Graz, Austria
e-mail: philipp.berghofer@uni-graz.at

H. A. Wiltsche (✉)
Department of Culture and Society (IKOS), Linköping University, Linköping, Sweden
e-mail: harald.wiltsche@liu.se

H. A. Wiltsche, P. Berghofer (eds.), *Phenomenological Approaches to Physics*,
Synthese Library 429, https://doi.org/10.1007/978-3-030-46973-3_1

It has been argued that the relative absence of continental influences on contemporary philosophy of science is a result of the historical contexts from which analytic and continental philosophy have emerged. According to Barry Smith, for instance, "post-Kantian philosophy in the German-speaking world [of the nineteenth and early twentieth century] ought properly to be divided into two distinct strands which we might refer to as the *German* and *Austrian* traditions" (Smith 1994, 1). Smith argues that the works of "Austrians" such as Bernard Bolzano, Franz Brentano, Ernst Mach, or Alexius Meinong are characterized by a sympathy towards British empiricism, by their anti-Kantianism, by the employment of a clear and concise language, and by a strong interest in the special sciences. It is from this tradition that logical empiricism and, by extension, contemporary analytic philosophy of science has emerged. German philosophy, on the other hand, is the tradition from which continental philosophy grew out. What unites thinkers as diverse as Johann Gottlieb Fichte, Georg Wilhelm Friedrich Hegel, Wilhelm Windelband, or Heinrich Rickert is, according to Smith, their idealist or even romantic leanings, their lack in linguistic clarity, and—most important for our purposes here—their neglect of the empirical sciences. It is, so the story goes, especially the ignorance of the sciences "which can be seen to have thwarted the development of a native German tradition in the philosophy of science" (Smith 1994, 4).

Although it would lead us too far afield to discuss Smith's account in detail, it is worth noting that recent years have seen a steady increase in studies contributing to a less Whiggish view of the historical context from which contemporary philosophy of science has emerged. In many of these studies, special emphasis has been put on the neo-Kantian tradition that dominated the German-speaking world at the end of the nineteenth and the beginning of the twentieth century. The renewed interest in neo-Kantianism has advanced our understanding of the history of philosophy of science in several important ways. First, it has become clear that the strict separation between an Austrian and a German tradition oversimplifies the complex personal, institutional, and intellectual interactions between seemingly incompatible philosophical cultures. Take logical empiricism—arguably the pinnacle and endpoint of what Smith refers to as the Austrian tradition—as an example: not only is it the case that many of its leading figures (such as Rudolf Carnap, Moritz Schlick, or Hans Reichenbach) started out as neo-Kantians; despite their sometimes violent anti-Kantian rhetoric, many logical empiricists sided with the neo-Kantians in their rejection of naturalism or their understanding of philosophy as a reflective, second-order discipline (cf. Glock, 2015). Second, the contention that a serious engagement with the special sciences has never been part of the German tradition is in fact a highly questionable one: As early as in the 1880s, several inner-scientific developments such as the introduction of non-Euclidean and non-metrical geometries as well as the rise of field and statistical theories in physics attracted the attention of systematically minded Neo-Kantians like Hermann Cohen or Paul Natorp (cf., for instance, Richardson, 2006). But even after physics had been revolutionized in 1905 and 1915, Neo-Kantians such as Ernst Cassirer forcefully countered the claim according to which Kantianism in all of its guises was proven untenable by Einstein's theories of special and general relativity.

The point of the previous remarks is that our understanding of the development of twentieth century philosophy of science is certainly more nuanced today than it was only a few years ago. However, since neo-Kantianism is, as we shall see, by no means the only influence from outside the well-established canon of analytic philosophy, still more work needs to be done in order to arrive at an adequate picture of how philosophical reflections on the sciences have evolved over the course of the previous century. The aim of this anthology is to address one of the remaining blind spots, namely the impact *phenomenology* had on the development of twentieth and twenty-first century philosophy of science. In particular, this anthology focuses on the role phenomenology plays in the ongoing attempts to understand the development and nature of physics from a philosophical point of view. What is more, we will also take a closer look at the ways in which phenomenology influenced the development of twentieth century physics.

The idea that phenomenological reflections can contribute to our understanding of physics, or even to the development of physics itself, may come as a surprise to some. After all, one might suspect that it is already due to methodological reasons that the relationship between phenomenology and physics is likely to be fraught with difficulties. For phenomenology, as it was conceived by its founding father Edmund Husserl, is an a priori science that proceeds from the first-person perspective and primarily aims at revealing essential structures of consciousness. Physics, on the other hand, is an a posteriori science that proceeds from the third-person perspective and aims at revealing contingent laws and facts about spatio-temporal entities. Why, one could ask, should an a priori study of consciousness contribute to our understanding of a cognitive enterprise that seeks to unveil the deep-structure of reality by empirical means, and, as it is often argued, through a systematic and methodologically regimented exclusion of everything subjective? The aim of this anthology is to give an answer to this (and related) questions and to present phenomenology as a useful framework for the philosophical interpretation of the physical sciences. As we shall see, phenomenological reflections on, for instance, the relationship between mathematics and physics, the role of experience in science, or the relationship between subjectivity and objective knowledge provide rich resources for addressing many of the most pressing issues in (philosophy of) physics.

The structure of this introductory chapter is as follows. Since we do not expect all readers to be familiar with phenomenology, we will start out with an overview of some of its characteristic features in Sect. 1.2. In Sect. 1.3 we will focus on the role physics plays in the works of Edmund Husserl, the founding father of the phenomenological movement. Four topics will be addressed: Husserl's formal philosophy of science; his conception of regional ontologies and its relation to Hermann Weyl's "world geometry"; Husserl's critique of the "mathematization of nature"; and London and Bauer's phenomenological interpretation of quantum mechanics. While Sect. 1.4 will be concerned with Martin Heidegger's and Merleau-Ponty's views on physics, we shall provide a brief overview of the subsequent chapters in Sect. 1.5.

1.2 Husserlian Phenomenology

Giving a brief overview of phenomenology is by no means an easy task. Just as it is hard to say what *the* defining characteristics of analytic philosophy are, there is no general agreement within the phenomenological community on what makes a particular approach truly phenomenological. It is hence mainly for pragmatic reasons that our focus in this section will be on the founder of phenomenology, the Austrian-German philosopher Edmund Husserl. Even though many aspects of Husserl's philosophy have been rejected by later phenomenologists, it is a generally accepted fact that Husserl's oeuvre has set the agenda for subsequent developments in the field of phenomenological philosophy.

Edmund Husserl is one of the most influential and substantial thinkers of the twentieth century. A mathematician by training, Husserl paid special attention to the formal sciences at the beginning of his philosophical career. By the turn of the twentieth century, however, Husserl had already widened his interests and turned phenomenology into a general method for analyzing the essential structures of consciousness and the role they play in virtually all areas of cognitive practice. Even though the majority of the works that have been published during his lifetime were rather programmatic in nature, Husserl's voluminous *Nachlass*, consisting of some 40,000 pages, contains detailed analyses and significant contributions to almost all philosophical sub-disciplines. In light of the complexity and breadth of his oeuvre, any attempt to break down Husserl's philosophy into a set of defining features will inevitably be incomplete.[1] This limitation notwithstanding, we still hope that the following ten themes give an initial sense of what phenomenology is and why it constitutes a useful framework for the analysis of scientific cognition.

1.2.1 Anti-psychologism

The publication of Husserl's *Logical Investigations* in 1900/1901 is widely considered to mark the birth of phenomenology. Husserl himself considered it the "breakthrough" to phenomenology (Husserl 2001b, 3). The first volume of the *Logical Investigations*, the *Prolegomena*, is devoted to a detailed refutation of psychologism, i.e. the thesis that logic is merely a branch of psychology such that logical laws can be reduced to psychological laws (cf., in particular, Husserl, 2001b, 40). One of Husserl's main arguments against psychologism is that it is ultimately self-refuting due to its relativistic and skeptical consequences. While it is controversial whether Husserl should be read as subscribing to platonism or some

[1] Cf., for more detailed introductions to phenomenological philosophy, e.g. Smith (2007), Luft and Overgaard (2012) or Zahavi (2012). The relations between phenomenology and (philosophy of) science are discussed in Kockelmans and Kisiel (1970), Feist (2004), Gutting (2005) or Hyder and Rheinberger (2010).

kind of truth-value realism, it is clear that for Husserl logical laws are not a posteriori laws about how we (must) think; the laws of logic are a priori and objective.

There is wide agreement that the *Prolegomena*, along with Gottlob Frege's attacks on psychologism, were instrumental for the anti-psychologistic climate that was characteristic for much of phenomenology and early analytic philosophy. However, in the subsection *On certain basic defects of empiricism* (Husserl 2001b, 59–61) Husserl broadens his criticism to include classical empiricism as an ultimately self-refuting position. One of Husserl's main arguments is that empiricism "*destroys the possibility of the rational justification of mediate knowledge*, and so destroys *its own possibility* as a scientifically proven theory" (Husserl 2001b, 59). Husserl's point here is that empiricism does not allow for the possibility of immediately grasping substantial epistemological principles, including principles that would govern any form of inferential reasoning. As a consequence, mediate (i.e. inferential) justification and knowledge would be impossible if empiricism were true. It is interesting to note that one of the most vocal contemporary critics of empiricism, Laurence BonJour, makes basically the same point when he accuses empiricism of amounting to "intellectual suicide" (BonJour, 1998; cf. also Berghofer, 2018d; Berghofer & Wiltsche, 2019).

1.2.2 Intentionality

The second volume of the *Logical Investigations* consists of six interrelated investigations in which Husserl expounds his early phenomenological project. Of particular significance are the fifth investigation that focuses on the *intentionality* of consciousness, and the sixth investigation in which Husserl lays out his vision of a genuinely phenomenological epistemology in which the conception of intentionality plays a pivotal role. Quite generally, the term "intentionality" denotes the "aboutness" or "directedness" that is the mark of the mental. Mental states such as perceptual experiences, wishes, or desires are essentially characterized by their being directed at something beyond themselves. What is more, intentionality comes in many different flavors. One can be intentionally directed towards the same object in many different ways, such as when one first believes that one's bike is in the office, and then sees that one's bike is in the office. For Husserl, the ways in which objects present themselves in different kinds of intentional acts are of utmost epistemological importance. Intuitive acts (such as, for instance, perceptual acts) are experiences in which the object is given in a presentive manner, i.e. in which the intended object is not only meant but also immediately present. These acts are contrasted with empty (or signitive) acts in which what is given is not the object in its actual presence, but the object as something that is only meant. While believing that one's bike is in the office is an empty act, the intuitive act of seeing the bike *fulfills* the empty act of believing. For Husserl, fulfillment, i.e. the congruence between the object as it is emptily intended and the object as it is intuitively given, is what distinguishes knowledge from mere belief.

An important achievement of Husserl's mature phenomenology is the discovery of the *horizontal structure* of intentionality. To make a long story short: As phenomenological descriptions reveal, the meaning an object has for an experiencing subject always goes beyond what is directly and immediately given. Consider, for instance, a veridical perception of a material object. At first glance, what presents itself to the experiencing subject is a three-dimensional object in space. However, a more accurate description shows that what is really sensuously given is not a three-dimensional object in space, but only one single profile of the object, its current frontside. To be sure, the experiencing subject could alter her position and make the current backside the new frontside, and vice versa. But this does not change the fact that the intended object is always given in perspectives and that, more generally, objects always and necessarily have more parts, functions, and properties than can be actualized in one single intentional act. What this shows is that there is a describable difference between what is meant through a particular act (a three-dimensional object in space) and what is sensuously given (the object's facing side with its momentarily visible features). Phenomenologically construed, this discrepancy does not represent a problem that must be somehow remedied, e.g. by proposing a theory that explains how a number of seemingly disconnected profiles add up to a homogeneous thing to which we then attribute these profiles. The fact that our intentions always transcend the sphere of direct givenness is rather to be treated as a phenomenologically discoverable feature of experience itself: Fulfilled intentions towards objects are always embedded in horizons of intentions that are momentarily unactualized, but that could be actualized in the course of further acts. Intending is, as Husserl puts it, always and necessarily an "*intending-beyond-itself*" (Husserl 1960, 46).

1.2.3 Description and Eidetics

As we have already indicated, phenomenology is a descriptive study of consciousness as experienced from the first-person point of view. Given the fact that in contemporary analytic parlance the term "phenomenology" is often restricted to denote a property of some mental states, namely their "what-it's-like-ness," it could be assumed that phenomenologists are in the business of offering more or less random descriptions of the qualitative characteristics of their own experiences. It is important to note, however, that this construal misses the point of Husserl's philosophy almost entirely. Instead of delivering collections of particular facts about one's own experiences, phenomenology in Husserl's sense is an eidetic science that seeks to generate intuitive a priori knowledge of the essential, i.e. non-contingent, features of consciousness as such. Examples for eidetic laws of consciousness are: "All experiential consciousness is intentional," "Intuitive acts can fulfill empty intentions," or "Physical objects can only be given in perspectives."

It is not unreasonable to suspect that there is an irresolvable tension between the methodological principles of phenomenology on the one hand and its lofty aspirations as an eidetic science of consciousness on the other. The worry, in a nutshell, is this: Doesn't phenomenology's self-understanding as a descriptive first-person method exclude the possibility of knowing *general* facts about consciousness as such? After all, while the requirement to proceed from a first-person perspective seems to restrict the phenomenologist to her own consciousness, the commitment to a purely descriptive approach seems to preclude the possibility of general insights. Husserl's answer to this problem is that phenomenologists are required to perform the *eidetic reduction* in order to tease out the invariant components of experience and thus to intuit the essential laws underlying it. In a similar sense in which we must "look through" the factual peculiarities of a series of circular objects in order to intuit an Euclidean circle in its pure ideality, the point of the eidetic reduction is to bracket any considerations concerning the accidental and contingent, and to direct one's attention to essential laws instead.

Following Husserl's remarks in *Experience and Judgment* and elsewhere (cf. Husserl, 1973a, 341–348), the intuition of essential laws is preceded by the method of *eidetic variation*: In systematically varying the idea of a material thing, for instance, one realizes that there are features, such as its givenness in perspectives, without which something would no longer count as an exemplar of the kind of thing under consideration. It is thus through the identification of invariants that we gain knowledge of essential laws. It should be noted, however, that this knowledge is not inferential in nature. Essential laws can and must be immediately grasped; like certain mathematical truths they present themselves not to sensory intuition, but to categorial or eidetic intuition.

1.2.4 The Epistemic Significance of Experience

For Husserl, the most fundamental question in epistemology is how subjectivity can be the source of objective knowledge, how "objectivity becomes'presented', 'apprehended' in knowledge, and so ends up by being subjective" (Husserl 2001c, 169). In contemporary terminology, Husserl is an epistemic internalist in a twofold sense. He states that, first, mental states are our justifiers, that "*subjective acts provide the reasons for everything*" (Husserl 2008, 120), and that, second, it is only internal factors that give subjective acts their justificatory force.[2] On Husserl's view, the kind of acts that play the role of justifiers for all sorts of beliefs are *originary presentive intuitions*. What makes this particular category of acts special is the fact that they present their objects as "bodily present," "actually present," or simply "self-given" (Husserl 1997, 12). Since all mediate justification leads back to immediate justification, and since originary presentive intuitions are the source of this kind

[2]For more details on Husserl's epistemology and his conception of experiential justification, cf. Berghofer (2018a, 2019). For how Husserl's approach can enrich current debates in epistemology, cf. Berghofer (2018c).

of justification, originary presentive intuitions also play the role of *ultimate* (albeit fallible) justifiers. The overall image that emerges from Husserl's detailed analyses of different kinds of intentional acts is summarized in the famous *principle of all principles*:

> No conceivable theory can make us err with respect to the *principle of all principles: that every originary presentive intuition is a legitimizing source of cognition*, that *everything originarily* (so to speak, in its "personal" actuality) *offered* to us *in "intuition" is to be accepted simply as what it is presented as being*, but also *only within the limits in which it is presented there*. (Husserl 1983, 44)

To say that an experience is an originary presentive intuition is to say that this experience presents its object in immediate givenness. However, what this means in concrete contexts depends on the type of experience one is having. Different types of experiences correspond to different types of (originary) givenness, and different types of originary givenness correspond to different types of evidence. Very roughly, one can distinguish between inadequate (perceptual), adequate (introspective), and apodictic (eidetic) evidence. What this means is that, on Husserl's view, perceptual experiences of material objects are just a subcategory of originary presentive intuitions, and that, say, introspective experiences of one's own mental states or eidetic experiences of ideal objects belong into this category as well. Since all these types of originary presentive intuitions can be regarded as experiences in a broad sense, Husserl claims that his phenomenological-epistemological system amounts to a "universal" form of empiricism (Husserl 1971, 89).

1.2.5 Phenomenology as First Philosophy

Husserlian phenomenology is an ambitious project, aiming at nothing less than realizing the venerable idea of a First Philosophy, the ultimate science. For Husserl, this means that for any science, indeed for any piece of knowledge, phenomenology must be capable of elucidating the legitimacy of this science or piece of knowledge. Here is how Husserl puts the basic idea:

> [I]t shall be shown that phenomenology encompasses the whole system of sources of knowledge from which all true sciences must draw their fundamental concepts and statements and the entire force of their ultimate justification [Rechtfertigung]. Precisely for this reason, phenomenology achieves the vocation to be "First Philosophy" in the true sense, the vocation, to confer to all other sciences unity due to ultimate grounding [Begründung] and a link to the ultimate principles and to reorganize all of these sciences as lively organs of a single, absolutely universal science, philosophy in its oldest sense. (Husserl, 2000, 200; our translation)

But how can phenomenology, a science of the essential structures of consciousness, serve as the ultimate science? The answer to this question, as indicated above, is to be found in Husserl's analyses of the variety, epistemic force, and systematic role of *experience*. The idea, roughly, is that every piece of knowledge can be traced back to epistemically foundational experiences. To be more precise, it is experiences that bear the mark of originary givenness that play this role. Investigating the sources of knowledge, then, means investigating modes of givenness—the ways experiences present the objects they are directed at. On Husserl's view, different sources of knowledge correspond to different types of experiences, which in turn correspond to different types of evidence. It is thus one of the most important tasks of phenomenology to clarify the different modes of originary givenness.

1.2.6 Husserl's Anti-naturalism

When Husserl elaborates on the epistemological significance of different types of experiences, his investigations do not make use of methods usually associated with the empirical sciences. Husserl does not classify experiences according to the bodily organs that produce them. He does not link justificatory force to causality or other external factors such as reliability or truth. What counts for him is how experiences present their contents, how objects are given within the respective experiences. The focus is on the phenomenal character of the experiences, not on any external factors. This emphasis on the internal can be regarded as a consequence of Husserl's anti-naturalism. Naturalism comes in ontological and methodological forms. Here we focus on methodological naturalism. Broadly speaking, in its methodological guise, naturalism states that only the methods of the natural sciences are acceptable forms of gaining knowledge. Accordingly, even philosophy must proceed like an empirical science.

Husserl's descriptive methodology, investigating experiences from the first-person perspective, as well as his eidetic methodology of gaining immediate a priori insights about necessary structures of consciousness are clearly opposed to methodological forms of naturalism. This is because the natural sciences are typically considered to proceed from the third-person perspective.[3] Here the basic idea is that we look at the world and then we quantify, generalize, and mathematize the data delivered by experience. Husserlian phenomenology, by contrast, is concerned with how we look at the world. What does it mean for a subject to undergo certain types of experiences, and what are the a priori correlations between modes of givenness, modes of evidence, and types of objects? Furthermore, Husserl stresses that phenomenological methods do not include inferential methods characteristic

[3]It should be mentioned, however, that such an apparently clear distinction between the empirical sciences and phenomenology would be blurred if the first-person perspective were incorporated to the natural sciences. For instance, there are trends in current experimental psychology that explicitly argue for incorporating the first-person perspective into science, emphasizing the significance of Husserlian phenomenology. One such proposal is Liliana Albertazzi's "experimental phenomenology" (cf. Albertazzi, 2013). As we will see, some phenomenologists, such as Merleau-Ponty, believe that the incorporation of subjectivity is even possible in physics.

of the natural sciences such as induction, deduction, or inferences to the best explanation. Instead, phenomenologists aim at immediately grasping a priori truths.

Finally, Husserl's above-mentioned ambition to introduce phenomenology as First Philosophy, the ultimate science, is also at odds with the spirit of naturalism. Phenomenology is not one science among the other individual sciences. Instead, according to Husserl, phenomenology is the science that clarifies the epistemological foundations of the individual sciences including mathematics and physics, thereby bestowing legitimacy on them. While the individual sciences make use of different types of experiences and different types of reasoning, phenomenology must investigate which types of experiences and reasoning are justification-conferring and why this is so.

1.2.7 The Life-World

One of the key concepts in Husserl's late philosophy, playing an important role not only in philosophy but also in other areas such as sociology or anthropology, is the *life-world*. Even though Husserl seems to use the term in different, sometimes even conflicting ways (cf. Moran, 2012, chapter 6), the life-world, broadly construed, is the world of ordinary objects, the world of tables and chairs, the world as it is immediately perceivable and familiar to us. However, the life-world is not only the pre-scientific world in which we all live. It is also the "meaning-fundament of natural science" (Husserl 1970, 48) and the "realm of original evidences" to which "[a]ll conceivable verification leads back" (Husserl, 1970, 127 f.; translation slightly modified).

The characterization of the life-world as both the meaning fundament and the epistemic basis of science makes clear why the life-world concept plays such a pivotal role in Husserl's late attempts to come to grips with the status of modern science in the wider context of human intellectual life. As the title of his last major publication, *The Crisis of European Sciences and Transcendental Phenomenology*, indicates, Husserl considers modern scientific culture to be haunted by a deeply-rooted crisis. However, it is important to note that the crisis diagnosed by Husserl does not concern the sciences themselves, but rather our philosophical understanding of science and thus the meaning science has for us as members of modern society. Husserl's argument, in a nutshell, is this: Since its inception in the seventeenth century, modern science is bewitched by an objectivist mindset according to which science, and only science, describes reality as it is in itself. As a consequence, the status of the life-world is degraded to that of a mere illusion (cf., in particular, Husserl, 1970, 48–53). For Husserl, the main problem with this view is that it is based on a mistaken construal of the relationship between scientific theorizing on the one hand and the realm of pre-scientific experience on the other. As Husserl seeks to show in quite some detail, the mathematical models that are used in science since the time of Galileo require the life-world as their unsurpassable meaning fundament. If this is correct—if the worldview that threatens to eliminate the life-world is necessarily grounded and thus presupposes the lifeworld—, objectivism indeed appears to be flawed: To substitute the scientific image for the life-world of pre-scientific experience would then be like sawing off the branch on which science is sitting. On Husserl's view, the only cure for the objectivist mindset is to engage in the project of a phenomenological clarification of the sciences:

> One must fully clarify, i.e., bring to ultimate evidence, how all the evidence of objective-logical accomplishments, through which objective theory (thus mathematical and natural-scientific theory) is grounded in respect of form and content, has its hidden sources of grounding in the ultimately accomplishing life, the life in which the evident givenness of the life-world forever has, has attained, and attains anew its prescientific ontic meaning. (Husserl, 1970, 128; translation slightly modified)

1.2.8 Historicity and Genetic Phenomenology

Husserlian phenomenology is *critical* in the sense that it seeks to unveil the implicit structures that are always already presupposed when we approach the world from within the natural attitude. When dealing with the realm of material things, for instance, the aim of phenomenological analysis is to identify the essential laws that govern the appearance of these things as well as the sorts of intentional acts in which these things are presented. In order to engage in this kind of analysis, it is sufficient to treat material things as objectivities that are already fully constituted. Experiencing subjects have an initial, implicit understanding of what these things are, and the goal of phenomenology is to unpack this initial understanding by descriptive means. By proceeding in this manner, one engages in what is commonly referred to as *static phenomenology*.

Yet, as the later Husserl came to realize, static phenomenology is but one possible approach, and a limited one at that. Instead of taking fully constituted objectivities as a starting point, one can also focus on the *becoming* of these objectivities, their "history of objectivation," as Husserl puts it (Husserl 2001a, 634), and thus on the sedimented layers of constitution that underlie our experience of objects. A particularly telling example of such a *genetic* approach is the late draft essay "The Origin of Geometry" (Husserl, 1970, 353–378; cf., also, da Silva, 2017). In it Husserl employs a method of regressive inquiry in order to elucidate how the original constitution of geometrical objects came about, and what this "history of objectivation" means for the ideal objectivity we ascribe to them. An important upshot of Husserl's analysis is that the original constitution of geometrical objects such as Euclidean planes crucially depends on life-world practices such as land surveying or the gradual smoothing of real surfaces.

Instead of proceeding from the finished products of constitution, genetic phenomenology attempts to grasp how entire communities of subjects engage in the building up of sense through time. Seen from this perspective, then, constitution is not so much an instantaneous event that is brought about by a solitary subject, but a communal process that is essentially temporal in nature. By emphasizing the temporal character of constitutional processes, Husserl opened an avenue for a more hermeneutically oriented phenomenology, as it is most prominently exemplified in the works of Martin Heidegger. According to Heidegger, historicity—the Heideggerian notion for temporality—is one of the ontological structures that describe Dasein and its being.

1.2.9 Embodiment and Intersubjectivity

Whereas in the natural attitude we take the objectivity of the world as a starting point, the aim of phenomenology is to give a detailed account of how objectivity is constituted in consciousness in the first place. Yet, to treat objectivity as an analysandum not only means to focus on the sense of transcendence we usually ascribe to certain (ideal or real) objects. It also means to account for the kind of transcendence we ascribe to other subjects, the interrelations between them, and the sense of sociality that characterizes how the life-world presents itself to us. Husserl's analyses of the phenomenon of intersubjectivity roughly fall into three categories. First, Husserl seeks to offer detailed descriptions of the kinds of acts through which a subject experiences other subjects as both similar but also irreducibly different from oneself. In this context, special emphasis is put on acts of *empathy* and the "analogizing appresentation" in which they are grounded. Second, Husserl studies acts of empathy as the basis of our practical, moral, aesthetic, and emotional evaluations. Third, and perhaps most important, the mature Husserl offers an account of how intersubjectivity figures as a necessary condition for the possibility of experiencing the world as something objectively existing, as something that is there "for us." Consider, for instance, the perceptual experience of a material thing. As we have mentioned earlier, to perceive a thing also always means to co-intend a horizon of aspects that are absent in the currently present perception, but that could be actualized in the course of a continued perceptual encounter with the thing. This, however, raises the question concerning the constitutive status of these co-intended aspects. Going through Husserl's writings, two answers seem to prevail: Husserl sometimes claims that absent but co-intended aspects are constituted as aspects of the thing that were or could be actualized through past or future experiences. On other occasions, Husserl writes as if co-intended aspects were constituted as actually perceivable possibilities. Yet, as Zahavi has shown in detail (Zahavi 2001), the mature Husserl rejects both earlier views and opts for an interpretation that emphasizes the role of intersubjectivity. On this interpretation, absent but co-intended aspects are not merely constituted as the contents of possible experiences *I* could have, but as the contents of possible experiences *every member of an open community of subjects (including both foreign subjects and myself) could have*. Or, to put it in Husserl's own words:

> Every appearance that I have is from the very beginning a member of an open, endless, but not actualized range of possible appearances of the same, *and the subjectivity of these appearances is the open intersubjectivity.* (Husserl, 1973d, 289; our translation)

Husserl's account of perceptual experience allows us to address another important topic, namely that of embodiment. On Husserl's view, perceptual episodes consist in the continuous "probing" of intentional horizons, i.e. in the attempt to harmonize new incoming sensuous data with the anticipated aspects that are co-intended through the horizon. However, in order to generate new sensory input, the perceiving subject must engage in several bodily activities such as ocular movements or the variation of the subject's bodily location in space. For Husserl, such kinaesthetic abilities not only shape our perceptual interactions with reality—the fact that consciousness must be embodied is indeed an eidetic law that governs how the world presents itself to us.

1.2.10 Epoché, Transcendental Reduction, and Transcendental Idealism

Many of Husserl's early followers—especially those in the "Munich Circle"—were attracted by what they saw as a strong commitment to realism in early phenomenology. It thus came as a surprise to many that Husserl transformed his phenomenological project into a form of transcendental philosophy, effectively claiming, for instance, that "*an object existing in itself is never one with which consciousness or the Ego pertaining to consciousness has nothing to do*" (Husserl 1983, 106). Even though Husserl's "transcendental turn" already took place around 1905, his transcendental project was first developed in print in *Ideas I*.

Crucial for Husserl's transcendental phenomenology are two interrelated methodological devices, the epoché and the transcendental reduction. The epoché is the *suspension* of the *general thesis of the natural attitude*, i.e. our naive, pre-reflective belief in the mind-independent existence of the world and its objects. The epoché, then, enables the transcendental reduction which introduces a particular reflective attitude towards the world (Husserl, 1960, 21; Husserl, 1970, 152). After the epoché and the transcendental reduction have been performed, our attention is redirected from the objects we experience to the experiences themselves, to the givenness of the objects within experience, to the appearing of the objects, to the phenomena. It is these phenomena, as they appear after the general thesis of the natural attitude has been suspended, that make up the field for transcendental-phenomenological research and description.

The mature Husserl leaves no doubt that, on his view, phenomenology and transcendental idealism necessarily go hand in hand. In the *Cartesian Meditations*, for instance, we read that "[o]nly someone who misunderstands either the deepest sense of intentional method, or that of transcendental reduction, or perhaps both, can attempt to separate phenomenology from transcendental idealism" (Husserl 1960,

86). However, how Husserl's peculiar brand of transcendental idealism ought to be interpreted is still one of the most controversial topics in Husserl scholarship. The main question, of course, concerns the relationship between consciousness and the external world: Does transcendental phenomenology only imply that the *meaning* or *sense* of the intended objects is constituted by consciousness? Or does transcendental phenomenology advance the more radical claim that the objects themselves are constituted by consciousness and that, consequently, there is no reality beyond the phenomena? Basically, there are three lines of interpretation. First, there are those who understand Husserl's transcendental idealism as a purely methodological endeavor that is consistent with both metaphysical realism and metaphysical idealism (e.g., Carr, 1999; Crowell, 2001). Second, there are those who argue that transcendental phenomenology inevitably culminates in a form of metaphysical idealism (e.g., Smith, 2003; Meixner, 2010). Third and finally, some commentators argue that transcendental phenomenology has "metaphysical implications" (Zahavi 2003, 11) in that it can be considered "a rejection of metaphysical realism" (Zahavi 2010, 85), however without thereby collapsing into some sort of metaphysical idealism (Zahavi 2010, 81). Here we do not wish to take sides. However, clearly, one's stance concerning the exact interpretation of Husserl's transcendental phenomenology has implications for the phenomenological interpretation of science, for instance with respect to the scientific realism debate.

1.3 Husserl and (Philosophy of) Physics

Browsing through his voluminous oeuvre, one's overall sense is that Husserl was a rather isolated thinker, a thinker who was so absorbed in tinkering with improvements of his philosophical system that he invested relatively little energy in a detailed engagement with the intellectual context of his time. For instance, when Moritz Schlick leveled a series of attacks at him, Husserl reacted only once in the foreword to the second edition of the *Logical Investigations*. While it would have been an easy task to respond to Schlick's rather questionable objections in a forceful and philosophically rewarding manner, Husserl simply rejects them as "nonsense" (Husserl 2001c, 179) without substantiating his verdict in any way. Matters do not seem to be different in regard to the wider scientific context in which phenomenology stands: The fact that the most productive decades of his career were also marked by several fundamental revolutions in physics and mathematics seems to receive next to no attention in Husserl's philosophical writings. For instance, the name "Einstein" is, to the best of our knowledge, mentioned less than ten times in all 42 volumes of the *Husserliana* edition (cf. Husserl, 1970, 4, 125–126, 295; Husserl, 1973b, 229; Husserl, 2002, 297).

In light of these circumstances, it seems natural to assume that Husserl did not participate in the intellectual developments of his day and that, consequently, the attempt to extract philosophically illuminating analyses of the physical sciences from Husserl's writings is a pointless exercise. In our view, however, such a conclusion would be premature. For one thing, there is historical evidence indicating that Husserl had a better understanding of the physics of his day than most of his writings may suggest. In a recently published article (Hartimo 2018), Mirja Hartimo has analyzed Husserl's private library and came to the conclusion that books and articles on the development of theoretical physics outweigh those on other scientific disciplines (including mathematics and psychology) both in number and in the intensity of Husserl's markings and annotations. As Hartimo points out, Husserl had not only familiarized himself with the special and general theory of relativity already before the 1920s. He was also aware of the developments in quantum mechanics as well as of the interpretational issues arising from these novel physical paradigms. To be sure, this basic familiarity with the physics of his day does not make Husserl a philosopher of physics in the present-day sense of the word. At a minimum, however, it should make us more confident that Husserl's oeuvre contains at least some clues indicating how a genuinely phenomenological framework for the interpretation of physics may look like.

1.3.1 Husserl's Formal Philosophy of Science

Even though physics, per se, does not play a major role in Husserl's early works, the *Logical Investigations* contain several remarks that are relevant from the perspective of a general philosophy of science. Judged by today's standards, the early Husserl seems to advocate a rather conservative construal of scientific methodology. Unlike phenomenology and other eidetic disciplines, the empirical sciences are said to rely on indirect methods which have "deduction, verification and [...] repeated modification" (Husserl 2001b, 160) as their main components. Furthermore, the early Husserl strongly emphasizes the role of demonstrative reasoning by arguing that "every explanatory interconnection is deductive" (Husserl 2001b, 147) and that every scientific explanation depends on "the explanatory ground of a law, from which a class of necessary truths follow" (Husserl 2001b, 146).

Readers familiar with the history of philosophy of science will not fail to notice the similarities between these remarks and the model of scientific method that was widely discussed until the 1960s under the label of *hypothetico-deductivism* (cf., e.g., Hempel, 1966). In its simplest form, the idea behind hypothetico-deductivism is that a theory is confirmed (or disconfirmed) by its true (or false) observable consequences. Consider, to use an example used by Popper (2002, 38), the general hypothesis that pieces of thread will break whenever they are loaded with weights exceeding the thread's tensile strength. This general hypothesis logically entails the singular-predictive statement that a thread with a tensile strength of 1 kg will break if it is loaded with a weight of 2 kg. If experimental data proves the singular-predictive statement to be true, then the general hypothesis is thereby confirmed (or, on Popper's account, corroborated). If, on the other hand, experimental data proves the singular-predictive statement to be false, the hypothesis must be rejected or at least modified.

It should also be noted that hypothetico-deductivism is isomorphic to one of the classical accounts of scientific explanation, the so-called *deductive-nomological model* (Hempel, 1965, 335–276; Popper, 2002, 38–40). In line with Husserl's aforementioned remarks about explanation, the point of this model is that an empirical occurrence is explained if it can be deduced from a set of premises that includes at least one law that is necessary to the deduction. On this view, then, the fact that a piece of thread is broken is explained by deducing the singular statement describing this occurrence from a general, law-like statement ("All pieces of thread will break whenever they are loaded with weights exceeding the thread's tensile strength") and certain singular statements specifying the initial conditions ("The tensile strength of the broken thread was 1 kilogram"; "The weight that was put on the broken thread was 2 kilograms").

Given his early remarks on the matter, it comes as no surprise that some commentators claim that "Husserl [subscribes] to something like the hypothetical-deductive model" (Hardy 2013, 29), and that, more generally, Husserl's vision of science "resembles that of the logical empiricists" (Gutting 1978, 47). Like his contemporaries in Vienna and Berlin, Husserl seems to be a proponent of what is nowadays called the *syntactic view* of scientific theories: On this view, theories are conceived of as linguistic entities, or, to be more precise, as axiomatized systems of sentences, analyzable in terms of predicate logic. This view, of course, fits well with hypothetico-deductivism: Roughly put, the idea is that the axioms of the system—the underived laws fundamental to the theory—allow for the deduction of general hypotheses. From these general hypotheses, singular-predictive statements are derived. And, finally, these singular-predictive statements are compared with corresponding experimental reports. Building on this general framework, proponents of the syntactic view such as, for instance, Rudolf Carnap, have advanced the radical idea that "*the logic of science takes the place of the inextricable tangle of problems which is known as philosophy*" (Carnap 2002, 279). Hence, all philosophy does—or, at least, ought to do—is to engage in the logical analysis of science by studying the linguistic features of scientific theories. On Carnap's view, then, philosophy is nothing but *logic of science*, or, to use the German expression, *Wissenschaftslogik*.

Now on the face of it, Husserl's position does not seem to be entirely at odds with Carnap's. To be sure, Husserl would have had little sympathy for the radical idea that all meaningful problems in philosophy are problems concerning the logical syntax of the language of science. But Husserl is very outspoken in his conviction that phenomenology must, first, provide a clarification of the natural sciences, and that, second, logic plays a crucial role in the realization of this task. Consider, for instance, the following passage from the *Logical Investigations*:

> Whether a science is truly a science [. . .] depends on whether it accords with the aims that it strives for. Logic seeks to search into what pertains to genuine, valid science as such, what constitutes the Ideal of Science, so as to be able to use the latter to measure the empirically given sciences as to their agreement with their Idea, the degree to which approach it, and where they offend against it. (Husserl 2001b, 25)

As a closer look reveals, however, there are fundamental differences between Carnap's *Wissenschaftslogik* on the one hand and Husserl's appreciation of logic as a "theory of science" on the other. These differences become readily apparent when one takes into account that what Husserl calls logic is a much broader discipline than it is for the proponents of the syntactic view.

For Husserl, the term "science" denotes any systematic discipline in which we rely on theories in order to represent a particular domain of objects. Furthermore, all theories share certain essential properties which, on Husserl's view, are logical in nature. Hence, there must also be a scientific discipline that studies these essential logical properties, and that, accordingly, specifies the ideal conditions under which a theory can be said to be truly scientific. In Husserl's terminology, this meta-discipline is called *pure logic*. Since it studies what makes scientific theories truly scientific, logic is, strictly speaking, the *theory of theories*.

As Husserl points out in *Formal and Transcendental Logic* (Husserl 1969, 33–36), classical formal logic is characterized by its two-sidedness: As *formal apophantics*, it studies the domain of judgments by fixing pure meaning categories such as "Concept, Proposition [or] Truth" as well as "*elementary connective forms* [. . .] e.g. the conjunctive, disjunctive, hypothetical linkage of propositions to form new propositions" (Husserl 2001b, 153). Since every science will crucially rely on judgment and argument, apophantic logic constrains the formal structure of any possible theory with respect to its language, vocabulary, and grammar.

The systematic study of all possible forms of judgments, arguments, and their components is, without doubt, an important task. Yet, since judgments and arguments are always about something—since pure meaning categories always have pure object categories as their correlates—, formal apophantics must be complemented by what Husserl calls *formal ontology*, i.e. the formal-mathematical "theory of *something in general* and of its derived forms, thus of concepts like "object," "property," "relation," "plurality," and the like" (Husserl 1973a, 11). One can think of formal ontology in terms of a theoretical account of all possible objects of whatever kind, or, alternatively, in terms of a science of possible being. And since Husserl claims that knowledge of possibilities precedes knowledge of the actual (Husserl 1983, 190), formal ontology constrains the formal structure of every actual theory with respect to its object domain: The domain of an actual theory must, of course, be possible and for this reason has to comply with the laws of formal ontology.

Husserl even went a step further by extending and generalizing his formal philosophy of science into what he calls a *pure theory of manifolds*, i.e. "a science of the conditions of the possibility of theory in general" (Husserl 2001b, 155). Loosely put, the basic idea is this: To every theory corresponds a field of knowledge, i.e. a domain of objects to which the theory applies. Within the theory of manifolds—a mathematical theory that grew out of Riemann's attempts to generalize the concept of space—, only the form or structure of these fields of knowledge is taken into account. A manifold is thus an objectively structured collection of objects bearing certain relations. A theory of theories in the highest sense would then be a purely formal account of the nature of manifolds as such. A formal-mathematical metatheory of this kind would allow us to define and to study the possible forms of all formally possible theories as well as the most general form of the world that science seeks to describe.

Even though much more could and should be said about Husserl's formal philosophy of science, these remarks suffice to bring the differences between Husserl and the syntactic view into sharper focus. As a result of the fixation on language that was prevalent among logical empiricists and their followers, proponents of the syntactic view took it for granted that theories are linguistic entities, that, consequently, the reconstruction of theories is achieved in the framework of first order predicate logic, and that theories are (partially) interpretable by connecting principles such as bridge laws or reduction sentences. For reasons that we cannot discuss in detail here (cf., for a general overview, Winther, 2016), this view has been superseded in the 1960s by a rival conception that sees scientific theories not primarily as linguistic entities (but as non-linguistic entities such as models), and according to which the right tool for the reconstruction of theories is not logic but mathematics. As Thomas Mormann has shown in detail, Husserl's formal philosophy of science is an early anticipation of this shift in attitude because "[f]or Husserl it is *not* sufficient for a philosophically adequate description of an empirical theory to describe only its linguistic features; what is needed as well is a *mathematical description* of its models or *formal ontology*" (Mormann 1991, 61). Husserl can thus be seen to be an early forerunner of the *semantic view* as it was later introduced by Patrick Suppes, Bas van Fraassen, and others.

1.3.2 Regional Ontologies and Weyl's "World-Geometry"

Pure logic in Husserl's sense is an a priori discipline that studies the most general form of possible theories independently from their material content. From the perspective of the pure theory of manifolds, for instance, " '+' is not the sign for numerical addition, but for any connection for which laws of the form $a+b=b+a$ etc., hold" (Husserl 2001b, 156–157). It is clear, however, that a phenomenological interpretation of the sciences cannot restrict itself to this purely formal level. Formal ontology, which studies the essence of anything whatsoever, must be complemented by *regional ontologies* that study the essential forms belonging to particular material domains. At the highest level of generality, Husserl recognizes three essentially distinct material domains (or, in Husserl's preferred terminology, *regions*): nature, consciousness, and culture (Husserl 1989). Furthermore, the material essences under which all possible individuals in a given region fall are hierarchically ordered: While regional categories (such as "thing" or "color") are on top of the hierarchy, eidetic singularities (such as particular shades ascertainable in individual objects) are at the bottom.

Since, as we have already mentioned, Husserl claims that "*the cognition of 'possibilities' must precede the cognition of actualities*" (Husserl 1983, 190), both formal and regional (or material) ontologies are indispensable to the foundations of all empirical sciences. While formal ontology develops a concept of form as applicable to any objectivity whatsoever, regional ontologies determine in eidetic universality what must belong to a particular entity in order to fall within the extension of a particular region. It is precisely in this sense that one of the main functions of regional ontologies is "*that of rationalizing the empirical*" (Husserl 1983, 19): The constitution of particulars within a certain realm always already refers back to formal-eidetic and material-eidetic laws that constrain how these particulars can in fact be constituted. For instance, subjects always and necessarily constitute material things as *spatial* entities *in time* without, however, having an explicit grasp of the material essences of space and time that determine the objective sense of spatio-temporal objecthood. In order to overcome this naïveté—in order to clarify the "posit of reality" made within a particular domain—, it is necessary to exhibit the essential characteristics and structures peculiar to each member of a certain class of entities. The "rationalization of the empirical" thus consists in the reflective endeavor to systematically study and explicate essential laws.

In the eyes of many, Husserl's goal to rationalize the empirical through a priori regional ontologies may appear as a remnant of a bygone era in which philosophers of science could still lose themselves in excessive system-building without paying attention to the realities of scientific practice. There are two important qualifications to be made, however: First, even though Husserl holds that regional ontologies are necessary for "*the interpretation, the ultimate interpretation, of the empirical sciences of reality*" (Husserl 2008, 98), this does not entail the subordination of science to armchair philosophizing. In *Ideas I*, for instance, Husserl explicitly states that *geometry* is the ontological discipline studying the essential laws pertaining to crucial aspects of material thinghood, and that the physical sciences made the first steps towards the goal of a "rational physics" when the revolutionaries of the seventeenth century amalgamated the empirical study of physical reality with mathematics.

Second, as Thomas Ryckman has shown in great detail and admirable clarity, Husserl's conception of regional ontologies as well as other parts of his methodological toolbox did in fact exert a decided impact on the development of contemporary physics. One of the main protagonists of Ryckman's book-length study is Hermann Weyl, one of the premier mathematicians and theoretical physicists of the twentieth century whose scientific and philosophical thinking was deeply influenced by Husserl.[4] Even though phenomenological traces can be found in many places of Weyl's oeuvre, the context of Ryckman's instructive case study is Weyl's critical-reflective analysis of Einstein's general theory of relativity. According to Ryckman, Weyl's reformulation of gravitational and electromagnetic theory within the framework of a "purely infinitesimal geometry" can be understood as the phenomenological attempt to fully *rationalize* the empirical, as it is constituted in

[4]Cf., for detailed information concerning the personal relationship between Husserl and Weyl, Ryckman (2005b, chapter 5).

the general theory of relativity (cf., for the following, Ryckman, 2005b and, for a less technical summary, Ryckman, 2005a).

The general theory of relativity, as it was presented by Albert Einstein in 1915, is formulated in the mathematical language of Riemannian geometry. A feature of this geometry is that it treats the magnitude and the directions of vectors quite differently: If we have two points p and q at finite separation in the manifold, then the metric of Riemannian geometry does not permit direct comparison of two vectors A at p and B at q with respect to their direction. What is permitted, however, is the direct comparison of the magnitude (or length) of A and B. It was this possibility of direct length comparisons between distant points of the manifold to which Weyl took exception. Instead of naively presupposing the global availability of a measuring rod, Weyl sought to recast general relativity in the framework of a "purely infinitesimal geometry" that only "recognize[s] the principle of the transference of a length from one point to another point infinitely near to the first" (Weyl 1923, 203). Weyl's non-Riemannian geometry thus permitted the unit of scale to vary (smoothly) from space-time point to space-time point; from this new degree of freedom, he was able to show that Maxwell's electromagnetism, in addition to Einstein's gravitation, could be incorporated into the metric of space-time. Hence was born the contemporary idea that a physical theory must be "gauge invariant," i.e. remain invariant under transformation of certain local degrees of freedom. As reinterpreted in the context of quantum mechanics by Weyl himself in 1929, the derivation of electromagnetism from gauge freedom pertains not to a factor of scale but to the arbitrary phase of the electron wave-function represented by the Abelian (i.e., commutative) group U(1). Yang and Mills in 1954 further generalized Weyl's idea of local gauge invariance to non-Abelian Lie groups (O'Raifeartaigh 1997); it is no overstatement to say that non-Abelian gauge fields are the very core of the Standard Model of contemporary particle physics of which the most recent triumph is the experimental detection of the Higgs boson at CERN in 2012.

What is particularly relevant in the context of this chapter is the rationale behind Weyl's line of thinking. Quite generally, Weyl engages in a reflective analysis of general relativity that is supposed to elucidate the very meaning of the "posit of reality" made in Einstein's theory. In order to do so, Weyl pays special attention to the regional ontology underlying general relativity, i.e. the supposed mathematical representation of the material essence of space-time. The question Weyl seeks to address is how such a mathematical representation can be constructed in a phenomenologically permissible way. The first step in Weyl's analysis is to identify an arbitrary point in the space-time manifold with an idealized cognizing subject. This cognizing subject is surrounded by a so-called tangent space, an infinitesimal Euclidean space associated with every point in the space-time manifold. From the viewpoint of the cognizing subject, only the tangent space is the locus of *Evidenz*, or originary presentive intuition—everything that lies beyond the tangent space cannot present itself in direct, originary givenness. This, of course, is also the reason for Weyl's rejection of direct length comparisons between distant points of the space-time manifold: The fact that this operation presupposes the global availability of an idealized measuring rod shows that direct length comparisons transcend the sphere of intuitive givenness and must thus be replaced by a phenomenologically permissible procedure. In order to live up to the phenomenological demand "*to work completely consciously, 'to trace back to Evidenz*'" (Husserl 2008, 440), Weyl's procedure of length comparison consists in the parallel transportation of a comparison vector in infinitesimal increments along the path between the points p and q. Since the unit of scale ("gauge") is re-configured at each point on the path between p and q, there is no longer any need for globally available measuring rods or any other intuition-transcending auxiliary tools. Weyl's "world-geometry" can thus be seen as a "remarkably sustained attempt to probe the 'darker depths' of the 'origins' of the objective physical world portrayed in relativity theory through mathematical construction guided by the phenomenological method of 'essential analysis' " (Ryckman 2005b, 117).

1.3.3 The Mathematization of Nature

While his early philosophy of science is largely constructive, the late Husserl strikes a more critical tone in his assessment of mathematized sciences and the role they play in the wider context of contemporary intellectual life. As we have already mentioned, the late Husserl considers *objectivism* to be the main reason for the deeply rooted crisis that, on his view, haunts modern scientific culture. Objectivism in Husserl's sense combines two claims that are familiar from contemporary forms of scientific realism and naturalism: first, that knowledge of the "world in itself" can only be acquired through the methods of the sciences, and that this aim is already achieved at least in some areas; second, that there is no perspective over and above the scientific perspective from which, in principle, all meaningful questions can be answered.

The attempt to reject objectivism in all of its guises is a unifying thread that runs through virtually all stages of Husserl's development. Yet, a variation of this topic that comes to the surface only in his last major publication, the *Crisis*, is that objectivism emerged as an unintended by-product during the scientific revolution of the seventeenth century. Husserl thus takes a historical approach to show how the objectivist mindset arose from a naive understanding of the methodological innovations that mark the birth of modern physics. Husserl's foray into the history of science serves a therapeutical purpose: Today objectivist tendencies are so deeply ingrained in the thinking of most philosophers and scientists that they find it difficult even to imagine any other way of looking at science. Once the historical roots of objectivism are exposed, however, it becomes easier to acknowledge its status as an unfounded metaphysical hypostatization of scientific methodology.

According to Husserl, the formative moment in the development of modern physics was Galileo's reformation of scientific method, which consisted in a complete amalgamation of mathematics and experimentation. To be sure, as Husserl clearly recognizes, Galileo's use of mathematics was not unprecedented in the history of the physical sciences. But what distinguished Galileo from the tradition before him is that he did not just make occasional use of geometrical models in order to "save the appearances" in this or that segment of reality. Husserl argues that Galileo was after something much more radical, namely the complete "mathematization of nature [through which] nature itself is idealized under the guidance of the new mathematics [and] becomes [. . .] a mathematical manifold" (Husserl 1970, 23). So, according to Husserl, the radicalism of the Galilean project lends itself to the thesis "that everything which manifests itself as real [. . .] must have its mathematical index" (Husserl 1970, 37) and must therefore be translatable into the language of geometry. Mathematizability thus becomes an ontological criterion: In order to be included among the primary qualities, a property must be amenable to quantification and geometric representation. Secondary qualities like color or odor, on the other hand, do not belong to the domain of what is objectively real.[5]

On Husserl's view, the conviction that mathematics is a reliable guide towards the one true description of physical reality makes up an important component of our modern scientific mindset. Nowadays, this conviction is backed up by reference to the immense predictive and practical success of modern mathematized science. During the first half of the seventeenth century, however, Galileo's call for a complete amalgamation of mathematics and physics was just a bold methodological conjecture that could only be substantiated by metaphysical means, i.e. by assuming that the deep-structure of reality is in fact mathematical in nature. What is more, since Galileo failed to inquire into the meaning and origin of geometry, his methodological revolution is also marked by a fundamental naïveté: In a similar sense in which Weyl criticized Einstein for what he saw as an uncritical adoption of the already existing framework of Riemannian geometry, Husserl takes exception to the fact that Galileo merely inherited Euclidean and Archimedean proportional geometry from the tradition before him. For Husserl, Galileo's unwillingness to deal with questions concerning the origin and meaning of geometry is a "fateful omission" (Husserl 1970, 49) that ultimately lies at the heart of modern objectivism. Yet, in order to understand the reasons for this verdict, it is necessary to say a word or two on Husserl's own take on the "primal establishment" (Husserl 1970, 362) of geometrical thinking (cf., for further details, Wiltsche, 2016; 2019).

Although they are ubiquitous in Galilean science, abstract objects such as ideal spheres or frictionless planes are nowhere to be found in the life-world of pre-scientific experience. These objects only come into existence through a special mental operation through which one generates a limiting case against which actual instances of spherical bodies and real planes can be projected. But how does this mental operation come about? Following the late Husserl of the *Crisis*, there are two preconditions for the original constitution of something like a frictionless

[5]Although there have been critical voices as well (Ihde 2011), Husserl's interpretation of Galilean science had a strong impact on several Galileo scholars, especially on the French historian of science Alexandre Koyré. Cf., for a discussion of the relationship between Husserl and Koyré, Parker (2017).

plane: first, the acquaintance with real surfaces of different degrees of flatness; and, second, the acquaintance with tools that give us the "technical [...] capacity to make [...] the flat flatter" (Husserl 1970, 25). Looking at a series of real surfaces with increasing degrees of flatness, one can either ponder over practical ways to push the limits of technological perfection. Or one can ignore questions of technological realizability and instead focus on the ideal limiting pole "towards which the particular series of perfectings tend" (Husserl 1970, 26), namely, the abstract, empirically unrealizable conception of a perfectly flat plane. However, what is needed in order to grasp this ideal limiting case in a distinct and self-conscious manner is a "peculiar sort of mental accomplishment: idealization" (Husserl 1970, 348). Idealization in Husserl's sense is the process through which the vague, imprecise, and morphological concepts with which we describe real things are replaced by exact, precise, and mathematical concepts. Hence, it is a progression of similarities between concrete things, and an additional act of idealization in which abstract objects such as frictionless planes find their "primal establishment."

The take-home message of Husserl's genetic inquiry is that the original constitution of abstract objects depends, first, on life-world experiences of real things, and, second, on higher-order acts of idealization. However, as Husserl also makes clear, these two preconditions are not yet sufficient to account for the "ideal objectivity" (Husserl 1970, 356) which we normally ascribe to abstract objects. According to Husserl, this kind of objectivity is only attained if the meaning of abstract objects is consolidated and stabilized by detaching it from the intellectual accomplishments of singular subjects. Husserl calls the process through which such a consolidation is achieved *sedimentation*. Crucial to this process of sedimentation is the externalization of original, intuitive thought by means of formal notations: Once abstract objects have been constituted in intuitive acts of idealization, these objects can be "liberated from all intuited actuality" (Husserl 1970, 44) through further acts of formalizing abstraction. One of the historical examples Husserl gives for this process is the algebraization of geometry (Husserl 1970, 43–48). Considering, for instance, the proportional geometry that operates at the heart of Galilean mechanics, it is clear that the concepts used by Galileo retain their reference to the material contexts that originally gave meaning to them. This is particularly obvious in the case of Galileo's graphical representations of levers, weights, or planes: Although the referents of these representations are without doubt abstract objects, the symbols used by Galileo are easily recognizable as idealizations of sensible shapes that can be found in the life-world of pre-mathematical experience. It is exactly this intuitive connection between geometric symbols and the underlying sensible shapes that is undermined when the materially determined concepts of proportional geometry are replaced with purely formal algebraic expressions. Innovations such as the Cartesian coordinate system allow for the direct translation of complex geometrical properties into the formal language of algebra. As a consequence, complex geometrical problems can be solved by means of materially undetermined algebraic equations.

The processes of sedimentation and formalization are of utmost importance for Husserl's overall argument as well as for his historical critique of objectivism. Once a field such as geometry is formalized, it can become a "calculating technique" in which strings of symbols are manipulated "according to technical rules" (Husserl 1970, 46) and without regard for the content to which these symbols correspond. This means not only that it becomes possible to solve geometrical problems without repeating the intuitive acts that were necessary for the original constitution of geometrical objects. It also means that one can solve equations in an almost game-like fashion, i.e. without even asking for what the purely formal symbols stand for or how they were bestowed with meaning in the first place. For the development of modern mathematized science, this "*technization of formal-mathematical thinking*" (Husserl, 1970, 48; our emphasis) is both a blessing and a curse. It is a blessing because science would be practically impossible if novices were under the constant pressure to think everything anew. However, as Husserl repeatedly stresses in the *Crisis*, formalization is also a curse because it harbors the danger of a dangerous forgetfulness with regard to science's roots in the life-world of pre-theoretical experience.

The kind of objectivism that originates in the works of Galileo considers the mathematical models that are built and applied in the physical sciences as the best candidates for delivering truthful representations of the "world in itself." And since the distance between the world, as it is allegedly represented in these models, and the life-world of pre-scientific experience dramatically increases the further science progresses, it becomes harder and harder to reconcile the "scientific image" with the "manifest image." Objectivism reacts to this problem in a very straightforward manner, namely by arguing "that the common sense world of physical objects [...] is *unreal*" (Sellars, 1991, 173; our emphasis). Yet, relegating the life-world to the status of an illusion not only produces the crisis which Husserl opposes so vehemently in the *Crisis*. If Husserl's genetic analysis of the origin of mathematics is correct, then the demotion of the life-world also leaves us in a quandary with respect to the unsurpassable foundation of scientific cognition: On the one hand, objectivism implies that the life-world is nothing but a veil that needs to be removed in order to catch a glimpse of the deep-structure of the "world in itself." At the same time, however, the methods through which this veil ought to be removed presuppose the life-world as their necessary "meaning-fundament." If this is true, then objectivism leaves us in a paradoxical situation indeed: To advocate objectivism is, as we have said earlier, to saw off the branch on which science is sitting.

In light of Husserl's rejection of objectivism, an obvious question arises: If, phenomenologically construed, scientific theories are not truthful representations of the "world in itself," what are they then? Or, to put the question differently, how should philosophers with phenomenological leanings react to the still ongoing disputes between different forms of scientific realism on the one hand and different forms of scientific anti-realism on the other? Even though this question has been widely discussed, there is no general consensus within the secondary literature: While there have been attempts to render phenomenology compatible with anti-realist lines of thought (Wiltsche, 2012; 2017; for critical reactions: Reynolds, 2018, chapter 3; Berghofer, 2017), others have argued that nothing prevents the phenomenologist from adopting a realist stance (Gutting, 1978; Harvey, 1986, 1989; Belousek, 1998; Soffer, 1990; Vallor, 2013). Still others have claimed that

Husserlian phenomenology lacks any particular impact on the scientific realism debate and thus resembles Arthur Fine's deflationist "NOA" (Rouse 1987).

1.3.4 Phenomenology and Quantum Mechanics

Although Husserl never publicly commented on the emerging quantum paradigm, his phenomenology had at least an indirect impact on quantum mechanics through the work of the German physicist Fritz London. While not widely known in philosophical circles, it is no overstatement to say that London is a truly remarkable figure who—like many other scientists during the first half of the twentieth century—transcended the disciplinary boundaries between philosophy and physics (cf., for an insightful biography of London, Gavroglu, 1995). Nowadays London is mainly remembered as the founder of quantum chemistry. However, a number of substantial contributions to theoretical physics and philosophy of physics as well as four nominations for the Nobel Prize in Chemistry and one nomination for the Nobel Prize in Physics attest to the wide scope and significance of his thinking. Yet, interestingly enough, London began his academic career not as a scientist, but as a philosopher. His doctoral dissertation *Über die Bedingungen der Möglichkeit einer deduktiven Theorie* was supervised by the Munich phenomenologist Alexander Pfänder and appeared in Husserl's *Jahrbuch für Philosophie und Phänomenologische Forschung* in 1923. As Mormann has noted, London's thesis can be regarded as a piece of Husserlian-style mathematical philosophy of science which deals with "a set theoretic concretization of Husserl's largely programmatic account of a *macrological philosophy of science*" (Mormann 1991, 70).

After graduating from the University of Munich at the age of 21, London's focus shifted to physics where he was mainly interested in the newly emerging field of quantum mechanics. After studying previous attempts to unify gravity and electromagnetism, London formed the idea that quantum mechanics could be the right framework for the task at hand. As it turned out, London's idea was immensely fruitful: Building on Weyl's work on unification, London was among the first to realize that the gauge invariance underlying electrodynamics is, other than Weyl had expected, not a scale invariance *but a phase invariance*.[6]

What is most relevant in the context of this chapter, however, is London's work on interpretational issues of quantum mechanics. In 1939 London published a monograph entitled *La Théorie de l'Observation en Mécanique Quantique* together with the French physicist Edmond Bauer. This work has two main objectives: First, London and Bauer seek to offer a "concise and simple" (London & Bauer 1983, 219) account of the measurement problem in the spirit of von Neumann's groundbreaking *Mathematische Grundlagen der Quantenmechanik* (1932). Providing the axiomatic

[6]For an excellent analysis of the historical origins of gauge theory as well as an overview of its role in string theory, cf. O'Raifeartaigh and Straumann (2000).

foundations of quantum mechanics, von Neumann's book was one of the most influential works of early quantum mechanics and his "conception of the measurement problem became the framework of almost all subsequent theories of measurement" (Jammer 1974, 474). Since London and Bauer were in broad agreement with von Neumann, their monograph was not intended as a counter project, but as a more accessible version of von Neumann's highly technical work which, to add insult to injury, was written in German.

Second, London and Bauer seek to shed more light on the relationship between the observed and the observer, thus aiming at clarifying the role of consciousness in quantum mechanics. Although it was clear for von Neumann "that it is impossible to formulate a complete and consistent theory of quantum mechanical measurement without reference to human consciousness" (Jammer 1974, 480), he said very little about what consciousness is or what role it plays in quantum mechanics. In fact, "it was the London and Bauer treatment that effectively cemented consciousness into the 'received view' " (French 2002, 470). We shall say more on London and Bauer's take on the role of consciousness in quantum mechanics in a moment. Before that, however, some brief remarks about the measurement problem are in order.

At the heart of quantum mechanics there are two seemingly conflicting principles: On the one hand, we have the Schrödinger equation that describes the evolution of the quantum state over time. The Schrödinger equation is a unitary, deterministic, and linear equation. Its linearity entails that the sum of two solutions is again a solution to the equation. This, then, is the principle of quantum superposition that highlights the wave character of quantum objects. The quantum state of a system is described by its wave function. The superposition principle entails that wave functions can be added together to form a new wave function.

On the other hand, there are the principles dealing with the *apparent collapse of the wave function*. The collapse postulate states that when a measurement takes place, the wave function collapses such that the quantum state is not in a state of superposition anymore, but now has a definite value. The necessary character of this postulate stems from the apparent fact that we never observe superposition states but only definite values. For instance, when we measure the spin of an electron, we never observe a superposition of spin-up and spin-down. What we observe is always the electron being in one of these states. Understanding the apparent collapse of the wave function is the core of the measurement problem.

Let us now turn to London and Bauer's approach to the problem. For our purpose, it is instructive to begin at the very end of their monograph. Here, they point out that the whole debate about the measurement problem relates to a much broader philosophical issue, namely "the determination of the necessary and sufficient conditions for an object of thought to possess objectivity and to be an object of science" (London & Bauer 1983, 259). They continue by adding that "[m]ore recently Husserl [. . .] has systematically studied such questions and has thus created a new method of investigation called 'Phenomenology' " (London & Bauer 1983, 259). Given this explicit reference to Husserl, and given Lodnon's background in phenomenology, it is easy to agree with commentators such as Gavroglu (1995) and French (2002) that the way London and Bauer set up the measurement problem has a distinctively phenomenological ring to it. As we have seen earlier, the most fundamental problem in Husserl's epistemology concerns the question as to how "objectivity becomes 'presented', 'apprehended' in knowledge, and so ends up by being subjective" (Husserl 2001c, 169). When London and Bauer raise the problem of how "an object of thought" can be objective and be scientifically investigated, then this is easily identifiable as a variation of Husserl's original question. What is more, London and Bauer also agree with Husserl's anti-naturalistic stance by claiming that "[p]hysics insofar as it is an empirical science cannot enter into such problems in their generality" (London & Bauer 1983, 259). Hence, although they insist that physics can at least lead to significant " 'negative' philosophical discoveries" (London & Bauer 1983, 259), London and Bauer consider the measurement problem primarily as a problem of (phenomenological) philosophy.

Concerning the specific problems surrounding quantum mechanics, London and Bauer make it clear that "[t]he heart of the matter is the difficulty of separating the object and the observer" (London & Bauer 1983, 220). On their view, modern physics reveals that "the idea of an observable world totally independent of the observer, was a vacuous idea" (London & Bauer 1983, 220). Here we see what they mean by saying that physics can lead to significant negative philosophical insights. According to London and Bauer, "the formalism of quantum mechanics already implies a well-defined theory of the relation between the object and the observer, a relation quite different from that implicit in naive realism, which had seemed, until then, one of the indispensable foundation stones of every natural science" (London & Bauer 1983, 220). As we shall see later, it is this very idea that had a tremendous impact on Maurice Merlau-Ponty; the idea that modern physics, and quantum mechanics in particular, undermines (naive) realism, and that our most sophisticated theories undermine the expectation that science could possibly offer an entirely objective account of the world.

Let us now turn to London and Bauer's proposed solution to the measurement problem and the role they ascribe to the observer's consciousness. The first thing to note is that, in their view, a measurement is only complete when the outcome "has been *observed*" (London & Bauer 1983, 251). The observer, then,

> possesses a characteristic and quite familiar faculty which we can call the "faculty of introspection". He can keep track from moment to moment of his own state. By virtue of this "immanent knowledge" he attributes to himself the right to create his own objectivity. (London & Bauer 1983, 252)

London and Bauer thus come to the conclusion that

> it is not a mysterious interaction between the apparatus and the object that produces a new ψ for the system during the measurement. It is only the consciousness of an "I" who can separate himself from the former function $\Psi(x, y, z)$ and, by virtue of his observation, set up a new objectivity in attributing to the object henceforward a new function $\psi(x) = u_k(x)$. (London & Bauer 1983, 252)

After rightly pointing out that terms such as *immanent knowledge* "clearly demand a phenomenological reading" (French 2002, 484), French interprets London and Bauer's take on the separation between the ego and the superposition as follows:

> This separation should not be thought of in terms of consciousness "causing", in whatever sense, the wave function to collapse, but rather in Husserlian terms, as that of a *mutual separation* of both an Ego-pole and an object-pole through a characteristic act of reflection. (French 2002, 484)

As French adds, this phenomenological reading of their solution to the measurement problem has the additional advantage of avoiding the main objections that have been brought forward against London and Bauer.

1.4 Beyond Husserl

Although the focus in this chapter is on Husserl, this should not be taken to suggest that other figures of the phenomenological movement did not engage with physics in novel and creative ways. In what follows, we will indicate some directions in which a genuinely phenomenological analysis of the physical science was taken by later phenomenologists. To be sure, space limitations prevent us from providing a comprehensive overview of the entire field of post-Husserlian phenomenology of science—discussions of, for instance, Becker (1973), Ströker (1997), Heelan (1983), Kockelmans (1966), or Ihde (1991) will have to wait for another occasion. In our view, however, there are two figures in particular who merit closer consideration: Martin Heidegger and Maurice Merlau-Ponty. Since both philosophers had a tremendous impact on the entire phenomenological movement, a brief discussion of some of their key insights will help to gain a better understanding of how phenomenological analyses of the physical sciences evolved in the second part of the twentieth century.

1.4.1 Martin Heidegger

Even to mention Heidegger in the context of a serious philosophical engagement with the sciences might be enough to raise some eyebrows. After all, in light of remarks such as that "[s]cience does not think" (Heidegger 1968, 8), or that "science's knowledge [...] already has annihilated things as things long before the atom bomb exploded" (Heidegger 1971, 168) it seems hard to deny that parts of Heidegger's oeuvre are characterized by a pessimistic, if not hostile attitude concerning the sciences. However, recent years have seen an increase in studies highlighting the constructive potential that lurks behind the seemingly anti-scientific façade of Heidegger's philosophy (cf., for a general orientation, Kockelmans, 1985; Glazebrook, 2000, 2012). Heidegger, who studied physics for two years, and kept close contact with leading physicists such as Werner Heisenberg or Carl von Weizsäcker, is not only said to have "had a remarkable knowledge of both physics and biology" (Kockelmans 1985, 17). Some commentators go so far as to argue "that philosophy of science was at the center of his project and its development throughout his career" (Rouse 2005, 124).

While Heidegger's philosophy is sometimes acclaimed as both a rejection and an advancement of Husserl's phenomenology, one cannot help but notice certain similarities between parts of their philosophies of science. To begin with, although Heidegger's stance towards naturalism can generally be seen as somewhat ambiguous (cf. Rouse, 2005), he agrees with the Husserlian sentiment that the natural sciences are in principle incapable of investigating themselves in a philosophically satisfactory manner: "The moment we talk 'about' a science and reflect upon it, all the means and methods of this science in which we are well versed fail us" (Heidegger 1967, 177). This is equally true of biology, where we "cannot put biology under the microscope" (Heidegger 1967, 177), and of physics, which "itself is no a possible object of a physical experiment" (Kockelmans 1970b, 170). If this view is correct, it not only follows that serious reflections on any particular science must transcend the standpoint and methodological repertoire of that science. For Heidegger, the limitations of any particular methodology also result in a pluralist image of science: Instead of absolutizing one particular discipline with its own specific methods and values, Heidegger seems to promote a vision of science in which different methodologies and sets of values can coexist without standing in a relation of super- or subordination. Joseph Kockelmans summarizes the pluralistic sentiment of Heidegger's philosophy of science as follows:

> The rigor of mathematical physics is exactness. An event can be considered as an event of nature if, and only if, it is determined beforehand as a kinematic magnitude. Such a determination can be effected by means of measurements and with the help of their resulting numbers and the calculations performed on them. However, [...] the exactness of mathematical physics is not due to the fact that it calculates exactly; it must calculate exactly, precisely because the mode in which it is bound to its own realm of objects by its *original project* has the character of exactness. That is why the humanistic sciences can be rigorous without for that matter being exact. (Kockelmans, 1970a, 189; our emphasis)

What is particularly noteworthy about this passage is Kockelman's remark that, on Heidegger's view, a particular science "is bound to its own realm of objects by its *original project*". "Project" or "Projection" (in German: *Entwurf*) is a technical term in Heidegger's philosophy, and clarifying its meaning will allow us to highlight another similarity between Husserl's and Heidegger's views on science.

Heidegger's interest in science—and in physics in particular—is already evidenced in one of his earliest works, his 1916 habilitation lecture "Der Zeitbegriff in der Geschichtswissenschaft" (Heidegger 1978, 413–433). Quite generally, Heidegger's aim is to distinguish the historical sciences from physics on the basis of the concepts of time that are operative in both disciplines. The approach he chooses to tackle this issue is in perfect agreement with the basic tenets of Husserlian phenomenology: Instead of presupposing certain pre-established conceptions of time, physics, or history, Heidegger employs a method of regressive inquiry in which one begins with a particular existing science and then works back to determine the formal and material conditions underlying it (Heidegger 1978, 417–418). The outcome of Heidegger's analysis is that the very essence of modern physics, as it was inaugurated by Galileo and Newton, lies in the *mathematical projection of nature*. What this means can be made clear by considering a crucial passage from *Being and Time*:

> What is decisive for its development [the development of mathematical physics] does not lie in its rather high esteem for the observation of "facts", nor in its "application" of mathematics in determining the character of natural processes; it lies rather in *the way in which Nature herself is mathematically projected*. In this projection something constantly present-at-hand (matter) is uncovered beforehand, and the horizon is opened so that one may be guided in looking at those constitutive items in it which are quantitatively determinable (motion, force, location, and time). Only "in the light" of a Nature which has been projected in this fashion can anything like a "fact" be found and set up for an experiment regulated and delimited in terms of this projection. The "grounding" of "factual science" was possible only because the researchers understood that in principle there are no "bare facts". In the mathematical projection of Nature, moreover, what is decisive [. . .] is that this projection *discloses something that is a priori*. (Heidegger 1962, 413–414)

This passage contains several insights that deserve closer attention. To begin with, by stating "that in principle there are no 'bare facts' ", Heidegger anticipates the debate over the *theory-ladenness of observation*, as it is discussed in "mainstream" philosophy of science since the 1960s. What is clear in light of the above-quoted passage is that Heidegger opposes the idea that scientific facts could ever be "neutral" in the sense that they can be disentangled from the theoretical framework in which they are situated. For Heidegger, however, theory-ladenness does not primarily occur on the level of scientific *theories* that are said to impinge on the perceptions of scientists. Heideggerian theory-ladenness is much more fundamental because it has to do with the a priori conditions that must already be in place in order for concrete scientific work to be possible. Before a scientist can even begin to collect data, to devise theories, to make calculations, or to design experiments, the "world" or "region" at which the scientist aims must already be constituted in a way that makes it amenable to a particular kind of scientific inquiry. In the case of modern physics, this primal constitution of the region is achieved through what Heidegger calls the *mathematical projection of nature*, which "maps out in advance the way in which the procedure of knowing is to bind itself to the region that is opened up" (Heidegger 2002, 50). Since it determines in advance what counts as a being and as experience, the mathematical projection itself is, on Heidegger's view, not grounded in experience of beings—it is a priori.

From these remarks it should be evident that there are clear affinities between the Heideggerian notion of a mathematical projection of nature and Husserl's conception of regional ontologies. Heidegger would most certainly agree with Husserl that, explicitly or implicitly, the special sciences are necessarily grounded in regional ontologies which are a priori, and which express the essence—or, to put it in Heideggerian terms, the "basic state of being" (Heidegger 1962, 246)—of the entities in their domain. What is more, Heidegger also agrees that these regional ontologies can only play their foundational role if they have been, first, explicated, and, secondly, critically examined:

> [A]ll ontology, no matter how rich and firmly compacted a system of categories it has at its disposal, remains blind and perverted from its ownmost aim, if it has not first adequately clarified the meaning of Being, and conceived this clarification as its fundamental task. (Heidegger 1962, 31)

Seen from this perspective, then, it is not surprising that Heidegger praises Hermann Weyl for his insight that in "the theory of relativity in contemporary physics [...] the notion of field is normative" (Heidegger, 1997, 81; cf., regarding the relationship between Heidegger and Weyl, Webb, 2009, chapter 5; Sieroka, this volume). Unfortunately, Heidegger does not go into any detail of Weyl's reformulation of general relativity theory. As one can suspect, however, Heidegger correctly identified Weyl as a philosophically-minded physicist who took up the hermeneutical task of critically engaging in an "interrogation of being," as it becomes manifest in Einstein's theory.

Although, as we have seen, there are interesting parallels between Husserl's and Heidegger's philosophies of science, there are also points of divergence. Perhaps the most fundamental difference concerns the starting point from which phenomenological analysis must proceed. As Henry Pietersma has aptly put it, the point of departure for Husserl is "that a human being is basically a knower [...,] that whatever engages a human being is (or at least should be) based on what she knows or justifiably believes" (Pietersma 2000, 86). As a consequence, the most fundamental task in Husserlian phenomenology is to spell out the conditions under which subjects may be said to have achieved the goal of knowledge, both in scientific and everyday contexts. While there is, of course, still ample room for ontology and metaphysics, Husserl's transcendental phenomenology is, first and foremost, an epistemological project. Heidegger, on the other hand, strongly opposed any claim to the primacy of knowledge. Quite the opposite, knowledge, according to Heidegger, is but a derivative mode of *being-in-the-world*, i.e. the fundamental ground upon which every further determination of Being rests. Heidegger's argument, in a nutshell, goes as follows: Traditionally, knowledge has always been characterized as some kind of relation between (at least) two relata, someone who knows, and something that is known. The task of epistemology, then, is to specify exactly how this relation must look like in order for genuine knowledge to occur. On Heidegger's view, however, any separation between two (or more) relata is itself the result of a particular projection, which—like any other projection—refers back to Dasein's essential state of "being-in-the-world." Consequently, the most fundamental question in philosophy is not epistemological in nature; it is rather the *ontological* question concerning the nature and understanding of *Being*.

Heidegger's "ontological turn" has far-reaching consequences for his philosophy of science: In *Being and Time*, Heidegger attacks what he calls the "logical conception of science" (Heidegger 1962, 408) that focuses on systems of statements or mathematical models as the finished product of research, and then raises the question of how these representational vehicles can be used to mirror particular empirical target systems. The point of Heidegger's argument is that this conception must be replaced with what he calls an "existential conception of science [that] understands science as a way of existence and thus as a mode of Being-in-the-world"

(Heidegger 1962, 408). If one accepts this existential conception, one no longer conceives of science as a project that is primarily geared towards the accumulation of truthful (mental, linguistic, or mathematical) representations of reality. Following the existential conception, science is, first and foremost, something human beings *do*—it is an activity that is essentially linked to tools and equipment, and that aims at a local manifestation of reality in experimental settings and instrumental work. In their everyday research, scientists are not primarily concerned with entities, their properties, or with the relations between them. What scientists are actually concerned with is technological equipment that must always already be understood as being useable for a particular purpose, and that is best understood in being so used. Hence, from the viewpoint of an existential conception of science, *practical understanding* has priority of *theoretical knowledge*.

By giving phenomenology an existential-hermeneutic twist, Heidegger was one of the main influences for the emergence of what later became known as *hermeneutic philosophy of science* (cf., for a general overview, Babich, 2016). Philosophers such as Theodore Kisiel, Joseph Kockelmans, Patrick Heelan, Babette Babich, Dimitri Ginev, Joseph Rouse, or Don Ihde employ methods and insights from phenomenology, hermeneutics, and post-positivist philosophy of science in order to gain a firmer grasp on science as an embodied, culturally, and historically situated practice that materializes itself in what Patrick Heelan has called "readable technologies" (Heelan 1983).

1.4.2 Maurice Merleau-Ponty

Among the three classical phenomenologists we discuss in this chapter, Maurice Merleau-Ponty is the one whose work contains the most detailed analysis of contemporary physics. It is well known that Merleau-Ponty cared deeply about the sciences.[7] His work on psychology in particular is a common starting point for many contemporary phenomenologists who are conducting research at the interface between philosophy of mind, psychology, and the cognitive sciences. What is less well known, however, is that Merleau-Ponty also explicitly addressed physics, aiming at a deeper understanding of how physics and philosophy can enrich each other, and of how a genuinely phenomenological philosophy of physics might look like. This is true, in particular, of the essay "Modern Science and Nature," which was part of a lecture course Merleau-Ponty held at the Collège de France (Merleau-Ponty, 2003, 81–122). Here, he carefully engages with quantum mechanics, outlining his more general phenomenological approach to physics. Unfortunately, it seems that Merleau-Ponty's interest in philosophy of physics came to the fore only quite late in his career: While the aforementioned lecture course *La*

[7]Cf., for discussions of Merleau-Ponty's views on the exact sciences, e.g. Kisiel (1970), Rouse (1986), Matherne (2018), and Romdenh-Romluc (2018).

Nature was held between 1956 and 1960, and thus one year before his death, his treatise *The Visible and the Invisible*, which also addresses physics in general and the "relations between the observer and the observed" (Merleau-Ponty 1968, 15) in particular, was published only posthumously.

In his philosophy of physics, Merleau-Ponty discusses the limits of objectivity and raises the question as to whether physics could ever deliver a picture of the world that also incorporates the physicist who observes and experiments. On his view, modern physics and quantum mechanics in particular exemplifies or at least leads to such a new kind of physics, which—unlike classical physics—not only "posits nature as an object spread out in front of us, [but rather] places its own object *and its relation to this object in question*" (Merleau-Ponty, 2003, 85; our emphasis). It is in this context that Merleau-Ponty discusses the aforementioned London and Bauer interpretation of quantum mechanics, adding that the emerging picture could be called a "participationist conception" or a "partial realism" Merleau-Ponty, 2003, 97–98.[8] This is a striking similarity both in content and in terminology to a recent interpretation of quantum mechanics that goes by the name *QBism* and is also often referred to as a kind of "participatory realism" (Fuchs 2017). Two contributions to this volume are explicitly dedicated to a discussion of this interpretation of quantum mechanics.

According to Merleau-Ponty, physics in its most sophisticated form abandons the goal of delivering an entirely objective picture of the world. Instead, it incorporates the physicist herself, and thereby accounts for the fact that the life-world is always and necessarily the meaning-fundament of all scientific endeavors. Consider the following passage from *The Visible and the Invisible*:

> Philosophy is not science, because science believes it can soar over its object and holds the correlation of knowledge with being as established, whereas philosophy is the set of questions wherein he who questions is himself implicated by the question. But a physics that has learned to situate the physicist physically, a psychology that has learned to situate the psychologist in the socio-historical world, have lost the illusion of the absolute view from above: they do not only tolerate, they enjoin a radical examination of our belongingness to the world before all science. (Merleau-Ponty 1968, 27)

Of crucial importance for Merleau-Ponty's overall position is the rejection of two traditional assumptions: first, that there is a "physical object in itself" (Merleau-Ponty 1968, 15) which exists prior to our theorizing, and thus "has an individual existence" (Merleau-Ponty 2003, 92); and, second, that determining the nature of this object is the main goal of physical research. Instead, Merleau-Ponty recognizes the "relations between the observer and the observed [as the] ultimate physical beings" (Merleau-Ponty 1968, 15).

Another topic that receives much attention in Merelau-Ponty's philosophy of physics concerns the question about the nature and role of measurements. This

[8]Merleau-Ponty adopts this terminology from the French physicist and logician Paulette Destouches-Février. Below we see in more detail how strongly Merleau-Ponty was influenced by Destouches-Février. For a portrayal and further development of the approaches of Destouches-Février and her husband Jean-Louis Destouches, cf. Bitbol (1998, 2001).

question is, to be sure, of special relevance in quantum mechanics where, as we have seen, the role of observations and measurements seems particularly mysterious. Contrasting the measuring apparatus in classical physics with the measuring apparatus in quantum mechanics, Merleau-Ponty comes to the conclusion that while classically "the apparatus is the prolongation of our senses," in quantum mechanics "[t]he apparatus does not present the object to us." Instead, "[i]t realizes a sampling of this phenomenon as well as a fixation. [...] Known nature is artificial nature" (Merleau-Ponty 2003, 93).

In light of the above, one may wonder how Merleau-Ponty's position relates to the contemporary scientific realism debate. Given his rejection of the idea that physical objects are things in themselves, and given his analysis of the measuring apparatus in quantum mechanics, it seems natural to consider him a scientific anti-realist. On closer inspection, however, things are not that simple. The first thing to note is that Merleau-Ponty is very outspoken in his negative assessment of one of the more popular versions of scientific anti-realism, namely instrumentalism:

> Physics should not be conceived as a search for the truth, it should give up determining a real physics: it would be only an ensemble of measurements linked to equations, allowing [us] to foresee the result of future measurements. Formalist physics receives all freedom, but it loses its ontological content. It signifies no mode of being, no reality. Like all radical nominalism, this nominalism cannot articulate itself. (Merleau-Ponty 2003, 95–96)

After dismissing instrumentalism without much argument, Merleau-Ponty goes on to say that it would also be a mistake to adopt an idealist position. Drawing on the work of the French physicist and logician Paulette Destouches-Février, Merleau-Ponty claims that the problem with idealism is that, just like realism, it amounts to a form of objectivism. To be more precise, idealism is an objectivism that "objectifies human representations" (Merleau-Ponty 2003, 96). Contrary to objectivism, Merleau-Ponty is convinced that "[t]he relations between reality and measurement must be conceived outside of the dichotomy of in-itself/representation" (Merleau-Ponty 2003, 96). As we have seen above, acknowledging that "[p]hysics cannot be realist in the classical sense" but "cannot be idealist, either," Merleau-Ponty chooses to call his position "a 'partial realism' or a 'participationist' conception" (Merleau-Ponty 2003, 97–98). This terminology, adopted from Paulette Destouches-Février, highlights the interrelatedness and inseparability of the observer and the observed. What is more, the term "partial realism" indicates that Merleau-Ponty seeks to find a middle ground between instrumentalism on the one hand and a full-blown realism on the other. Hence, returning to our initial question, we need to ask: What are the specific features of the kind of realism Merleau-Ponty endorses?

In this context, Merleau-Ponty says, we must begin with "distinguish[ing] several meanings [of reality]" (Merleau-Ponty 2003, 98): first, a "plane of reality, where objects exist in themselves and where the properties that we attribute to them are intrinsic" (Merleau-Ponty 2003, 98). After dismissing this realist notion of reality, Merleau-Ponty goes on to mention, second, an intersubjective plane of reality, "where reality is constituted uniquely by the 'results of measurement' " (Merleau-Ponty 2003, 98). Since this notion is too instrumentalist for Merleau-Ponty, he finally introduces "a third plane, the structural plane" (Merleau-Ponty 2003, 98). After doing so, Merleau-Ponty reproduces a long passage from Destouches-Février which we will quote in full[9]:

> From the fact that this plane transcends the subjective-objective duality, the structural relations dress an absolute character up in the framework of theory. In effect they are independent of the results and of the process of measurement. They are however relative to the species of the system studied. By their independence from the results of observations, they dress up a certain objectivity, comparable to the Platonic objectivity of the Idea vis-à-vis its sensible realizations. But on the other hand, this independence which detaches them from all sensible contact with the object could make them refuse objectivity. In effect they refer not to an object, but to certain mathematical forms necessary for the description of the relation of the subject to the object. They present the same ambiguity if we envisage them under the angle of reality; to the extent that they appear completely detached from the results of measurement—that is, from the immediate meeting with the objects studied—they lose all reality, and their nature approaches mathematical being; but we just saw that the whole critique of knowledge withdrawn into modern physics consisted exactly in unmasking the illusory character of the phenomenal reality as just as sensible as rational. Of such kind that the character of reality seems to have to take refuge, preferably in the structural plane, relatively more independent, permanent, and coherent than the two preceding planes. Moreover, the fact that structures are determined by the theory in which they intervene—since they schematize the general conditions on the observers in their relations with the objects—confers unto them a reality that purely mathematical beings independent of all sensible signification do not possess. (Destouches-Février, as quoted in Merleau-Ponty 2003, 98)

Although Merleau-Ponty does not do much to clarify or go beyond these remarks, it seems clear that the partial realism of Merleau-Ponty and Destouches-Février is in many ways similar to a currently popular version of realism, namely structural realism (for more details, cf. Ladyman, 2016; Berghofer, 2018b). Instead of taking a realist stance concerning the unobservable objects posited by our best theories, structural realists claim that we should limit our epistemic and/or ontological commitments to the mathematical or structural content of theories. In light of Merleau-Ponty's pronouncement to take the "mathematical forms necessary for the description of the relation of the subject to the object" as the fundamental entities of physical theorizing, his partial realism has indeed much common ground with contemporary structural realism. Of course, the vast majority of contemporary structural realists take the observer-independence of physical theories for granted, and would not, consequently, regard structural relations as relations between subject and object. One might even be tempted to suspect that Merleau-Ponty's position collapses into a very peculiar form of structural *idealism*. However, since for Merleau-Ponty the structural relations between observer and the observed cannot be reduced to anything subjective or mental, his position clearly has a realist flavor to it. In particular, and as mentioned earlier, there are interesting parallels between Merleau-Ponty and QBism, according to which quantum mechanics tells

[9]Unfortunately, the fact that Merleau-Ponty is quoting Destouches-Février here is easily overlooked in the English translation because the quotation marks are missing (Merleau-Ponty 2003, 98).

us something very important about reality, namely "that reality is *more* than any third-person perspective can capture" (Fuchs 2017, 113).

It is crucial to note that Merleau-Ponty's position is by far the strongest we have discussed so far. In order to see why, let us return to a claim that is widely endorsed by virtually all phenomenologists, viz. the claim that physics, at best, can only yield a perspectival image of reality. Typically, this perspectivity is said to arise due to the role subjectivity plays in our cognitive interactions with reality. An argument to this effect can be found, for instance, in Hermann Weyl's *Philosophy of Mathematics and Natural Science* (Weyl 1949, 110–113): For Weyl, the driving force behind modern physics is the attempt to objectify reality through a systematic exclusion of everything subjective. Historically, the first steps in this direction were made when Galileo and others introduced the distinction between primary and secondary qualities, and moreover argued that mathematizability is a reliable criterion for what can count as objectively real. On this view, then, only primary qualities belong to the inventory of objective reality because secondary qualities like color or odor are not amenable to direct mathematizability due to their subjective character. As Weyl observes, the development of physics culminated in purely symbolic representations of the world where everything that is granted physical significance must find its expression in mathematical symbols.

Following Weyl's historical narrative, the systematic exclusion of everything subjective appears to be a regulative idea in the Kantian sense, i.e. a prescriptive telos that gives physical research its normative direction. As with all regulative ideas, however, the focal point towards which the movement of symbolization strives will never be fully realizable. The reason for this is, according to Weyl, that even the most abstract mathematical tools still carry the trace of transcendental subjectivity. Consider the following passage from *Philosophy of Mathematics and Natural Science*:

> How is it possible to assign to the points of a point-field marks or labels which could serve for their identification or distinction? The labels are supposed to be self-created, distinctive and always reproducible symbols, such as names, numbers (or number triples x, y, z, etc.). Only after this has been accomplished can one think of representing the spectacle of the actually given world by construction in a field of symbols. All knowledge, while it starts with intuitive description, tends toward symbolic construction. No serious difficulty is encountered as long as one deals with a domain consisting of a finite number of points only [...]. The problem becomes a serious one when the point-field is infinite, in particular when it is a continuum. A conceptual fixation of points by labels of the above-described nature that would enable one to reconstruct any point when it has been lost, is here possible only in relation to a *coordinate system*, or frame of reference, that has to be exhibited by an individual demonstrative act. The objectivation, by elimination of the ego and its immediate life of intuition, does not fully succeed, and the coordinate system remains as the necessary residue of the ego-extinction. (Weyl 1949, 75)

Much could be said about this telling passage (cf. Ryckman, 2005b, 128–136).[10] In our view, however, the take-home message is this: Whenever we seek to establish a link between the mathematical formalism and observational data—for instance, when we carry out measurements—it is necessary to introduce a coordinate system in order to single out individual objects from a continuously extended object domain. Yet on Weyl's view, it is precisely this implementation of a coordinate system that reintroduces subjectivity into the purely symbolic representation of reality. The reason for this claim is straightforward: For Weyl the origin of the coordinate system is the most formal representation of the physicist's lived body, her "zero point of orientation" (Husserl 1989, 166); the axes of the coordinate system, on the other hand, determine the physicist's orientation in space. On this interpretation, then, the perspectivity of every symbolic representation of reality is indeed mandated by physics itself: Whenever we seek to establish a link between the mathematical formalism and reality, a coordinate system must be introduced. Yet whenever a coordinate system is introduced, subjectivity creeps back into our purely symbolic representation of the world.

In light of what has just been said, it is clear that Weyl fully embraces the claim according to which every symbolic representation will necessarily be perspectival in nature. Yet for Weyl this is a meta-theoretical claim that tells us something about how to understand physics and how to interpret its results: Physics is fine as it is, but we need to keep in mind that its purportedly objective methodology is essentially limited in scope. Now there can be no doubt that Merleau-Ponty also accepts the perspectivity of our scientific image of reality. For Merleau-Ponty, however, this claim is not the result of a reflective analysis from outside of physics. Quite the opposite, on Merleau-Ponty's reading, quantum mechanics itself implies the strong ontological claim that the classical picture of a purely objective, observer-independent physical reality is untenable, and that every complete physical description of reality must incorporate the physicist as well as her experience. Seen from this perspective, then, quantum mechanics has the potential to live up to the ideal of a fully rationalized, critical, and ultimately *phenomenological* physics.

1.5 Summaries of the Chapters

In this section, we provide summaries of all chapters in this volume, so as to permit the reader to identify those most likely to be of interest to her.

[10]For instance, readers familiar with Husserl's oeuvre will not fail to notice Weyl's allusion to section 49 of *Ideas I*: The notion of the "coordinate system as the necessary residue of the ego-extinction" is, of course, a reference to Husserl's thought experiment of the "absolute consciousness as the residue of the annihilation of the world" (Husserl 1983, section 49). Moreover, it interesting to note that the trained mathematician Husserl also explicitly refers to the "origin of the coordinate system" (Husserl, 1973c, 116; our translation) in order to elucidate the role of the embodied subject in our cognitive engagements with the world.

In his chapter "Explaining Phenomenology to Physicists," Robert Crease sheds light on the objective and significance of philosophy of physics, distinguishes different traditions within the field, and focuses on one in particular, namely the phenomenological tradition. Crease starts out by emphasizing "the unavoidability of philosophical commitments in science." Recently, physicists such as Stephen Hawking, Leonard Mlodinow, or Sander Bias have dismissed philosophical reflections on physics as useless, thus echoing Richard Feynman's dictum that "philosophy of science is about as useful to scientists as ornithology is to birds." As Crease rightly observes, however, physicists are already engaging in "amateur philosophizing" when they reflect on physics or its relationship to philosophy. Since philosophy seems inevitable whenever we raise questions concerning, for instance, the aim of physics or its place in the ensemble of intellectual practices, it might be better, as Crease argues, to construe the difference between physicists and philosophers as a difference of stances. The scientist, in her scientific stance, objectifies what is being studied. The philosopher, in his philosophical/phenomenological stance, is interested in the relationship between the scientist and her object of study. Concerning philosophy of science, Crease distinguishes between analytic, pragmatic, and phenomenological traditions. For the phenomenologist, the individual sciences need to be epistemically grounded in a more fundamental science, namely phenomenology. This means that the aim of phenomenology is "to reflectively justify scientific activity, and describe how it arises out of the grounds of human experience." In the final section of his chapter, Crease addresses the project of a *phenomenology of physics*. The phenomenology of physics thematizes the framing in which physical research is conducted and recognizes this framing as a human product. In this context, Crease discusses three different paths the phenomenology of physics can take.

Mirja Hartimo's chapter "Husserl's Phenomenology of Scientific Practice" addresses Husserl's approach to the natural sciences, arguing that his goal is to *describe scientific practice*, rather than to impose norms or restrictions on the sciences. Hartimo comes to the bold conclusion that "Husserl is the first philosopher who took seriously the importance of concrete and diverse scientific practices." After discussing the relationship between mathematics and physics, Hartimo sheds light on the development of Husserl's position on this matter. Her focus is on *Ideas I*, *Formal and Transcendental Logic*, and *Crisis*. The aim of her analysis is to show that while Husserl in *Ideas I* subscribes to the idea of a pre-established harmony between mathematics and physics—an idea shared by his Göttingen mathematical colleagues—, he eventually emphasizes the difference between the two disciplines. In *Formal and Transcendental Logic* Husserl introduces what he calls the method of "Besinnung" that seeks to investigate the intentional genesis of the sciences. Yet the goal of "Besinnung" is not to examine and judge the sciences from above, but to be as close to scientific practice as possible. Ultimately, the outcome of Husserl's transcendental analysis is that mathematics and mathematical physics need to be clearly separated. In this context, Hartimo points to interesting similarities between Husserl's intentional analysis and Penelope Maddy's study of the development of applied mathematics. The take-home message of Hartimo's chapter is that, although

Husserl's views concerning the sciences change, the goal of his phenomenological analysis remains the same: *to provide a phenomenological clarification of scientific practice*.

In his chapter, Paolo Palmieri addresses a question raised by Husserl in the *Crisis*: Why do the deductive methods employed in mathematical physics yield so much clarity although the axioms are anything but self-evident? Palmieri approaches this problem through three case studies that exemplify three different stages of the development of modern physics: Galileo's mathematical natural philosophy as the birth of modern physics; Helmholtz' analysis of human sound perception in terms of an infinite series of anharmonic oscillators; and the birth of quantum mechanics as brought forth by Heisenberg's paper "Über quantentheoretische Umdeutung kinematischer und mechanischer Beziehungen" in 1925. There is an interesting connection between Palmieri' case study based approach and Mirja Hartimo's chapter. This is because, as we have seen, the ambition of any phenomenological analysis is to be as close to the object of study as possible, in our case scientific practice and its historical development. Phenomenologically construed, there is no royal road to the practice of physics. Although phenomenology can, as in the case of Weyl's world-geometry, serve as a guide, phenomenologists are not in the business of imposing a priori rules on physics. Instead, physics is, at least to some degree, what physicists do. Or, to put it in Palmieri's own words: "A phenomenologically oriented philosophy of physics is grounded in a non-colonialist, subdued appreciation of legitimately autonomous and ethnically diverse mathematical and empirical styles that manifest themselves through history and are ultimately rooted in natural languages and in the life-worlds of the physicists."

Norman Sieroka's chapter addresses a blind spot in Weyl scholarship. While some ink has been spilled on Weyl's earlier writings and on his relationship to Husserl's phenomenology, not much has been done to elucidate the connections between Weyl and Heidegger. As Sieroka observes, this is surprising since there are quite a few places in Weyl's later writings where he discusses Heideggerian phenomenology in detail. Heidegger's influence is perhaps most obvious in the case of Weyl's claim that modern physics and mathematics develop towards an *existential standpoint*. Sieroka seeks to clarify the background of this claim by focussing on Weyl's notion of *symbolic constructions*. Although Weyl relies on several Heideggerian concepts in order to understand the role of symbols in physics, he does so in ways that would not have found Heidegger's approval. According to Sieroka, a particularly telling example is Heidegger's distinction between being *ready-to-hand* and being *present-at-hand*. While Heidegger insists that symbols can only be present-at-hand, Weyl rejects this on the basis that symbols—such as concrete strokes on a blackboard—are physical objects that can be manipulated. The gist of Sieroka's chapter is that in his later writings Weyl aims at establishing some middle ground between Heidegger and Cassirer—a middle ground that avoids both Heidegger's separation of science from human existence and Cassirer's scientism.

In his chapter, Matthias Egg aims at revealing an unexpected parallel between Husserl's late philosophy of science and a currently popular version of scientific metaphysics defended by James Ladyman and Don Ross. At first glance, the prospects of finding any common ground between the argument in Husserl's *Crisis* and contemporary scientific metaphysics seem dim, to say the least: As the late Husserl is at pains to show, the success of modern mathematized physics has misled scientists and philosophers to accept a dangerous metaphysical hypostatization of scientific methodology. By buying into what Husserl calls *objectivism*, they commit the mistake of confusing highly idealized mathematical models with truthful representations of reality, thereby demoting the life-world of pre-scientific experience to the status of a mere illusion. Although, as Egg admits, Husserl's criticism of objectivism also affects parts of Ladyman's and Ross' naturalistic program, there are at least two points of convergence: First, there is agreement that we need to abstract away from our naive encounter with the world as well as from our practical interests in order to make a truly fundamental science possible. Second, there is agreement that the sciences must be interpreted in a way that makes them relevant for the practicalities of everyday life. Egg argues that these two points of convergence imply a commitment to *weak metaphysics*, i.e. "the articulation of a world-view based on a certain stance." Despite the differences between them, Egg manages to bring Husserl's phenomenology of science and scientific metaphysics into a fruitful dialogue that is likely to spark further debates in the future.

Lee Hardy addresses the question of whether Husserl's phenomenological reflections on modern physics are consistent with scientific realism. In its most basic form, scientific realism is the view that scientific theories aim at a literal description of reality, and that we have reasons to believe that at least some of the claims of our best theories are (approximately) true. Scientific anti-realists, on the other hand, either hold that science does not aim at truth at all, or that we should restrict our epistemic and/or ontological commitments to what theories say about the observable world. As this initial characterization suggests, the main object of dispute between realists and anti-realists are the so-called *unobservable entities*, i.e. entities like atoms, quarks, fields or forces. Concerning such unobservable entities, Husserl seems to take up a straightforward position: For instance, Hardy draws our attention to section 20 of *Ideas I* where Husserl proclaims that "[i]f *'positivism'* is tantamount to an absolutely unprejudiced grounding of all sciences on the 'positive,' [. . .] on what can be seized upon originaliter, then *we* are the genuine positivists" (Husserl 1983, 38). As Hardy notes, Husserl's self-identification as a positivist is potentially problematic in light of the current scientific realism debate. This is because "an ontology restricted to perceivable physical objects [. . .] and an instrumentalist interpretation of scientific theories [seems] incapable of doing justice to the rapid and impressive advance of the physical sciences into the hidden regions of the unobservable." However, based on a careful analysis of the distinction between physical things, ideal objects, and theoretical entities on the one hand and between scientific theories and scientific laws on the other, Hardy argues that Husserl's phenomenology of physics is nevertheless "entirely compatible with a realistic construal of scientific theories." This means that, according to Hardy, Husserl's approach to physics remains a viable option for scientific realists.

In their chapter, Arezoo Islami and Harald A. Wiltsche address one of the most important problems at the interface between philosophy of physics and philosophy of mathematics, namely the so-called *applicability problem*. The problem, in a nutshell, is this: Why is it that mathematical methods and models are so successful in physics? Most notably, this problem has been raised by Eugene Wigner in his essay "The Unreasonable Effectiveness of Mathematics in the Natural Sciences." There exist a number of different approaches to this problem, none of which enjoys general consent among philosophers. Islami and Wiltsche thus aim at a distinctively transcendental-phenomenological approach. For them, this implies a shift from the *why-question*—Why is mathematics is so successfully applied in physics?—to the *how-question*—How do physicists apply mathematical methods and models? As it is typical for phenomenological approaches, the focus is on scientific practice. More precisely, Islami and Wiltsche claim that "[a]ddressing the how-question from a first-person perspective puts us in a position to gain a firmer grip of the intentional structures that are operative in concrete cases of mathematical-physical theorizing." Building on the distinction between a *synchronic* and a *diachronic* analysis of the ways in which mathematics is applied in physics, Islami and Wiltsche argue that the applicability problem, as it is traditionally viewed, disappears as soon as we realize that the objects of modern physics are the result of a quite peculiar form of constitution which transcends the strict separation between an abstract and an empirical sector of reality.

Thomas Ryckman's chapter addresses one of the pillars of modern physics: the gauge principle. Given its crucial role in, for instance, the standard model of particle physics, in string theory or in general relativity, it is not surprising that prominent voices have called for an elucidation of the gauge principle as one of the most important tasks of philosophy of physics. Ryckman argues that revisiting the philosophical-phenomenological motifs that led Weyl to introduce the gauge principle in the first place can contribute to this task. Ryckman begins his discussion with Weyl's thesis that in order to understand the world, physics must proceed "by bottom-up symbolic construction starting from mathematical relations in the infinitely small." For Weyl, physical laws must be grounded in *Evidenz*, and *Evidenz*, in fundamental physics, is only to be found in the infinitely small, since the range of intuition of the cognizing subject, i.e., the ego-center, is regarded to be limited to its immediate spatial-temporal neighborhood. Ryckman identifies this line of thought as a commitment to transcendental phenomenological idealism in a Husserlian spirit. Towards the end of his chapter, after shedding light on the transcendental-phenomenological origins of the gauge principle and the role it plays in modern physics, Ryckman brings up an issue that puzzles many contemporary physicists and philosophers of physics: Since gauge transformations lead to new degrees of freedom that appear to be redundancies, it seems that the gauge symmetries do not correspond to symmetries of nature, but only to symmetries of our symbolic representations of nature. Concerning these arbitrary purely mathematical degrees of freedom, Ryckman argues that "the arbitrariness can be understood phenomenologically, as each point indifferently can be considered the locus of an experiencing, constructing ego." In our view, Ryckman's phenomenological clarification of (the origins of) the gauge principle is not only of utmost historical importance, but paves the way towards at a better understanding of the mathematical structure of modern physical theories as well as of the world these theories purport to describe.

Steven French aims at paving the way for a phenomenological approach to the notorious measurement problem in quantum mechanics. The interpretation offered by London and Bauer serves as the starting point of his chapter. By shedding light on Fritz London's phenomenological background, French argues convincingly that the interpretation of London and Bauer goes beyond von Neumann's interpretation in that it aims at a *phenomenological clarification of the role of consciousness* in the apparent collapse of the wave function. This is not to say that the observer and her consciousness are to be placed outside of a quantum mechanical description, thus mysteriously causing the wave function to collapse. Rather, the observer must be included into the quantum mechanical description: When observation takes place, a separation occurs in the sense that the "object and subject poles of the relationship between the knower and the world emerge." It is this embeddedness of the observer and the observed within a common theoretical structure that is of particular interest to French. On his view, a phenomenological reconstruction of the London and Bauer interpretation results in a position that avoids the objections that have been traditionally raised against von Neumann as well as against London and Bauer. What is more, in the final section of his chapter, French sketches how a phenomenological interpretation of quantum mechanics could be introduced into current debates about the interpretation of quantum mechanics. In particular, French highlights how a phenomenological approach could adopt plausible elements of rival interpretations such as Dieks' perspectivalism, Rovelli's relationalism, or Everett's many-worlds interpretation without being obliged to postulate branching worlds or branching minds. Without a doubt, French's chapter will be an important stepping stone for further attempts to position phenomenology as a fruitful framework for the interpretation of quantum mechanics.

In his chapter, Michel Bitbol addresses certain systematic similarities between a novel interpretation of quantum mechanics, QBism, and phenomenological motifs, particularly as we find them in Husserl and Merleau-Ponty. Bitbol identifies three features of QBism that are shared by the phenomenological tradition. The first concerns the first-person approach. The idea, in a nutshell is this: Phenomenology is well-known for its demand that philosophy, as the ultimate science, must proceed from the first-person perspective in order to account for objective knowledge in terms of the underlying structures of (transcendental) subjectivity. QBism seems to follow a similar trajectory by aiming "to reconstruct a new, self-conscious, type of objective knowledge, [and by] starting everything afresh from the first-person standpoint of knowers and agents." The second common feature is the demand to direct our attention away from the external objects of physical theorizing, and focus on the mental acts that present these objects instead. According to Bitbol, many proponents of QBism "proceed phenomenologically" by suspending their judgments concerning the existence of external objects and by redirecting their attention "towards the epistemic function and the practical use of the symbols of quantum mechanics." The third common feature concerns the claim, popular among defenders of QBism, that quantum mechanics only tells us something about the *expectations* we should have concerning the outcomes of experiments. Bitbol argues that this line of interpretation is in many ways similar to Husserl's conception of horizontal intentionality. This third common feature takes center stage in the final chapter of this volume.

In the final chapter of this volume, Laura de la Tremblaye discusses similarities between a QBist and a phenomenological epistemology, particularly addressing the conception of horizontal intentionality in Husserl's theory of perception. One distinctive feature that is shared by QBists and phenomenologists is the recognition of the central role of the subject and her experiences. The QBist slogan "experience first" is identified as a basic phenomenological principle. What is more, Tremblaye argues for a phenomenological reading of QBism that links the Husserlian notions of anticipation and fulfillment to the QBist understanding of the measuring process in an astonishingly straightforward way. In this picture, "the perceptual horizon parallels the QBist quantum state, the perceptual act corresponds to the physicist's measurement and the modification of my possible horizon corresponds to the modification of the state vector after the measurement." Finally, Tremblaye addresses one important distinction between QBism and the Copenhagen interpretation. While QBism and the Copenhagen interpretation share many important similarities, the Copenhagen interpretation does not single out the subject and her experiences. It ascribes a central role to measurements but not to the subject conducting the measurement. Accordingly, the Copenhagen interpretation remains within the third-person perspective. QBism, on the other hand, aims at a first-person interpretation of quantum mechanics. In this sense, Tremblaye concludes that "if interpreted phenomenologically, QBism reveals the special relations that unite the physicist and her experience, as well as the nature of the knowledge of which theoretical physics is a special case."

Acknowledgments The majority of the chapters in this volume were initially presented at the conference "Phenomenological Approaches to Physics," which we organized at the University of Graz in June 2018. We would like to thank the participants who made the conference a truly memorable event. Special thanks are also due to Otávio Bueno, the editor of the Synthese Library series, to Sonja Rinofner-Kreidl, the head of the working unit "Phenomenology" at the University of Graz, and to Springer's project coordinator Palani Murugesan. Finally, we would like to thank our families for emotional support, and the Austrian Research Fund (FWF) for a generous grant to carry out a research project on Hermann Weyl's phenomenological philosophy of physics (project number: P31758).

References

Albertazzi, L. (2013). Experimental phenomenology: An introduction. In L. Albertazzi (Ed.), *Handbook of experimental phenomenology. Visual perception of shape, space and appearance* (pp. 1–36). Malden: Wiley-Blackwell.

Babich, B. (2016). Hermeneutic philosophy of science: Interpreting nature, reading laboratory science. In N. Keane & C. Lawn (Eds.), *The Blackwell companion to hermeneutics* (pp. 492–504). Malden: Blackwell.

Becker, O. (1973). *Beiträge zur phänomenologischen Begründung der Geometrie und ihrer physikalischen Anwendung*. Tübingen: Max Niemeyer.

Belousek, D. W. (1998). Husserl on scientific method and conceptual change: A realist appraisal. *Synthese, 115*, 71–98.

Berghofer, P. (2017). Transcendental phenomenology and unobservable entities. *Perspectives, 7*(1), 1–13.

Berghofer, P. (2018a). Husserl's conception of experiential justification: What it is and why it matters. *Husserl Studies, 34*(2), 145–170.

Berghofer, P. (2018b). Ontic structural realism and quantum field theory: Are there intrinsic properties at the most fundamental level of reality? *Studies in History and Philosophy of Modern Physics, 62*, 176–188.

Berghofer, P. (2018c). Towards a phenomenological conception of experiential justfication. *Synthese*. https://doi.org/10.1007/s11229-018-1744-5.

Berghofer, P. (2018d). Why Husserl's universal empiricism is a moderate rationalism. *Axiomathes, 28*, 539–563.

Berghofer, P. (2019). On the nature and systematic role of evidence: Husserl as a proponent of metalist evidentialism? *European Journal of Philosophy, 27*(1), 8–117.

Berghofer, P., & Wiltsche, H. (2019). Phänomenologie. In M. Grajner & G. Melchior (Eds.), *Handbuch Erkenntnistheorie* (pp. 35–42). J.B. Stuttgart: Metzler.

Bitbol, M. (1998). Some steps towards a transcendental deduction of quantum mechanics. *Philosophia Naturalis, 35*, 253–280.

Bitbol, M. (2001). Jean-Louis destouches: théories de le prévision et individualité. *Philosophia Scientiae, 5*(1), 1–30.

BonJour, L. (1998). *In defense of pure reason. A rationalist account of a priori justification*. Cambridge: Cambridge University Press.

Carnap, R. (2002). *The logical syntax of language*. Chicago: Open Court.

Carr, D. (1999). *The paradox of subjectivity*. Oxford: Oxford University Press.

Crowell, S. (2001). *Husserl, Heidegger, and the space of meaning*. Evanston: Northwestern University Press.

da Silva, J. J. (Ed.) (2017). *Mathematics and its applications. A transcendental-idealist perspective*. Cham: Springer.

Feist, R. (Ed.) (2004). *Husserl and the sciences*. Ottawa: University of Ottawa Press.

French, S. (2002). A phenomenological solution to the measurement problem? Husserl and foundations of quantum mechanics. *Studies in the History and Philosophy of Science, 33*, 467–491.

Fuchs, C. (2017). On participatory realism. In I. Durham & D. Rickels (Eds.), *Information and interaction* (pp. 113–134). Cham: Springer.

Gavroglu, K. (1995). *Fritz London: A scientific biography*. Cambridge: Cambridge University Press.

Glazebrook, T. (2000). *Heidegger's philosophy of science*. New York: Fordham University Press.

Glazebrook, T. (Ed.) (2012). *Heidegger on science*. Albany: State University of New York Press.

Glock, H.-J. (2015). Neo-Kantianism and analytic philosophy. In N. D. Warren & A. Staiti (Eds.), *New approaches to Neo-Kantianism* (pp. 59–81). Cambridge: Cambridge University Press.

Gutting, G. (1978). Husserl and scientific realism. *Philosophy and Phenomenological Research, 39*(1), 42–56.

Gutting, G. (Ed.) (2005). *Continental philosophy of science*. Malden: Blackwell.
Hardy, L. (2013). *Nature's suit. Husserl's phenomenological philosophy and the physical sciences*. Athens: Ohio University Press.
Hartimo, M. (2018). Husserl's scientific context 1917–1938, a look into Husserl's private library. *The new yearbook for phenomenology and phenomenological philosophy, 16* (pp. 335–355).
Harvey, C. W. (1986). Husserl and the problem of theoretical entities. *Synthese, 66*, 291–309.
Harvey, C. W. (1989). *Husserl's phenomenology and the foundations of natural science*. Athens: Ohio University Press.
Heelan, P. (1983). *Space-perception and the philosophy of science*. Berkeley: University of California Press.
Heidegger, M. (1962). *Being and time*. Oxford: Basil Blackwell.
Heidegger, M. (1967). *What is a thing?* South Bend: Gateway Editions.
Heidegger, M. (1968). *What is called thinking?* New York: Harper and Row.
Heidegger, M. (1971). *Poetry, language, thought*. New York: Harper and Row.
Heidegger, M. (1978). *Frühe Schriften. Gesamtausgabe Band 1*. Frankfurt am Main: Vittorio Klostermann.
Heidegger, M. (1997). *Plato's sophist*. Bloomington: Indiana University Press.
Heidegger, M. (2002). *Off the beaten track*. Cambridge: Cambridge University Press.
Hempel, C. G. (1965). *Aspects of scientific explanation*. New York: Free Press.
Hempel, C. G. (1966). *Philosophy of natural science*. Englewood Cliffs: Prentice-Hall.
Husserl, E. (1960). *Cartesian meditations. An introduction to phenomenology*. The Hague: Martinus Nijhoff.
Husserl, E. (1969). *Formal and transcendental logic*. The Hague: Martinus Nijhoff.
Husserl, E. (1970). *The crisis of the European sciences and transcendental phenomenology. An introduction to phenomenological philosophy*. Evanston: Northwestern University Press.
Husserl, E. (1971). "Phenomenology", Edmund Husserl's article for the encyclopedia Brittanica (1927): New complete translation by Richard E. Palmer. *Journal of the British Society for Phenomenology, 2*, 77–90.
Husserl, E. (1973a). *Experience and judgment. Investigations in a genealogy of logic*. Evanston: Northwestern University Press.
Husserl, E. (1973b). *Zur Phänomenologie der Intersubjektivität. Texte aus dem Nachlass. Dritter Teil: 1925–1935*. Den Haag: Martinus Nijhoff.
Husserl, E. (1973c). *Zur Phänomenologie der Intersubjektivität. Texte aus dem Nachlass. Erster Teil: 1905–1920*. Den Haag: Martinus Nijhoff.
Husserl, E. (1973d). *Zur Phänomenologie der Intersubjektivität. Texte aus dem Nachlass. Zweiter Teil: 1921–1928*. Den Haag: Martinus Nijhoff.
Husserl, E. (1983). *Ideas pertaining to a pure phenomenology and to a phenomenological philosophy. First book: General introduction to a pure phenomenology*. The Hague: Martinus Nijhoff.
Husserl, E. (1989). *Ideas pertaining to a pure phenomenology and to a phenomenological philosophy. Second book: Studies in the phenomenology of constitution*. Dordrecht: Kluwer.
Husserl, E. (1997). *Thing and space: Lectures of 1907*. Dordrecht: Kluwer.
Husserl, E. (2000). Londoner Vorträge. *Husserl Studies, 16*, 183–254.
Husserl, E. (2001a). *Analyses concerning passive and active synthesis: Lectures on transcendental logic*. Dordrecht: Kluwer.
Husserl, E. (2001b). *Logical investigations. Volume 1*. London/New York: Routledge.
Husserl, E. (2001c). *Logical investigations. Volume 2*. London/New York: Routledge.
Husserl, E. (2002). *Einleitung in die Philosophie. Vorlesungen 1922/23*. Den Haag: Martinus Nijhoff.
Husserl, E. (2008). *Introduction to logic and theory of knowledge. Lectures 1906/07*. Dordrecht: Springer.
Hyder, D., & Rheinberger, H. (Eds.) (2010). *Science and the life-world. Essays on Husserl's crisis of the European sciences*. Stanford: Stanford University Press.

Ihde, D. (1991). *Instrumental realism. The interface between philosophy of science and philosophy of technology*. Bloomington: Indiana University Press.
Ihde, D. (2011). Husserl's Galileo needed a telescope! *Philosophy of Technology, 24*, 69–82.
Jammer, M. (1974). *The philosophy of quantum mechanics*. New York: Wiley.
Kisiel, T. J. (1970). Merelau-Ponty on philosophy and science. In J. J. Kockelmans & T. J. Kisiel (Eds.), *Phenomenology and the natural sciences* (pp. 251–273). Evanston: Northwestern University Press.
Kockelmans, J. J. (1966). *Phenomenology of the physical sciences*. Pittsburgh: Duquesne University Press.
Kockelmans, J. J. (1970a). The era of the world-as-picture. In J. J. Kockelmans & T. J. Kisiel (Eds.), *Phenomenology and the natural sciences* (pp. 184–201). Evanston: Northwestern University Press.
Kockelmans, J. J. (1970b). Heidegger on the essential difference and necessary relationship between philosophy and science. In J. J. Kockelmans & T. J. Kisiel (Eds.), *Phenomenology and the natural sciences* (pp. 147–166). Evanston: Northwestern University Press.
Kockelmans, J. J. (1985). *Heidegger and science*. Washington: University Press of America.
Kockelmans, J. J., & Kisiel, T. J. (Eds.) (1970). *Phenomenology and the natural sciences*. Evanston: Northwestern University Press.
Ladyman, J. (2016). Structural realism. In E. N. Zalta (Ed.), *The stanford encyclopedia of philosophy*. Metaphysics Research Lab, Stanford University, winter 2016 ed.
London, F., & Bauer, E. (1983). The theory of observation in quantum mechanics. In J. A. Wheeler & W. H. Zurek (Eds.), *Quantum theory and measurement* (pp. 217–259). Princeton: Princeton University Press.
Luft, S., & Overgaard, S. (Eds.) (2012). *The Routledge companion to phenomenology*. New York/London: Routledge.
Matherne, S. (2018). Merleau-Ponty on abstract thought in mathematics and natural science. *European Journal of Philosophy, 26*(2), 780–797.
Meixner, U. (2010). Husserls transzendentaler Idealismus als Supervenienzthese. Ein interner Realismus. In M. Frank & N. Weidtmann (Eds.), *Husserl und die Philosophie des Geistes* (pp. 178–208). Frankfurt am Main: Surhkamp.
Merleau-Ponty, M. (1968). *The visible and the invisible*. Evanston: Northwestern University Press.
Merleau-Ponty, M. (2003). *Nature. Course notes from the Collège de France*. Evanston: Northwestern University Press.
Moran, D. (2012). *Husserl's crisis of the European sciences and transcendental phenomenology: An introduction*. Oxford: Oxford University Press.
Mormann, T. (1991). Husserl's philosophy of science and the semantic approach. *Philosophy of Science, 58*(1), 61–83.
O'Raifeartaigh, L. (1997). *The dawning of Gauge theory*. Princeton: Princeton University Press.
O'Raifeartaigh, L., & Straumann, N. (2000). Gauge theory: Historical origins and some modern developments. *Review of Modern Physics, 72*(1), 1–23.
Parker, R. K. B. (2017). The history between Koyré and Husserl. In R. Pisano, J. Agassi, & D. Drozdova (Eds.), *Hypotheses and perspectives in the history and philosophy of science* (pp. 243–276). Cham: Springer.
Pietersma, H. (2000). *Phenomenological epistemology*. Oxford: Oxford University Press.
Popper, K. R. (2002). *The logic of scientific discovery*. London/New York: Routledge.
Reynolds, J. (2018). *Phenomenology, naturalism and science: A hybrid and heretical proposal*. New York/London: Routledge.
Richardson, A. (2006). "The Fact of Science" and critique of knowledge: Exact science as problem and resource in Marburg Neo-Kantianism. In M. Friedman & A. Nordmann (Eds.), *The Kantian legacy in nineteenth-century science* (pp. 211–226). Cambridge: MIT Press.
Romdenh-Romluc, K. (2018). Science in Merleau-Ponty's phenomenology. In D. Zahavi (Ed.), *The Oxford handbook of the history of phenomenology* (pp. 340–359). Oxford: Oxford University Press.

Rouse, J. (1986). Merleau-Ponty and the existential conception of science. *Synthese, 66*(2), 249–272.
Rouse, J. (1987). Husserlian phenomenology and scientific realism. *Philosophy of Science, 54*, 222–232.
Rouse, J. (2005). Heidegger on science and naturalism. In G. Gutting (Ed.), *Continental philosophy of science* (pp. 123–141). Malden: Blackwell.
Ryckman, T. (2005a). Recovering first philosophy in philosophy of physics. *Philosophy Today, 49*, 13–22.
Ryckman, T. (2005b). *The reign of relativity. Philosophy in physics 1915–1925*. New York: Oxford University Press.
Sellars, W. (1991). Philosophy and the scientific image of man. In *Science, perception and reality* (pp. 1–37). Ridgeview: Alascadero.
Smith, B. (1994). *Austrian philosophy. The legacy of Franz Brentano*. Chicago: Open Court.
Smith, A. D. (2003). *Husserl and the cartesian meditations*. London/New York: Routledge.
Smith, D. W. (2007). *Husserl*. New York/London: Routledge.
Soffer, G. (1990). Phenomenology and scientific realism: Husserl's critique of Galileo. *Review of Metaphysics, 44*, 67–94.
Ströker, E. (1997). *The Husserlian foundations of science*. Dordrecht: Kluwer.
Vallor, S. (2009). The pregnancy of the real: A phenomenological defense of experimental realism. *Inquiry, 52*(1), 1–25.
Webb, D. (2009). *Heidegger, ethics, and the practice of ontology*. London/New York: Continuum.
Weyl, H. (1923). Gravitation and electricity. In A. Einstein, H. A. Lorentz, H. Weyl, & H. Minkowski (Eds.), *The principle of relativity* (pp. 199–216). New York: Dover.
Weyl, H. (1949). *Philosophy of mathematics and natural science*. Princeton: Princeton University Press.
Wiltsche, H. A. (2012). What is wrong With Husserl's scientific anti-realism? *Inquiry, 55*(2), 105–130.
Wiltsche, H. A. (2016). Mechanics lost: Husserl's Galileo and Ihde's telescope. *Husserl Studies, 32*(2), 149–173.
Wiltsche, H. A. (2017). Science, realism, and correlationism. A phenomenological critique of Meillassoux' argument from ancestrality. *European Journal for Philosophy, 25*(3), 808–832.
Wiltsche, H. A. (2019). Models, science, and intersubjectivity. In F. Kjosavik, C. Beyer, & C. Fricke (Eds.), *Husserl's phenomenology of intersubjectivity. Historical interpretations and contemporary applications* (pp. 339–358). New York/London: Routledge.
Winther, R. G. (2016). The structure of scientific theories. In E. N. Zalta (Ed.), *The stanford encyclopedia of philosophy*. Metaphysics Research Lab, Stanford University, winter 2016 edition.
Zahavi, D. (2001). *Husserl and transcendental intersubjectivity. A response to the linguistic-pragmatic critique*. Athens: Ohio University Press.
Zahavi, D. (2003). Phenomenology and metaphysics. In D. Zahavi, S. Heinämaa, & H. Ruin (Eds.), *Metaphysics, facticity, interpretation* (pp. 3–22). Dordrecht: Kluwer.
Zahavi, D. (2010). Husserl and the "Absolute". In C. Ierna, H. Jacobs, & F. Mattens (Eds.), *Metaphysics, facticity, interpretation* (pp. 71–92). Dordrecht: Springer.
Zahavi, D. (Ed.) (2012). *The Oxford handbook of contemporary phenomenology*. Oxford: Oxford University Press.

Part I
On the Origins and Systematic Value of Phenomenological Approaches to Physics

Chapter 2
Explaining Phenomenology to Physicists

Robert P. Crease

Abstract This essay attempts to outline how one might present phenomenology of physics in a way that might engage reductive scientific accounts. The basic strategy is to point out that quantum mechanics in particular forces recognition of the scientific workshop frame as not a given for scientific activity, but the product of a certain way of being as well as a certain method of framing – and recognition that the appearing of objects in that frame depends on how such framing is carried out.

2.1 Introduction

Philosophy is the systematic practice of critical reflection to examine assumptions and practices usually taken for granted in ordinary human life. Philosophy of science is the use of such critical reflection to examine conflicting assumptions and practices that arise in science, viewing and analyzing in the light of similar cases elsewhere in human activity.[1] It is often difficult to explain to scientists. One reason is that its language, like the discourse of science itself, often takes a narrow focus and is preoccupied with special topics and technical issues whose value understandably may not be obvious to outsiders. Another reason is that many scientists share the attitude that only the measurable is meaningful. The phenomenal, qualitative world that philosophers typically address seems less tangible, concrete, and even less interesting than the grandeur of things like Newtonian physics and the intricate beauty of quantum mechanics.

[1] I am greatly indebted to Delicia Kamins, Paul Rubery and James Sares – the members of the Phenomenology of Science Research Group at Stony Brook – for comments on and help with this article.

R. P. Crease (✉)
Stony Brook University, Stony Brook, NY, USA
e-mail: robert.crease@stonybrook.edu

H. A. Wiltsche, P. Berghofer (eds.), *Phenomenological Approaches to Physics*, Synthese Library 429, https://doi.org/10.1007/978-3-030-46973-3_2

Philosophy of science is particularly difficult to explain to physicists, who have often dismissed or ridiculed it. Such reactions are not insignificant, and reveal specific misunderstandings that provide philosophers with clues for how to respond. "Philosophy is dead" write Stephen Hawking and Leonard Mlodinow on the first page of their book *The Grand Design*, an international best-seller praised by physicists and physics students. Just a few pages later, they proceed to engage in amateur philosophizing by championing a form of idealism they call "model-dependent reality," of a sort whose conceptual limitations were exposed by philosophers long ago (Hawking & Mlodinow, 2010, 1). This case of philosophy-dismissal can be used to illustrate the unavoidability of philosophical commitments in science, to point out the overlap between issues encountered by physicists and philosophers, and to indicate that such issues cannot be addressed without contradiction if they are not set in a context broader than physics itself.

To choose another example, the theoretical physicist Sander Bias, in his book *In Praise of Science: Curiosity, Understanding, and Progress*, likens philosophers' discussions of science to doctors who diagnose patients before considering symptoms (Bias, 2010). The metaphor, said in a joking tone but clearly meant seriously, proposes that philosophers are in the business of trying to find meanings in science without being able to read relevant evidence. A philosopher might use the occasion to point out how and why the meanings sought by physicists and philosophers differ, as do their methods and evidence.

It is curious to find physicists so confidently and vehemently condemning a field that is not their own when their technical training is to be inquisitive, resist overstepping what they know, withhold judgment until certain, and accompany claims with error bars. This, too, is significant. To convey the purpose of philosophy to physicists it is not enough to point out conceptual flaws in the informally expressed but nevertheless passionately felt convictions of Hawking, Mlodinow, Bias, and others.

If explaining philosophy of science to physicists is challenging, explaining phenomenology is still more so, given that its approach can be misread as involving appeal to subjective feelings. In the following brief discussion, I will not attempt to do phenomenology of physics or report its findings. Rather, I will try to outline a path by which phenomenologists might respond to anti-phenomenological stances on the part of physicists.

In general, these stances spring from the assumption that philosophers are essentially looking for the same things as physicists rather than at how physicists are encountering and engaging the objects they study. The physicists' objections, that is, are produced by a particular, naturalist stance. That stance leads physicists to assume that any difficulty they have in grasping what philosophers say must be due to the failure of philosophers to have properly understood the subject-matter. The natural response to such an assumption, then, is either to break down the subject as if explaining it to a novice, or to dismiss philosophers altogether.

The most effective way, I think, to begin to respond is to call attention to the fact that such scientists have that scientific, naturalist stance, and the difference between it and the phenomenological stance. The scientific stance, in brief, involves objectifying what is being studied, while the philosophical/phenomenological stance is to examine the engagement between the scientist and what is being studied. The philosopher/phenomenologist is engaging science differently. To appropriate Bais's analogy, the philosopher/phenomenologist is not seeking to comment on the disease but rather to comment on how scientists seek the disease, or what kind of engagement is involved in disease-seeking. The challenge is then to find ways of conveying that this is the calling of philosophers/phenomenologists, as well as its scholarly legitimacy and value.

2.2 The Workshop and Its Frame

The first step, I think, in bringing to light the difference between the scientific and the phenomenological perspectives is to appeal to the metaphor of science as a vast "workshop." This workshop is a specialized and regulated environment in which special things can be created and studied – subatomic particles and their interactions, chemical elements and reactions, the effect of nutrients and toxins on plants, and so forth. In the surrounding world, these things do not show themselves directly and clearly in a way that we can study them. But they can be made to show themselves in a way so that they can be measured thanks to the experimental staging possible in the workshop. In the workshop, one can be in near-complete control of the things and events we stage in preparing to measure them. One can therefore make reasonably sure that the results are general and do not depend on features of the world outside the workshop. The workshop is not a place where nature is seen "as it is," but a controlled and supervised environment in which suitably trained individuals can "frame" how nature reveals itself experimentally. In the workshop, researchers can put questions to nature, in Galileo's words. One can make sure that the results are general and do not depend on features of the world outside the workshop. These results can then be used to help understand the surrounding world and to effect changes in it.

The "frame" of the workshop is the set of assumptions that determine how objects reveal themselves in that workshop. In the classical Newtonian workshop, for instance, the frame is provided by the restriction of objects of study to things with measurable properties, along with a set of other assumptions – localization in space and time, for instance – about such objects. In the Newtonian frame, the only things that can appear as true objects of study are masses, the only thing that these masses can do is move about, and the only things that start or stop or otherwise affect these motions are forces. Thomas Seebohm called this framing the "first abstraction" that constitutes the objects of modern science, one that introduces and establishes the split between primary and secondary properties (Seebohm, 2015, 222).

Using the workshop and frame images can be a first step towards clarifying for physicists what the phenomenology of physics investigates. Philosophers, one can then say, are interested in the ongoing interactive activity of the workshop. But they pay attention to the process differently than physicists, for what philosophers seek to understand is not what physicists know but how they are engaged with what they know. This is not just an epistemological insight, for how objects of study reveal themselves depends on the workshop's frame, how its frame compares with others, and the way of life which finds it important to frame the encounter with nature this way.

Werner Heisenberg once remarked that "what we observe is not nature in itself but nature exposed to our method of questioning" (Heisenberg, 1962, 32). This observation applies not only to physics but as well to philosophy. Another key step in explaining phenomenology to physicists and other scientists is to point out the existence of different philosophical traditions that go about questioning their objects of study in different ways. This step is crucial, for otherwise scientists may think of philosophers as a motley crew of would-be inquirers who do not even agree on what they are inquiring into or what they find, and are constantly talking past each other. For a crude comparison, one might ask scientists to imagine a panel of physicists, linguists, and ecologists discussing how to do science. Each of these disciplines puts different features of the objects they study center-stage, and questions these features in different vocabularies for different ends. Only when it is made clear how each group questions will the discussion become easy to follow (Crease, 2017). Their approaches to science are a function of how their respective traditions understand its nature and importance.

Three philosophical traditions have paid particular attention to science: the analytic, pragmatic, and phenomenological traditions. They are stylistically and methodologically different, though ingredients of each are blended together (with the history of science) in what is known as science studies, which approaches science from the start as a cultural and historical product. Though practitioners may blend elements of other traditions, most tend to be able to be placed in one. Each tradition, one would explain, begins with a different understanding of scientific practice that guides the questions they pose to it.

The analytic and pragmatic traditions can be discussed briefly. The *analytic* tradition, crudely put, mainly focuses on the logic of science and the meaning of its basic concepts. Analytic perspectives on quantum mechanics have often begun by seeking a logic for its formalism. Peter Gibbins summarizes the core of the analytic approach to the philosophy of quantum mechanics as follows: Because we cannot picture the quantum realm – that is, make a graphic or iconic representation of it – "[t]herefore understanding quantum mechanics must be a matter of understanding the logic of the words and the mathematics of quantum mechanics" (Gibbins, 1987, 127). He characterizes the approach as "quantum mechanics baffles us because we misunderstand its logic." The founding figures of analytic philosophy, in fact, included logicians, physicists and mathematicians such as Rudolf Carnap, Hans Reichenbach and Bertrand Russell. Analysts position themselves in the workshop, so to speak, as onlookers to the workshop frame alongside scientists, sharing their naturalistic orientation.

Several founders of the pragmatic approach, too, were scientists, including John Dewey (psychology), William James (psychology), George Herbert Mead (psychology), Chauncey Wright (mathematics), and Charles Peirce (logician, mathematician, and metrologist). Most conducted scientific experiments at some point in their careers. Oriented by the model of science, they focused on how scientists approach and solve puzzles – on how what is in the frame was arrived at. Pragmatic philosophers of science tend to position themselves outside the workshop while respecting its process of inquiry. Like analysts, pragmatists tend to be naturalists, and while aware of the changing contents of the frame do not raise the question of the framing itself. Pragmatists are also aware that humans do not spring into being as scientists but apprentice to become them. They also assume, however, that the puzzle-solving process that takes place in the workshop is essentially the same activity that drives everyday life – and are interested in its puzzle-solving activity insofar as the solutions make a difference to science and the world. True ideas are the ones that make a difference, and the meaning of such ideas is the totality of its effects. "The truth is what works," said James in what is surely his most famous pronouncement.

The phenomenological tradition is different. It focuses on the *framing*; on the way of being that motivates individuals to construct and use that frame at all, and the reciprocal impact of that frame and what appears in it on their way of being. The way of being inside the workshop – which discloses things in a framed way – is only one possible mode of being, and not the default setting of human beings. All modes of being, one may say to non-philosophers, arise through modifications of a matrix of ways by which human beings practically connect to the world that precedes any cognitive understanding. This matrix is not cognitive, and not articulable in terms of propositions. The technical term that phenomenologists use for this matrix or horizon is the "lifeworld". Human beings have different modes of being in the lifeworld, with the world appearing differently in each. Each mode involves an implicit understanding of the world and what matters in it, and "discloses" the world in a different way. The phenomenological approach to science focuses on the way of life that finds it important to frame the encounter with nature in this scientific way in the first place, and how it shapes what scientists study. The desire to frame, that is, is not a universal human trait but essential to what science is all about. Examining this dimension of science is part of the phenomenological task.

Many contemporary philosophers of science combine elements of one or more of these approaches. But these three remain the key ingredients, and it is important to understand the basic positions. In shorthand to ourselves, for instance, we might say that, while pragmatism and science studies stop at the historical, phenomenology raises the question of the transcendental. It is not necessary to describe how and why they emerged, or exactly how they relate to each other, though their differences and value is best appreciated by seeing how they approach specific controversies that erupt within physics, such as the nature of the fundamental, the scientific status of string theory, and the interpretation of quantum mechanics (Crease, 2017).

The language of analytic philosophy tends to dominate discussion of philosophy of physics – partly a function of the fact that so many of its founders were logicians and scientists. But phenomenology now can be described as an alternate method of questioning that philosophers have, one that has distinctive approaches and results.

2.3 Phenomenology

Now one can move to phenomenology, and describe in more detail how its mode of questioning is different from others. To broach the subject to an outsider one might start with Husserl's work. Husserl's phenomenology began with the insight called intentionality – that "all consciousness is consciousness of something" – and proceeded by exploring how what we experience is always coordinated with a particular manner of experiencing. We experience things like sticks and stones, triangles, moods, and the breeze in different ways that can be described if one pays careful – phenomenological – attention. A phenomenological approach thus recognizes from the outset the activity of the subject in grasping different kinds of objects. Experience is not a mosaic of sensations, but is structured in a way that gives us objects that transcend – give themselves as being more than – those sensations. This structure provides a horizon thanks to which our everyday activities can be conducted. In what Husserl calls the "natural attitude," we usually accept things that appears on the horizon as simply existing on their own. Phenomenology is the attempt to explore what it is for such objects to appear on the horizon at all. But the lifeworld, like any horizon, tends to drop out of view. It is overlooked in favor of the objects, plans and goals that appear in and thanks to that horizon. Scientists go directly to that object, in what we might call the naturalistic attitude. Still, we have a relationship with whatever we encounter; what appears always does so from somewhere to someone on a horizon.

Husserl's approach superficially resembles that of Immanuel Kant, who regarded the experienced world as shaped by a set of rules that make it possible – the "transcendental", in his vocabulary. But Husserl's approach is not Kantian in several respects. First, the way objects are given cannot be systematized through rules (i.e., categories and judgments); second, how objects are given cannot be deduced transcendentally but only examined through experience; third, the modes of givenness of objects are not static but can evolve (though this point is emphasized more by later phenomenologists).

This general approach, one might continue, is useful for understanding natural science. The natural sciences tend to adopt a naturalistic attitude that assumes the existence of the objects they study apart from the lifeworld. What's different about scientific activity as a way of being is that it seeks to objectify the world by framing the experience of it so that it can be measured. Measurement allows properties of objects to appear in a way that seems to eliminate human intentions and desires in achieving an "objective" picture. But this is only one mode among others of experiencing nature, and involves a human role in the framing itself, including the application of measurement tools, mathematical models, and so forth. If modern science set in motion the dream of an objectivity that would let nature speak, Husserl sought to achieve this, not by seeking to expunge subjectivity, but by assuming a human role in the speaking.

In a sense, though, phenomenology only brings to bear what happens in ordinary natural science. The data of experiments consists, not of free-floating numbers, but numbers produced by a particular group of scientists in a particular way by particular instruments. Different or more sensitive instruments will result in different numbers, but this difference does not imply that a different phenomenon is being studied, or that science is "relative." For the workshop environment aims to create conditions so that these numbers cannot be written off as reducible to cultural, historical, or psychological factors.

In Husserl's last work, *The Crisis of European Sciences*, he elaborated these insights into an account of how scientists frame what they study, and how this framing can be useful but also dangerously misconstrued in the lifeworld. Scientists – and he was thinking of natural scientists and particularly physicists – use ideal mathematical objects as their basic conceptual tools for their descriptions of the world. These ideal mathematical objects comprise a world of their own. Husserl then outlines what amounts to a distinction between laws and theories (Hardy, 2013, 34). Laws are expressed in a mathematical formalism and concern relations between ideal objects. Examples are $P \propto 1/V$, $G = m_1m_2/r^2$, $\Delta x \Delta p \geq h/4\pi$. By themselves, these equations don't "say" anything about the world, but merely state ideal relationships in an ideal world. When I drop a ball or squeeze a balloon, it does not affect these formulae. Theories, however, connect these ideal relationships – the concepts and formalism – with the world: P with pressure, m with masses, and so on. Theories are interpretations of these formula. They do attempt to say something about the world we experience – experience understood as including what we encounter mediated by instruments in experimentation. In classical mechanics, it appears relatively easy to connect in a rough and ready way the formalism with familiar concepts such force, mass, acceleration. One can then pretend that phenomena actually possess these as properties. Mathematics and mathematical models can then be more or less straightforwardly identified with the fundamental framework of the world. A good mathematical theory gives us explanations that seem to be justified; connected with our experience.

Mathematical formulae, for Husserl, are tools by which scientists approach the real via the ideal. Ideal gas laws are for ideal gases, but help us understand real gases. Our experience of real phenomena may change – one can always make more sensitive instruments to get more data points – and the ideal law may still hold or be replaced by another. Using these ideal laws, Husserl wrote, scientists cloak the experienced world in "a well-fitting garb of ideas." The "clothing" is mathematical formalism, the "clothed" the phenomena being described, and Husserl found an absolute separation between the two. The distinction between laws (formalisms) and theories is why one can say "Shut up and calculate!" while disagreeing about what one is calculating about.

The motivation to build scientific workshops, Husserl thought, arises from the lifeworld; from our recognition of the importance of finding ways to better understand or control nature. So does the building of equipment and the development and use of ideal mathematical objects as tools. But even inside workshops the work of scientists springs from the lifeworld. Albert Michelson, Husserl writes, may well be trying to describe the behavior of light as it shows up in his instruments – but his motives for doing so, his ambitions and planning and discussing and all his other work in the laboratory – his human experience – are grounded in something else besides the scientific frame into which he is peering.

Husserl also concluded that it was easy to mistake the ideal world for the real world. The world – "reality" at the deepest level – is readily conceived as a vast geometrical space whose objects are mapped by mathematical formula. This creates the ground for the "crisis" referred to in Husserl's book. In a dramatic ontological reversal, the ideal world is substituted for the world itself, formulae for the meaning of being, the suit for the body. The connection with the lifeworld is lost. The reality in which we – and even scientists like Michelson – are bathed when we wake up in the morning, share our family life, make friendships, play and work, and hope and fear, fades into the background as something subjective in comparison to the objective maps provided by the workshops. The maps become more important, more "real," than the landscape they were invented to help us navigate. This would not be a problem if reality were itself ideal. But the world is richer than the concepts we use to try to capture it; we experience the world as containing "more" than the ideal terms by we represent it. Phenomenology, it has been said, is the study of the invisible; all those things around objects that are not part of a naturalistic account that pertain to the founding role of the lifeworld.

Husserl, who started his career as a mathematician, did not think that framing the world was useless. It is immensely powerful, and its products are indispensable to the modern world. These products can be known and applied without talking about their origins. The particular character of the modern world is that the very successes of science lead us to think that only the quantifiable aspects of objects are meaningful. Yet it is a dangerous mistake to regard the real as only what is quantifiable and mathematically expressible. In bypassing the lifeworld, one produces an illusory picture of what it is to be human.

The perspective of the workshop alone, Husserl concluded, is insufficient for understanding the whole of science – how it springs from the lifeworld and how its knowledge is ultimately justified. Its "accomplishments can be understood only in terms of the activity that accomplishes them" (Husserl, 1970, 117). The sciences, he said, "are in principle incapable of solving the basic epistemological problem" (Hardy, 2013, 63), namely, the origins of their own validity. The sciences cannot be validated if they are simply realms of knowing "facts;" one would have to examine how facts appear, what counts as a fact, why facts can appear to change, and so forth. What is required is another science of a broader scope that is able to reflectively justify scientific activity, and describe how it arises out of the grounds of human experience. For Husserl, that science is phenomenology. "Phenomenology is to keep the entire superstructure of the sciences in touch with its generative base in experience" (Hardy, 2013, 43).

Later phenomenologists, including Martin Heidegger, Maurice Merleau-Ponty, Patrick Heelan, and Thomas Seebohm, carried on Husserl's insights in different ways. They also elaborated aspects like the nature of inquiry – the back and forth cycle of interpretation and reinterpretation that takes place in the workshop, known technically as the hermeneutic circle – as well as the mode of being by which nature is objectified. They also study the difference between how such things as space and time appear in the workshop frame and that of ordinary human experience; the lifeworld.

2.4 Phenomenology of Physics

Phenomenology, then, starts with the recognition that the framing that allows scientific research to happen is already a human product. There is no dualism between what is measured in the frame and something beyond or behind it; there is no need to figure out how to relate these two realms. The frame, along with everything in it, is already as it were an engagement between consciousness and world. The known is already the product of such an engagement. One does not need to "introduce" subjectivity in order to account for it; the subject is there from the beginning. To take the subject as an "add-on" to science renders it impossible to analyze science with philosophical rigor. Explaining phenomenology can proceed by noting what is left out in scientific practice without a recognition of such engagement.

From here, phenomenological research into physics can take several paths. One is to carry out a phenomenology of the scientific attitude and how it is grounded in the lifeworld. What is the kind of attitude that makes Michelson a scientist? How does he approach his objects of study? What is the attitude that makes one think it important to frame at all? This would study the scientific attitude as one way of being among others.

Another path is to carry out a phenomenology of the lifeworld as it pertains to physics research. What features of the lifeworld are present but unthematized in the workshop, but nevertheless necessary to its activity? What, for instance, allowed Michelson to imagine, plan, discuss, collaborate, communicate, discover, and interpret his work in the first place? In explaining phenomenology to a physicist, here is the place to contrast the physicist's professional understanding of time with lived time – the time of measurement versus the time of the measuring. In ordinary experience time is permanently present in the world. Continuous and flowing, it moves in one direction from past to future, their border being a momentary now. This is not a confused or blurred way of experiencing measured time. Thanks to temporality, humans remember, perceive, plan, and act consciously and deliberately. Humans do so as individuals and in groups, transforming themselves and the world, creating culture, history, and science. Even practicing physics, in which you creatively use what you already know to find out something you want to discover, requires experiencing time this way. Yet many physicists declare this a mirage. "For we convinced physicists," Einstein wrote in one such declaration, "the distinction between past, present, and future is only an illusion, however persistent." But experienced time comes first, even before the distinction between experienced and measured time. The world is disclosed in and thanks to experienced time, which therefore has a kind of priority over what appears. Physicists don't live through time as a sequence of discrete moments.

Yet a third path is to carry out a phenomenology of the concepts and objects of physics itself. Here the case of quantum mechanics provides a useful illustration. While in classical physics mathematics can be virtually identified with the fundamental framework of the world in conjunction with a set of assumptions including localization in space and time, quantum mechanics disrupts this possibility. The culprit is the uncertainly relations, which specify which properties and values can be known simultaneously and where they cannot. Quantum mechanics therefore forces the recognition that mathematics cannot be treated as providing the truth about the phenomena being measured but as an abstraction, an interpretation. It forces recognition of the engagement between those who would understand, and not merely practice, science. It forces, therefore, a recognition of the role of the framing. Ignoring or failing to recognize this engagement can only be achieved by introducing paradoxes and insoluble puzzles. The result gives rise to interpretations, like the Copenhagen Interpretation, that are content to write off parts of quantum mechanics as mysteries in order to preserve classical assumptions like space-time localization. Paradoxically, the attempt to preserve realism and reject the engagement of scientist and objects of scientific study requires the introduction of subjectivity, and the idea of limitations to the knowledge of nature. Quantum mechanics, in short, forces recognition of the need for a phenomenology of the invisible.

Phenomenologists have approached this in several ways. One is through a mereological approach. The challenge of quantum mechanics is that phenomena show themselves in such a way that the mathematics can't be treated either as identical to the phenomenon or as hiding it. How then to describe how the phenomena show themselves? Husserl, for instance, engaged in a phenomenology of morphological forms in nature, though calling them pre-scientific and in the service of practical life interests. Other phenomenologists, including Pedro Alves, propose to ignore this restriction. "In my opinion," he writes, "a phenomenological Philosophy of Nature should be centered on natural wholes (they have a peculiar formal ontology), instead of on the phenomenology of the constitution of the *res extensa*, *temporalis,* and *materialis*, that is to say, of a simple bulk of matter in time and space" (Alves, 2020). That is, we don't have to ground a study of nature on specific assumptions about space and time. Alves, in short, proposes carrying forward the phenomenological philosophy of nature through a "formal ontology of morphological unities," that is, "wholes sustained by complex relations of interdependency and not analyzable in ultimate units according to relations of foundation." Turning to quantum mechanics, he says "The eigenvectors, the Hermitian operators, its underlying Algebra and the Hilbert spaces belong to the logical-mathematical structure of the noematic content, not to the object which is intended by it." The mathematical formalism is "not a horizon from which the object detaches itself, in the way the co-givenness of the world furnishes an external horizon for the givenness of each individual object." This, he continues "is the right place for the intervention of a phenomenological philosophy of nature." Quantum mechanics "is a physics of the 'invisible,' in the sense that its theoretical objects cannot eventually be referred to morphological entities in the perceptual world." But this does not prevent us from conceiving "of a morphological system and posit it as the *independent reality* we are searching for." The philosophical approach able to do this, he concludes, is not positivism nor analytic but phenomenological.

2.5 Conclusion

The above account is, of course, sketchy. What I have attempted to provide is not an overview of phenomenology of physics nor have I attempted to engage in it. Rather, I have attempted to outline how one might present phenomenology of physics in a way that might engage reductive scientific accounts. The basic strategy is to point out that quantum mechanics in particular forces recognition of the workshop frame as not a given for scientific activity, but as the product of a certain way of being as well as a certain method of framing – and recognition that the appearing of objects depends on how such framing is carried out.

References

Alves, P. M. S. (2020). From the epistemology of physics to the phenomenology of nature: Some reflections in the wake of Seebohm's theses. In T. Nenon (Ed.), *Thomas Seebohm on the foundations of the sciences. An analysis and critical appraisal*. New York: Springer.

Bias, S. (2010). *In praise of science: Curiosity, understanding, and progress*. Cambridge, MA: MIT Press.

Crease, R. P. (2017). *Philosophy of physics*. Institute of Physics. https://iopscience.iop.org/book/978-0-7503-1542-5

Gibbins, P. (1987). *Particles and paradoxes: The limits of quantum logic*. Cambridge, UK: Cambridge University Press.

Hardy, L. (2013). *Nature's suit: Husserl's phenomenological philosophy of the physical sciences*. Athens, Greece: Ohio University Press.

Hawking, S., & Mlodinow, L. (2010). *The grand design* New York: Bantam.

Heisenberg, W. (1962). *Physics and philosophy: The revolution in modern physics*. New York: Harper.

Husserl, E. (1970). *The crisis of European sciences and transcendental phenomenology*. Evanston, IL: Northwestern University Press.

Seebohm, T. M. (2015). *History as a science and the system of the sciences: Phenomenological investigations*. New York: Springer.

Chapter 3
Husserl's Phenomenology of Scientific Practice

Mirja Hartimo

Abstract In this paper I will interpret and discuss Husserl's approach to exact sciences focusing especially on *Ideas I* (1913), *Formal and Transcendental Logic* (1929), and *Crisis* (the 1930s). This development shows that: (1) Husserl's phenomenology is primarily a method (rather than a metaphysical thesis); (2) the method is context-dependent and hence it is not tied to any particular philosophical approach to mathematics or physics; (3) it emphasizes practice in a manner that anticipates more recent philosophical analyses of the scientific practice; and finally (4) its aim is to reveal the metaphysical commitments of scientists, rather than to formulate an argument for any particular metaphysical position. All this conforms to the views of contemporary *naturalists in philosophy of science*. They hold that philosophers should approach sciences as they are, and hence take the scientific practices as the starting point of the philosophical investigations (as opposed to earlier a priori reflection of what sciences should be like). Accordingly, the paper argues that Husserl's approach anticipates the naturalistic turn in philosophy of science: he did not engage in building models about what science should be like, instead he described the scientific practice and the normative goals that guide it. However, the task of transcendental phenomenology is to provide a critique of scientific practice as it is. Looked at from the Husserlian point of view, this is what contemporary naturalists are missing, and hence their approach remains philosophically naïve. The paper thus argues that phenomenology provides tools that allow naturalist philosophers of science to make their approach critical and critically *philosophical*, while retaining the basic naturalist commitments not to accept appeals to the mysterious and to approach sciences as they are.

I wish to thank Sara Heinämaa, Frode Kjosavik, and Philipp Berghofer for their feedback on earlier versions of this paper. Thanks are also due to Jaakko Kuorikoski and Petri Ylikoski for their help in my quest to find out about the recent developments in philosophy of science.

M. Hartimo (✉)
Tampere University, Tampere, Finland
e-mail: mirjahartimo@gmail.com

H. A. Wiltsche, P. Berghofer (eds.), *Phenomenological Approaches to Physics*, Synthese Library 429, https://doi.org/10.1007/978-3-030-46973-3_3

3.1 Introduction

My main aim in this paper is to explain the way in which Husserl's phenomenology can be interpreted and further developed as philosophy of scientific practice. I will further claim that, due to its interest in and orientation to sciences as they are practiced, it anticipates the naturalistic turn in philosophy of science. The term 'naturalism' in this context refers to the conviction that philosophers should approach sciences as they are, and hence take the scientific practices as the starting point of their philosophical investigations. Such approaches have developed since the 1980s as a reaction to the logical positivist and falsificationist heritage to view philosophy of science as an a priori normative discipline.[1]Opposed to the reconstructive attempts of the earlier generations, contemporary naturalists in philosophy of science engage and cooperate closely with special scientific practitioners and incorporate developments from history, psychology, sociology and science studies. Instead of making sweeping claims about the nature of scientific explanation in general or the a priori structures of scientific theories, the naturalists emphasize the importance of local questions that rise in individual disciplines. While 'naturalism in philosophy of science' is a term applicable to a variety of diverse views, they all are connected by two commitments identified by Joseph Rouse (2002): all naturalists share a Nietzschean commitment to reject appeals to the mysterious or supernatural, and they share a Quinean commitment to avoid arbitrary impositions on the development of science (Rouse, 2002, esp. 4, 302–306). In this paper this minimal sense of the term 'naturalism' is what is meant by it. Rouse, too, emphasizes the role of scientific practices to account for normativity within naturalist philosophy of science.

Rouse's approach is pertinent for the present paper also because Husserl's view plays a rather central role in it. However, for him, Husserl's importance – together with Carnap's – consists in providing a foil against which putatively the more interesting approaches of Heidegger and Neurath can be understood and developed. For Rouse, the source of the normativity in Husserl's and Carnap's views is in atemporal a priori structures, whereas Heidegger and Neurath located the origin of normativity in contingently situated futural temporality (Rouse, 2002, esp. 72 and still in 2015, esp. 59). Rouse thus likens Husserl's phenomenology to the a priori prescriptive approach of the logical positivists. Given the apparent possibility of such an interpretation, I will first explain how to read Husserl as a philosopher of the embedded, embodied, and historically situated practices. On this view, Husserl is the first philosopher who took seriously the importance of concrete and diverse scientific practices in the manner now commonplace in contemporary naturalist

[1]I am thinking of the views influenced by, e.g., Arthur Fine's (1986) natural ontological attitude, *new experimentalism*, to which Ian Hacking's *Representing and Intervening* (Hacking, 1983) gave rise, and the consequent *naturalistic localism* in philosophy of science. Nowadays it is rather common to think that science is a social institution and a collective process (Giere, 1988). (The development is helpfully described in Ylikoski, 1996, see also Callebaut, 1993).

philosophy of science. Note that naturalism in philosophy of science should not be confused with ontological naturalism (that denies non-material substances), or methodological naturalism that holds that the methods of the natural sciences should be used in all disciplines, or with any other kind of reductionist naturalism. As discussed by Philipp Berghofer and Harald Wiltsche in the introductory chapter of this book, Husserl is not naturalist in these senses, and I am not trying to align his views with anything like it. However, due to its commitment to describe various phenomena as they are presented, Husserl's view ultimately aims at a realistic description of scientific practices, which is in his approach intertwined with transcendental phenomenological reflection. Whereas Husserl shares the former aim (i.e., the description of the practices) with the naturalists in philosophy of science, the latter adds to his approach an irreducible dimension that is not continuous with science but is not anything supernatural or mystical either.

I could base my initial introduction to phenomenology on any of Husserl's texts, but here I will draw mainly on the lecture course *Basic Problems of Phenomenology* (GP) which Husserl held repeatedly from 1910–1911 onwards and in which he explained the underlying motivation of phenomenological philosophy for the beginners. The same view is expressed in much more detail in Husserl's later published writings, but due to many details and technicalities, the overall view is easily lost. So, here goes:

In the opening paragraph of the lecture course, Husserl first points out that phenomenological investigations "require a completely different attitude than the natural one within which natural-scientific and psychological knowledge is attained" (GP, §1). In order to substantiate this fundamental methodological thesis, he then explains in detail the nature of the *natural attitude*, starting by explaining how each ego finds herself as an embodied agent in her surrounding world, where there are also other people to whom she is related by way of empathy and communicative acts.[2] The natural attitude is the naïve, non-philosophical attitude in which we do not question the realities that we posit in our experiences, even if we also know that experience can deceive (GP, §7). Nevertheless, in the natural attitude, experience is the source of legitimacy for our judgments:

> In general, we firmly maintain that experience has its legitimacy; more precisely, that the judgment in the natural attitude, 'on the basis of experience,' has its legitimacy as a matter of course; namely, on the most basic level, the sheer descriptive judgment, and then also, on a higher level, the inductive scientific judgment in the descriptive sciences; and finally, the judgment in the exact, objective sciences, which, in going beyond what is immediately experienced, arrives at conclusions about what is not experienced, but which, in doing just that, always relies on its ultimate legitimating ground, i.e., the immediate experiential givens. (GP, §7)

The whole of scientific research is also carried out in the natural attitude. Sciences also rely on experiences, even though they do so in a more systematic and self-

[2] Husserl discusses the role of communication, not in the main text of these lectures, but in the Appendix to §39 of the text, written probably in 1912.

critical manner than what we do in our everyday life. By the concept "natural scientific attitude," Husserl aims to capture the attitude that scientists have to their subject matters.

The *phenomenological attitude*, as opposed to the natural attitude, also the natural scientific attitude, is the one in which the natural naïve positing of realities is put in brackets, so that "we rather turn to the act itself and make it itself, plus what in it may present itself to us, an object" (GP, §15). The object – the real thing, process or event – is then no longer natural but becomes in Husserl's terminology *pure* (ibid.) because the naïve positing of existence of the object is excluded from it. This, however, does not mean that any features, or "predicates" of the experience, such as our embodiment or worldliness as the experiencing persons or the legitimacy of the experience would be removed or abstracted away. To put it bluntly, in epoché only existence is bracketed. Bracketing leaves all the predicates of our experiences intact, and existence is not a predicate. Instead of "going along" with the natural attitude, we pause to reflect on our activities and their objects. The experiences and the world as it is given in these experiences can now be examined in detail from the phenomenological point of view. Transcendental phenomenology aims to describe invariant structures of this realm of givenness. Husserl calls this realm 'absolute' because givenness, in its various modes (including uncertain and confused givenness), is indubitable. From this, it does not follow that our knowledge claims in the natural attitude are infallible. In 1910 Husserl describes our natural attitude beliefs as follows:

> Each I not only perceives, has not only experiences that posit intuitive existence, but also it has a more or less clear or confused knowledge; it thinks, it predicates and, as a scientific person, each I does science. Thereby, the I knows itself as one which sometimes judges correctly, one which sometimes falls into error, as one which occasionally succumbs to doubts and confusions, and also as one which occasionally presses on to clear conviction. But the I knows also, or is certain, in spite of all this, that the world is and that it, the I itself, is in the middle of this world, etc., just as we have thoroughly described it earlier. (GP, §2).

The change of attitude from natural to phenomenological takes place also when we start to reflect on scientific practices: we do not then do science but pause our experiments, reasonings or analyses in order to examine the scientific practice itself and its correlate, the objective world. The reduction thus does not mean abandoning the natural scientific attitude, but it enables thematizing it from the phenomenological point of view. As Husserl puts it, he is interested in "science (*die Wissenschaft*) within the phenomenological attitude" (GP, §16). In addition, Husserl argues that science *has to be* examined in the phenomenological attitude in order to be self-critical in all its aspects and hence worthy of the title "science" and to provide critically reflected and clarified knowledge.

The question then arises what is the science that is to be investigated within the phenomenological attitude? In what follows, I will argue that Husserl's view of science is largely in conformity with the views of most of the scientists that he had around him. He occasionally departs from mainstream views. When he does so, this is due to transcendental phenomenological reflection. Thus, Husserl's own view of sciences is a combination of an approximation of scientists' naïve view of the sciences and transcendental reflection of it. These two attitudes develop in parallel, fuelling each other so that at the end, in *Crisis*, Husserl's view has become a genuine *phenomenology of the scientific practice*: In *Ideas I*, Husserl shares the so-called "Göttingen [view of] pre-established harmony between mathematics and physics" with his colleagues. His phenomenological reflection demonstrates that axiomatic rationality should not be taken as an ideal outside of the exact sciences. In *Formal and Transcendental Logic*, Husserl's transcendental investigations reveal the separation of mathematics, on the one hand, and mathematical physics, on the other, as two independent and autonomous fields. His most original claim about the sciences in *Crisis* is also an outcome of his transcendental reflection on concrete scientific work: in Husserl's view, all sciences are practices carried out by embodied researchers in the life-world. Thus Husserl's conception of science evolves in tandem with the development of the sciences, and in the quest of radicality, transcendental phenomenology of science brings to light presuppositions of science, such as the Euclidean ideal, and ultimately the presupposition that the sciences are practices that are carried out in the lifeworld. This approach does not impose arbitrary philosophical restrictions on the sciences, nor does it appeal to speculations or to supernatural forces. It is thus compatible with the Rousean form of naturalism. This metaphysically neutral stance towards the sciences is complemented by critical reflection carried out in the transcendental attitude. Phenomenology thus offers tools to bring philosophical reflection to naturalistically conceived philosophy of science.

3.2 *Ideas I* and Göttingen Pre-established Harmony

Already before his Göttingen period,[3]Husserl had developed the notion of definite manifolds, which he held to be the guiding ideal of mathematics (and as such analogous to Hilbert's notion of completeness).[4]During his Göttingen period, as witnessed in *Ideas I* (1913), Husserl took this idea also to be a guiding ideal of physics. He thus shared with other Göttingen mathematicians, Hilbert, Minkowski, Klein, and later Weyl, a notion that has been called the "Göttingen idea of preestablished harmony between mathematics and physics." This meant, roughly, that the axiomatic ideal of mathematics served for Husserl, as well as for his colleagues, as an ideal of scientific rationality, as a device that was taken to guide empirical physical investigations "regulatively." The notion indicated a desire to impose mathematical order on physical phenomena. The core of the idea was expressed by Minkowski in a lecture course given in 1904:

> [T] hrough a peculiar, preestablished harmony, it has been shown that, by trying logically to elaborate the existing edifice of mathematics, one is directed on exactly the same path as by

[3]Husserl worked in Göttingen from 1901 until 1916 when he moved to Freiburg.

[4]This is a much discussed notion in the secondary literature. My view of it and the references for the discussion can be found in Hartimo (2018a).

> having responded to questions arising from the facts of physics and astronomy. (Pyenson, 1982, 145, Corry, 2004, 186)

For Hilbert, mathematics, geometry, and mathematical physics developed in an interplay, which he described using the metaphor of an edifice whose foundations were mathematics. He, too, held on to some version of the idea of preestablished harmony. For him, it boiled down to an insistent faith in the axiomatic method:

> [A] nything at all that can be the object of scientific thought becomes dependent on the axiomatic method, and thereby indirectly on mathematics, as soon as it is ripe for the formation of a theory. By pushing ahead to ever deeper layers of axioms in the sense explained above we also win ever-deeper insights into the essence of scientific thought itself, and we become ever more conscious of the unity of our knowledge. In the sign of the axiomatic method, mathematics is summoned to a leading role in science. (Hilbert, 1918, 1115)

And later, in 1925, Hilbert explicitly referred to preestablished harmony when explaining his usage of logical calculus in his formalist project:

> Fortunately that same preestablished harmony which we have so often observed operative in the history of the development of science, the same preestablished harmony which aided Einstein by giving him the general invariant calculus already fully developed for his gravitational theory, comes also to our aid: we find the logical calculus already worked out in advance. (Hilbert, 1925, 197)

The idea was that the *creative* part of the work in physics takes place in mathematics, while the task of experiments was simply to verify the existence of the truths that mathematicians had already devised. Preestablished harmony came to denote a conviction that mathematics is not only a useful tool for physics but that physics is fundamentally mathematical in nature (Kragh, 2015, 518, Pyenson, 1982, 147).[5] In *Crisis*, Husserl refers to this notion as the modern ideal of rationality:

> This, then, is for philosophy truly a realizable, through infinitely distant, goal – not for the individual or a given community of researchers but certainly for the infinite progression of the generations and their systematic researches. The world is in itself a rational systematic unity – this is thought to be a matter of apodictic insight – in which each and every singular detail must be rationally determined. Its systematic form (the universal structure of its essence) can be attained, is indeed known and ready for us in advance, at least insofar as it is purely mathematical. Only its particularity remains to be determined; and unfortunately this is possible only through induction. (Crisis, §12)

The Göttingen mathematicians' reference to a preestablished harmony between mathematics and physics expresses the modern ideal of sciences, the original invention or formulation of which is attributed to Galileo. This is the ideal of

[5] Among philosophers the idea of a preestablished harmony was discussed and defended by, for example, Ernst Cassirer, first in a monograph on Leibniz (*Leibniz' System in seinen wissenschaftlichen Grundlagen*, Marburg, 1902), and later in his *Substanzbegriff und Funktionsbegriff: Untersuchungen über die Grundfragen der Erkenntniskritik* (1910). In contrast, Paul Natorp objected to Minkowski's interpretation of Lorentz and Einstein on the grounds that he could not accept Minkowski's idea of a preestablished harmony between mathematical and empirical nature. For more on the various philosophers' views on the matter, see Pyenson (1982, 148–152).

rationality that Husserl's colleagues in Göttingen, during Husserl's years there, held quite universally. The Göttingen mathematicians were thus a living evidence of the way in which a goal-idea, established by Galileo, binds together a chain of generations (cf. Crisis, §15).

Given that the point of the natural attitude, as described in further detail in *Ideas I*, is to describe the scientists' straightforward beliefs, prior to any philosophizing, it is no surprise that in that work Husserl shares his colleagues' assumption of a preestablished harmony between mathematics and physics, that is an isomorphism between their axiomatic systems. Accordingly, he thinks that mathematics is the source of formal ontology.

> If we fashion *the idea of a perfectly rationalized experiential science* of Nature, i.e. one so far advanced in its theorization that every particular included in it has been traced back to that particular's most universal and essential grounds, then it is clear *that the realization of that idea essentially depends on the elaboration of the corresponding eidetic sciences*; that is to say, it depends not only on the elaboration of *formal mathesis*, which is related in one and the same manner to all sciences taken universally, but especially on the elaboration of those *disciplines of material ontology* which explicate with rational purity, i.e. eidetically, the *essence* of Nature and therefore the essences of all essential sorts of natural objectivities as such. And obviously that holds for any other region.
>
> Also with regard to *cognitive practice* it is to be expected beforehand that the closer an experiential science comes to the "rational" level, the level of "exact," of nomological science—thus the higher the degree to which an experiential science is provided with developed eidetic disciplines as its fundamentals and utilizes them for its [cognitive] groundings—the greater will become the scope and power of its cognitive-practical performance. (Ideas I, §9, 22/19)

In Husserl's view, the closer to an axiomatic science an exact discipline is, the more rational it is and the more explanatory power it has. This formal structure is defined by definite axiom systems; they define what Husserl calls a formal definite manifold (Ideas I, §72). It is the ideal of rational nature that guides the physicist (§73), and thus forms the "practical ideal of exact eidetic science" (§7, 22/17). For him, the definite manifold thus prescribes a structure that has a normative role for conceptualization in the empirical sciences of nature. This may sound like an attempt to build a normative or prescriptive view of science, but it is not: Husserl is describing an actual scientific ideal shared by him and his colleagues in Göttingen. He is describing and clarifying a normative goal that guides scientific practices around him.

However, Husserl's view of sciences differs from the mainstream Göttingen view at two points. First, he thinks that the scientists should develop also material ontologies. The material ontologies are subordinated under formal ontology which "prescribes for material ontologies a formal structure common to them all" (Ideas I, §10, 27/21). Formal ontology, together with the various material ontologies, defines the normative ideal that guides physicists in their attempt to "mathematize" nature, as Husserl puts it later in the *Crisis* (in fact, he mentions the ideal of mathematization in passing in *Ideas I*, §7, 22/17). Husserl thus searches for more stringent ties with which to relate the axiomatic ideal to intuition.

The other divergence from the views of Göttingen mathematicians is in Husserl's insistence that the normativity of the exact sciences does not extend to all scientific (*wissenschaftlich*) domains. He writes, for example, that

> [i]t is only a misleading prejudice to believe that the methods of historically given *a priori* sciences, all of which are exclusively exact sciences of ideal objects, must serve forthwith as models for every new science, particularly for our transcendental phenomenology—as though there could be eidetic sciences of but one single methodic type, that of "exactness." (Ideas I, §75, 173/169)

In *Ideas I*, Husserl thus argues that the nature of the investigated domain must be well-understood before the establishment of the guiding ideals and the choice of methods. If its essences are not exact, but are morphological, axiomatic ideal should not be applied. This is the case with the sciences related to persons, personal wholes and consciousness itself, hence with cultural sciences and especially phenomenology itself. Husserl continues to hold on to this view. In FTL, he goes even further, writing: "[It] is high time that people got over being dazzled, particularly in philosophy and logic, by the ideal and regulative ideas and methods of the 'exact' sciences – as though the In-itself of such sciences were actually an absolute norm for objective being and for truth "(FTL, §105). Accordingly, he argues in *Crisis* that the origin of the pernicious dualism between mind and body is, likewise, in the ideal of rationality modeled after exact sciences that lead to increasing specialization in the sciences (esp. Crisis, §11).

The task of phenomenology is to clarify how precisely the objects of scientific research are given, and thereby to reveal unexamined presuppositions of the scientists. In *Ideas I*, Husserl does this by examining how exact sciences are legitimized and related to intuition, that is, how things and the essential structures in which they belong are given to us (i.e. constituted by us). The two above described divergences from the general Göttingen line of thought, the addition of material ontologies and establishment of a limit to the applicability of the axiomatic ideal, are results of Husserl's critical phenomenological clarifications. *Ideas I* is thus a critical clarification of the unclear "Göttingen presupposition of preestablished harmony" that held sway of the scientists as an implicit norm guiding the theoretical research as well as practical formation of concepts.

3.3 *Formal and Transcendental Logic* and Separation of Mathematics and Logic

In the introduction to FTL, Husserl states that his aim is "an intentional explication of the proper sense of formal logic" (FTL, 10). "Formal logic" for him refers to logic as a theory of science. More specifically, formal logic is for him a theory of the epistemic norms of the sciences. "Intentional explication," in turn, refers to the philosopher's task of clarifying and renewing the "final sense" or purpose of logic which has been the guiding idea of the sciences and towards which scientists have been continually aiming. The general aim of the book is thus to clarify the ultimate goals or purposes of the sciences. This will be carried out by a method named "*Besinnung*," with which Husserl aims to capture the scientists' intentional senses [*intendierende Sinne*], i.e., the purposes that guide the scientists in their endeavors, even if often only implicitly. He even declares that *Besinnung* has to be carried out in a "community of empathy" [*Eingefühlsgemeinschaft*] with the scientists (Hartimo, 2018b, 2019a), which is an explicit invitation to read his discourse as a reflection of the goal-directed practices of the scientists of his time.

Husserl's subsequent examination of the intentional history of the exact sciences makes it clear that it has developed in two parallel strands: as logic and as mathematics. The guiding concept of logic (i.e., theory of judgment) is that of truth. A closer inspection shows that truth presupposes non-contradiction and grammaticality. Thus, logic can be divided into three goals and accordingly into three 'layers': grammar, logic of non-contradiction, and logic of truth. While logic, and thus the empirical sciences, are directed at truth, the sense guiding formal mathematics is the Euclidean ideal, concretely captured by the notion of "definite manifold" (FTL, §31).

The transcendental phenomenological examination carried out in the second part of the book (i.e., transcendental logic), reveals that there are three different evidences corresponding to the three layers of logic: grammaticality, distinctness, and clarity. Clarity is obtained by attaining verification by means of the facts themselves. Mathematicians, however, do not need to care about facts. Thus they do not strive for clarity at all. In this way Husserl's phenomenological analysis shows mathematics and logic to be separate, and eventually that mathematics is an autonomous discipline that develops without a concern for empirical verification (ibid., §51). Husserl writes that for

> a "pure" formal mathematics, there can be no cognitional considerations other than those of "non-contradiction," of immediate or mediate analytic consequence or inconsistency which manifestly include all questions of mathematical "existence" ... One must see that a formal mathematics, reduced to the above-described purity, has its own legitimacy and that, for mathematics, there is in any case no necessity to go beyond that purity. (FTL, §52)

In the introduction to FTL, Husserl admits that it is his transcendental analyses that led him to thus distinguish between logic and mathematics. This distinction was increasingly emphasized in mathematics since the nineteenth century. Penelope Maddy, who in my view engages in many respects similar analyses of the goal-directed developments of the exact sciences, has identified several reasons for this: Mathematicians started to pursue specifically mathematical goals with no immediate connection to empirical applications; Euclidean geometry became a study of one particular space among many abstract mathematical spaces; and third, the best mathematical accounts of physical phenomena could not anymore be taken as literal truths (2008, 33). Mathematics developed into an independent discipline that was not taken to trace the structure of the world as was still assumed by the Göttingen mathematicians. This is what Husserl's *Besinnungen*, together with transcendental phenomenological reflection revealed to him so that he, too, distinguished mathematics from applied physics and abandoned his earlier assumption about a preestablished harmony between mathematics and physics. As Maddy puts it, the consequent view of the relationship between pure and applied mathematics became one in which "we are constructing abstract mathematical models and trying our best to make true assertions about the ways in which they do and do not correspond to the physical facts." (2008, 33). This view is implied also by Husserl's view of the relationship between the two as seen in FTL.[6]

3.4 Crisis and Mathematization

Analogously to *Ideas I* and FTL, in *Crisis*, too, Husserl first elaborates on the historically given scientific world view and only after this (beginning in §28) inquires back to the transcendental presuppositions of this view. By the time of *Crisis*, written in the 1930s, the modern view of the scientific ideal is breaking down partly due to the internal development in the sciences. Emphasizing the historically sedimented givenness of the scientific world view, Husserl discusses the formation and development of the scientific idea of rationality from Galileo onwards. He studies in detail the development that he calls "mathematization." The notion of definite manifold is the culmination of this development. Whereas in *Ideas I*, this ideal was conceived as a realizable, although infinitely distant, goal, this time Husserl looks at mathematics and the mathematical sciences as an ideology with which we construct our view of the "objectively actual and true" nature:

> Mathematics and mathematical science, as a garb of ideas, or the garb of symbols of the symbolic mathematical theories, encompasses everything which, for scientists and the educated generally, *represents* the life-world, *dresses it up* as "objectively actual and true" nature. It is through the garb of ideas that we take for *true being* what is actually a *method*—a method which is designed for the purpose of progressively improving, *in infinitum*, through "scientific" predictions, those rough predictions which are the only ones originally possible within the sphere of what is actually experienced and experienceable in the life-world. (Crisis, §9h, 51–52)

Mathematization then leads to "technization" – to calculating, rather than understanding the ways of the world. Instead of thinking axiomatic systems as models of the real structure of the world (*Ideas I*), Husserl now develops an instrumentalist take on axiomatization. Applied mathematics is characterized accordingly as a theoretical construction that has lost its tight relationship to the world. As Maddy puts it, "[p]aradoxical as it may sound, it now appears that even applied mathematics is pure" (2008, 33). This turn, however, is not a turn from realism to instrumentalism – Husserl's claims in the *Origin of Geometry* written around the same time shows that he does not give up on the idea of truth: invariant structures of the life-world are the source for eternal ideas for him. His aim is to describe metaphysical views to which

[6]See (Hartimo, 2018c, and Forthcoming a) for more detail about *Besinnung* and its use in FTL.

the scientists are committed in their practices. This view that takes as its starting point the science as it is, is then subjected to transcendental phenomenological clarification.

In *Crisis*, Husserl starts the transcendental questioning by first discussing the positions of the empiricists from Berkeley and Locke onwards to Kant, from where he takes it into his own hands. For various reasons,[7] Kant's approach left crucial presuppositions unexamined. Namely the fact that we live in the life-world, and also the fact that the sciences exist as cultural accomplishments in our life-world. We are embodied beings that constitute objects primarily in perception and the objects-in-themselves primarily in communicative intersubjectivity. We take the world to exist essentially prior to our questioning. The same is true of science, which Husserl now sees clearly as a praxis that takes place in the life-world:

> Objective science, too, asks questions only on the ground of this world's existing in advance through prescientific life. Like all praxis, objective science presupposes the being of this world, but it sets itself the task of transposing knowledge which is imperfect and prescientific in respect of scope and constancy into perfect knowledge – in accord with an idea of a correlative which is, to be sure, infinitely distant, i.e., of a world which itself is fixed and determined and of truths which are idealiter scientific ("truths-in-themselves") and which predicatively interpret this world. To realize this in a systematic process, in stages of perfection, through a method which makes possible a constant advance: this is the task.
>
> For the human being in his surrounding world there are many types of praxis, and among them is this peculiar and historically late one, theoretical praxis. It has its own professional methods; it is the art of theories, of discovering and securing truths with a certain new ideal sense which is foreign to prescientific life, the sense of a certain "final validity," "universal validity". (Crisis, §28)

The task of the scientific praxis is to obtain true knowledge instead of entertaining prescientific "subjective-relative" doxa. The scientific experimentation takes place among the people in the everyday world and in communicative relations. These experiments are then the basis for theoretical constructions, such as the one by Einstein, when usings Michelson's experiments for his theoretical purposes (§34b):

> It is, of course, the one world of experience, common to all, that Einstein and every other researcher knows he is in as a human being, even throughout all his activity of research. [But] precisely this world and everything that happens in it, used as needed for scientific and other ends, bears, on the other hand, for every natural scientist in his thematic orientation toward its "objective truth," the stamp "merely subjective and relative." The contrast to this determines, as we said, the sense of the "objective" task. This "subjective-relative" is supposed to be "overcome"; one can and should correlate with it a hypothetical being-in-itself, a substrate for logical-mathematical "truths-in-themselves" that one can approximate through ever newer and better hypothetical approaches, always justifying them through experiential verification. This is the one side. But while the natural scientist is thus interested in the objective and is involved in his activity, the subjective-relative is on the other hand still functioning for him, not as something irrelevant that must be passed through but as that which ultimately grounds the theoretical-logical ontic validity for all objective verification, i.e., as the source of self-evidence, the source of verification. The visible measuring scales,

[7] I explain some of these reasons in detail in Hartimo, 2019b.

> scale-markings, etc., are used as actually existing things, not as illusions; thus that which actually exists in the life-world, as something valid, is a premise. (Crisis, §34b)

The natural scientific theories are verified by means of experiences. These experiences are carried out in the life-world, and it is the observations (and inferences) in the life-world that serve as the source of validity of the construction. The task of describing how the scientific knowledge is then constructed is the task of describing the scientific practice. Its transcendental examination requires a further inquiry into the transcendental conditions of possibility of such a practice. This means inquiring into the normative goals of the scientific inquiry, into the basic principles and concepts used in the sciences and into their sources in the life-world. Last but not least, this entails the task of critically studying the possible ideological commitments that hold sway of people. One such commitment was the axiomatic ideal of rationality that turned out to be more limited than expected.

3.5 Morals

Husserl's view of mathematics and applied science starts from the Göttingen ideal of pre-established harmony, i.e., from the idea that mathematics is about structures that are isomorphic with the structure of nature, and that physics thus is about uncovering underlying mathematical structures. The Husserl of *Crisis*, ascribes the view to Galileo, instead of Leibniz, as what Göttingen mathematicians did. The texts discussed above show how Husserl gradually changes his analysis and at the end gives up this view, so that in *Formal and Transcendental Logic*, in 1929, he thinks that the goals of mathematicians are independent of those of exact sciences. The tasks of the physicists is to find those mathematical models that are empirically verifiable. *Crisis* also describes how even applied natural science becomes pure of material content: The mathematical accounts of physical phenomena are not literal truths, but they become hypothetical, mathematized constructs of what could be the case. Husserl's three works thus proceed in line with the development of mathematics, from the Göttingen ideal to a modern view of applied mathematics. The transcendental examination of this development eventually shows that science as a form of praxis has to be carried out in the life-world between communicating and embodied persons.

Husserl's development is not shifting from idealism to realism and then to instrumentalism, or something of the sort. He does not make a priori, metaphysical claims about the sciences. Nor is he aiming at a philosophical view of what the sciences *should* be or become like. Instead he is describing the scientific practices and their normative goals as he finds them at each point of time. During his long life, his conception of science undergoes changes from the modernistic ideal to the practical-philosophical emphasis on the scientific practice. Most of the time and for the most part, he agrees with the mainstream scientists' conception of science. When he departs from this conception, it is because his transcendental criticism has brought to light reasons to revise this view. Most importantly, his late view of science as life-world practice, as explicated most systematically in *Crisis,* is an important outcome of his transcendental questioning. It is a novel idea – such views started to be more commonly entertained only in the 1980s as explained in the introduction.

My interpretation of Husserl's contributions to philosophy of science demonstrates how his phenomenology is able to add critical philosophical reflection to localist naturalistic accounts of scientific practices. Taken by themselves these accounts easily remain philosophically unsatisfactory or naïve surveys of this or that. A phenomenologically strengthened naturalism of science would add to them a critical reflection on the presuppositions of the scientific practices at issue, including presuppositions of the purposes of sciences. It could identify and clearly distinguish between different kinds of natural attitudes and examine their relations to each other (e.g. so-called personalistic attitude vs. reductionist naturalistic attitude). Ultimately it clarifies the basic principles and concepts of the theories developed in the sciences. This inquiry is critical in two related senses: it is critical in the Kantian sense of revealing conditions of possibility of the scientific practice. It is also critical in the sense that it seeks to clarify the practice so that the scientific community can carry it out knowingly, aiming at genuine evidences instead of just engaging in habitual exchanges or confused activities handed down by previous generations of scientists.

The phenomenological reflection remains roughly the same throughout Husserl's three works examined above. Thus, there are no shifts, no turns in Husserl's views about metaphysics. His philosophizing remains metaphysically neutral throughout his life. The goal is always the same in each new scientific situation: to clarify the scientists' metaphysical or ontological and normative commitments. In a Rousian, minimalist naturalist manner, Husserl does not build models in order to argue how sciences should work, but he describes the scientists' goal-directed practices. His approach, which should be characterized more as a method than as a theory, is "philosophized" critical naturalism, where the term 'naturalism' is understood in the contemporary non-reductionistic sense that takes as its starting point metaphysically neutral description of scientific practices.

3.6 Abbreviations and the References to Husserl's Works

FTL Husserl, Edmund (1974). Formale und transzendentale Logik: Versuch einer Kritik der logischen Vernunft (1929). Husserliana XVII. Ed. P. Janssen. The Hague: Martinus Nijhoff. English translation: Formal and Transcendental Logic, translated by Dorion Cairns. Martinus Nijhoff. The Hague. 1969.

GP Husserl, Edmund (1973). "Aus den Vorlesungen, Grundprobleme der Phänomenologie, Wintersemester 1910/1911", in *Zur Phänomenologie der Intersubjektivität, Husserliana XIII*, edited by Iso Kern. Martinus Nijhoff: 111–193. English translation: *The Basic Problems of Phenomenology. From the Lectures, Winter Semester, 1910–1911. Collected Works XII.* Translated by Ingo Farin and James G. Hart. Springer. 2006.

Ideas I Edmund Husserl (1976). Ideen zu einer reinen Phänomenologie und phänomenologischen Philosophie. Erstes Buch: Allgemeine Einführung in die reine Phänomenologie. Husserliana Band III. Ed., K. Schuhmann. The Hague: Martinus Nijhoff. English Translation: Ideas pertaining to a pure phenomenology and to a phenomenological philosophy, translated by F. Kersten. Kluwer, Dordrecht, Boston, London. 1982.

Crisis Husserl, Edmund (1976). Die Krisis der europäischen Wissenschaften und die transzendentale Phänomenologie: Eine Einleitung in die phänomenologische Philosophie (1936). Husserliana VI. Ed. W. Biemel. The Hague: Martinus Nijhoff. English translation: The Crisis of European Sciences and Transcendental Phenomenology. An Introduction to Phenomenological Philosophy. Translated by David Carr. Evanston: Northwestern University Press, 1970.

References

Callebaut, W. (1993). *Taking the naturalistic turn or How real philosophy of science is done*. Chicago: The University of Chicago Press.

Corry, L. (2004). *David Hilbert and the axiomatization of physics (1898–1918): From Grundlagen der Geometrie to Grundlagen der Physik*. Amsterdam: Kluwer Academic.

Fine, A. (1986). The shaky game. In *Einstein, realism and the quantum theory*. Chicago: The University of Chicago Press.

Giere, R. (1988). *Explaining science. A cognitive approach*. Chicago: The University of Chicago Press.

Hacking, I. (1983). *Representing and intervening*. Cambridge, UK: Cambridge University Press.

Hartimo, M. (2018a). Husserl on completeness, definitely. *Synthese, 195*(2018), 1509–1527. https://doi.org/10.1007/s11229-016-1278-7

Hartimo, M. (2018b). On the origins of scientific objectivity. In F. Kjosavik, C. Beyer, & C. Fricke (Eds.), *Husserl's phenomenology of intersubjectivity: Historical interpretations and contemporary applications* (Vol. 2018, pp. 302–321). London: Routledge.

Hartimo, M. (2018c). Radical Besinnung in Formale und transzendentale Logik (1929). *Husserl Studies, 34*, 247–266. https://doi.org/10.1007/s10743-018-9228-5

Hartimo, M. (2019a). Husserl on '*Besinnung*' and formal ontology. In F. Kjosavik & C. Serck-Hanssen (Eds.), *Metametaphysics and the sciences: Historical and philosophical perspectives* (Vol. 2019, pp. 200–215). New York: Routledge.

Hartimo, M. (2019b). Husserl on Kant, and the critical view of logic *Inquiry. An Interdisciplinary Journal of Philosophy*. https://doi.org/10.1080/0020174X.2019.1651089

Hilbert, D. (1918). Axiomatic thought. In W. Ewald (Ed.), *(1996). From Kant to Hilbert: A source book in the foundations of mathematics* (Vol. II, pp. 1107–1115). Oxford, UK: Clarendon Press.

Hilbert, D. (1925). Über das Unendliche. Mathematische Annalen, 95. English translation: "On the infinite". In E. Putnam, G. J. Massey, P. in Benacerraf, & H. Putnam (Eds.), *(1983) Philosophy of mathematics. Selected readings* (2nd ed., pp. 183–201). Cambridge, UK: Cambridge University Press. [1964].

Kragh, H. (2015). Mathematics and physics: The idea of a pre-established harmony. *Science and Education, 24*, 515–527.

Maddy, P. (2008). How applied mathematics became pure. *The Review of Symbolic Logic, 1*(1), 16–41.

Pyenson, L. (1982). Relativity in late Wilhelmian Germany: The appeal to a pre established harmony between mathematics and physics. *Archive for History of Exact Sciences, 27*, 137–155.

Rouse, J. (2002). *How scientific practices matter reclaiming philosophical naturalism.* Chicago/London: The University of Chicago Press.

Rouse, J. (2015). *Articulating the world.* Chicago: The University of Chicago Press.

Ylikoski, P. (1996). Tieteenfilosofian Naturalistinen Käänne. [The Naturalistic Turn in Philosophy of Science]. *Niin&Näin, 3*, 20–26.

Chapter 4
Physics as a Form of Life

Paolo Palmieri

"We are products of the past and we live immersed in the past which encompasses us. How can we move towards the new life, create new activities without getting out of the past — without placing ourselves above it? (And how can we place ourselves above the past if we are in it and it is in us?) There is no other way out except through thought which does not break off relations with the past but rises ideally above it and converts it into knowledge ... Only historical judgment liberates the spirit from the pressure of the past; (it is pure and extraneous to conflicting parties, and guarding itself against their fury, their lures, and their insidiousness,) it maintains its neutrality and seeks only to furnish light — it alone makes possible the fixing of a practical purpose; opens a way to the development of action (and, in the process of action, to the struggle of good against bad, useful against harmful, beautiful against ugly, true against false, in a word, value against non-value)." Benedetto Croce.[1]

"Noi siamo prodotto del passato, e viviamo immersi nel passato, che tutt'intorno ci preme. Come muovere a nuova vita, come creare la nostra nuova azione senza uscire dal passato, senza metterci di sopra di esso? E come metterci disopra del passato, se vi siamo dentro, ed esso è in noi? Non v'ha che una sola via d'uscita, quella del pensiero, che non rompe il rapporto col passato ma sovr'esso s'innalza idealmente e lo converte in conoscenza.

[...] Solo il giudizio storico, che libera lo spirito dalla stretta del passato e, puro qual è ed estraneo alle parti in contrasto, guardingo contro i loro impeti ed i loro allettamenti e le loro insidie, mantiene la sua neutralità, ed attende unicamente a

[1]Quoted in: Kuhn, Heilbron, Forman, Allen, 1967, p. v. Missing sentences that diffract the original text and which were not signaled by ellipsis, have been added in parentheses (quoted from Croce, 1941, pp. 43–44, 48).

P. Palmieri (✉)
University of Pittsburgh, Pittsburgh, PA, USA
e-mail: pap7@pitt.edu

H. A. Wiltsche, P. Berghofer (eds.), *Phenomenological Approaches to Physics*, Synthese Library 429, https://doi.org/10.1007/978-3-030-46973-3_4

fornire al luce che gli si chiede, sol esso rende possibile il formarsi del pratico proposito e apre la via allo svolgersi dell'azione e, col processo dell'azione, alle opposizioni, tra le quali questa si deve travagliare, di bene contro male, di utile contro dannoso, di bello contro brutto, di vero contro falso, del valore insomma, contro il disvalore."

—Benedetto Croce. (Croce, 1937, pp. 21, 24).

Abstract In the *Crisis of the European Sciences* Husserl raised a fascinating question, namely (broadly paraphrasing*), why is it that the axioms of mathematical physics are not self-evident despite the evidence and clarity that is gained through the deductive processes that flow from them*? In this chapter, I hope to illuminate Husserl's foundational question by pursuing the idea that physics is a form of life. This idea should not be taken in a naive metaphorical sense but quite literally. The meaning of life must not be restricted to a biological definition but should be construed broadly as a manifold phenomenon appearing in historical contexts and linguistic frameworks. I will argue that nature manifests certain of her aspects to us, but that in her totality (including ourselves as observers of nature), crucially, she resists our insight. This being hidden of the totality nature, or her desire or necessity to hide herself, explains why the axioms of mathematical physics must appear to our intuition as obscure, according as Husserl noted. It is because they point us to nature as a totality, or put in another way, because nature cannot know herself in her totality. A phenomenologically oriented physics is grounded in diverse mathematical styles that evolve in history and are ultimately rooted in natural languages and in the life-worlds of the physicists. From this phenomenological viewpoint physics is not concerned with truth in the sense of a psychophysical parallelism (the conformity of mind and reality). Indeed, axioms cannot be true in this psychophysical sense given their unintelligibility and unobservability. Rather physics is a form of life coming to be in history and language.

4.1 Introduction

In the *Crisis of the European sciences* Husserl raised a fascinating question, namely (somewhat paraphrasing*), why is it that the axioms of mathematical physics are not self-evident despite the evidence and clarity that is gained through the deductive processes that flow from them*? In this chapter, I hope to illuminate and revitalize Husserl's foundational question by pursuing the idea that physics is a *form of life*.

This idea should not be taken in a metaphorical sense but quite literally. The meaning of life cannot be restricted to a biological definition but should be construed broadly as a manifold phenomenon appearing in historical contexts and linguistic frameworks (let this general meaning be notated as Life). Specifically, I will focus on three epochs of physics as a form of Life that are well suited to exploring Husserl's question, namely, the youth of physics, as manifested in Galileo's axiomatic "new science" (the deduction of the laws of motion through Euclidean proportional reasoning), the senescence of physics, as manifested in Helmholtz's struggle with the quadratic anharmonic oscillator, in search of the magic formula for combination tones, and finally the 'posthumous maturity' of physics (see Sect. 4.5 for clarification of this expression), as manifested in Heisenberg's translation of the classical anharmonic oscillator into the language of quanta and spectral lines — the mother tongue of quantum physics in the twentieth century. I will be thinking about the idea that nature makes certain of her aspects accessible to our knowledge, and hence to herself, but in her totality, which includes ourselves as living observers of nature, she crucially resists our efforts at unveiling *her* and *our* innermost secrets, and hence she resists self-knowledge.

This being hidden of nature as a totality, or her desire or necessity to hide herself from further scrutiny, which I would be tempted to qualify as nature's vow of virginity, explains why the axioms of mathematical physics must appear to our intuition as obscure, as Husserl noted in the *Crisis*. It is because, I suggest, they point us to the totality of nature, or put in another way, because nature cannot become wholly transparent to herself through humans. A phenomenologically oriented philosophy of physics is grounded in a non-colonialist, subdued appreciation of legitimately autonomous and ethnically diverse mathematical and empirical styles that manifest themselves through history and are ultimately rooted in natural languages and in the life-worlds of the physicists. Against Benedetto Croce's dictum (casually reported and sported by the historians who laid the foundations of an archive for the history of quantum mechanics), historical judgment does not liberate the spirit from the pressure of the past. It is not pure and extraneous to conflicting parties, and is impotent against their fury, their lures, and their insidiousness. Rather, it is spirit, Life as history, that marshals passions and conjures their myriad reflections in historical judgment, bringing them to maturity in forms of Life that engage in the struggle of good against bad, useful against harmful, beautiful against ugly, true against false, in a word, value against non-value. Indeed, as a manifestation of Life, historical judgment must absolve and redeem the past, despite all its wretchedness, unconditionally and joyously.

From the phenomenological viewpoint that I assume in this chapter physics does not appear to be concerned with truth in the sense of a psychophysical parallelism (or, in other words, in the sense of the conformity of mind and reality, *adequatio intellectus et rei*, according to Scholastic philosophical parlance). Indeed, the first principles of physics cannot be true in this psychophysical sense given their unintelligibility and unobservability (according to the intuition that nature loves hiding herself, or must remain hidden). Rather physics is a form of Life coming to be in history and language and undergoing processes of growth and maturation and decay, and eventually posthumous ripening. Thus, in this phenomenological perspective, physics is not to be regarded as a science in some classical sense (i.e., for example, as an episteme, or as technology of which humans are in possession) but as living nature emerging into consciousness—not fully awake but forever in a sleepy mood—together with the ascendance of the human species.

4.2 Husserl's Question

In her true innermost being nature is mathematical. The pure mathematics of space and time exposes to humans, with absolute certainty, a stratum of unconditionally valid laws in the deepest layers of the being of nature. The most elementary laws are revealed immediately while the others come to light only mediately. Humans possess an innate capacity for knowing the fine grain structure of the spatial temporal stratum of nature as mathematical idealizations that precede experience. However, in regard to the more superficial strata, where nature still appears to be obeying mathematical yet more complex laws, certainty about these laws is no longer absolute. Their knowledge must be gathered from experience by way of induction. Whereas the relationship of the ground for mathematical deduction to the deductive processes by which the elementary laws of nature are formulate appears to be unproblematic, in regard to the more complex inductive laws that govern the superficial strata of the being of nature, the relationship of ground and consequence cannot simply be understood as a deductive process but must be referred to causal processes in nature that as such are not a priori accessible to humans. A feeling of obscurity emerges when humans draw their attention to the relationship between the mathematics of the superficial strata of nature and the mathematics of the deepest strata.[2]

As Harald Wiltsche suggested to me, in this paragraph we do not have Husserl's view but rather what he would take to be a self-interpretation of the physicists. What is more, the Husserl of the *Crisis* would strongly disagree with the claim that the innermost being of nature is mathematical. On Husserl reading, this is exactly the metaphysical position Galileo holds, and which Husserl rejects as a naïve hypostatization of Galileo's mathematico-physical methodology. Husserl is no Platonist (especially not in the "mathematical monism"-sense of the word)

[2]I have paraphrased and interpreted Husserl's text from the *Crisis*. The original is as follows. "Die Natur ist in ihrem ‚wahren Sein an sich" mathematisch. Von diesem An-sich bringt die Reine Mathematik der Raumzeitlichkeit eine Gesetzesschicht in apodiktischer Evidenz als *unbedingt allgemein gültige*, zur Erkenntnis: unmittelbar die axiomatischen Elementargesetze der apriorischen Konstruktionen, in unendlichen Mittelbarkeiten die übrigen Gesetze. Hinsichtlich der Raumzeitform der Natur besitzen wir eben das uns (wie es später heißt) ‚eingeborene" Vermögen, wahres Ansichsein als Sein in mathematischer Idealität (vor aller wirklichen Erfahrung) bestimmt zu erkennen. Implizite ist sie selbst uns also eingeboren. Anders steht es mit der konkreteren universalen Naturgesetzlichkeit, obwohl auch sie durch und durch mathematisch ist. Sie ist,a posteriori", von den faktischen Erfahrungsgegebenheiten aus induktiv zugänglich. Vermeintlich voll verständlich stehen sich scharf unterschieden gegenüber: apriorische Mathematik der raumzeitlichen Gestalten und induktive — obschon reine Mathematik anwendende — Naturwissenschaft. Oder auch: Scharf unterscheidet sich das rein mathematische Verhältnis von Grund und Folge von dem des realen Grundes und der realen Folge, also dem der Naturkausalität. Und doch macht sich allmählich ein unbehagliches Gefühl der Unklarheit über das Verhältnis zwischen der Naturmathematik und der ihr doch zugehörigen Mathematik der Raumzeitform, zwischen dieser ‚eingeborenen" und jener nicht eingeborenen Mathematik geltend."(Husserl, 1954, pp. 54–55).

because for him all mathematics is constructed out of the life-world of everyday experience.[3] The chapter does not focus on the question whether, to what extent and in what sense a Platonism can be attributed to Husserl throughout his works.

The feeling of obscurity mentioned above appears uncanny, as the innate mathematics of humans is included in the inductive mathematics of nature — like the innermost crystalline shells are nested within the outermost ones in a Renaissance spherical universe. This belonging to nature herself of the pure mathematics of space and time, which governs the innermost strata of nature and comes to the fore in the consciousness of humans — appearing to them as an innate endowment sanctioning their capacity for knowing nature — is also the phenomenon which causes the feeling of obscurity. It is nature herself that precludes herself from knowing reflexively her own totality of laws. The clarity in the continuum of laws that are accessible to herself through human consciousness immediately only as spatial temporal laws, gradually fades into obscurity as they become more complex and their structure must be gathered by humans in the course of laborious empirical investigations which, however, never reach the originary clarity and apodictic certainty of pure a priori laws. Yet, is nature not thoroughly mathematical in her true being, as she appears to the scientific mind of the physicists? Must nature in the end not be reducible to a system of mathematical laws governed by fundamental axioms that in turn must be clear and accessible to humans? Why is it, would Husserl ask the physics community, that they remain beyond the intellectual grasp of humans after all, and never become absolutely certain to them? Is it because the innate cognitive endowment of humans precludes a complete access to the first principles that underpin the continuum of the laws of nature?[4]

In the remainder of the chapter, I explore Husserl's question and its emotional connotation about the obscurity of the fundamental principles governing the true mathematical being of nature by attending to the idea of physics as a form of Life, in which such persisting obscurity signals the desire or the necessity of nature to hide herself — from the self-scrutiny inflicted through the emergence of human consciousness in history.

[3] Many thanks to Harald Wiltsche for alerting me to the nuances of Husserl's views on Platonism in mathematics.

[4] "Aber ist nicht die Natur an sich durchaus mathematisch, muß nicht auch sie als einheitliches mathematisches System gedacht werden, also wirklich darstellbar sein in einer einheitlichen Naturmathematik: eben jener, die die Naturwissenschaft immer nur sucht, sucht als umgriffen von einem der Form nach ‚axiomatischen" Geesetzessystem, dessen Axiomatik immer nur Hypothese ist, also nie wirklich erreichbar? Warum eigentlich nicht, warum haben wir keine Aussicht, das der Natur eigene Axiomensystem als ein solches echter apodiktisch evidenter Axiome zu entdecken? Weil uns hier faktisch das eingeborene Vermögen fehlt?" (Husserl, 1954, pp. 55–56).

4.3 Youth

Galileo's mathematical natural philosophy created the conditions for the birth of mathematical physics in the early seventeenth century. It was the aesthetic ideal of the imitation of nature through refined geometrical means of representation — grounded in Galileo's aesthetic theory of painting and sculpture — that framed his work in mathematical physics and controlled experiment. Yet Galileo did not overcome the fatal obscurity that took center stage in his mathematical natural philosophy, and which he had inherited from classical Greek geometry, namely, the obscurity of Euclid's definition of proportionality for continuous magnitudes. Both Galileo and his predecessors had been consciously attentive to the menacing presence of this obscurity.[5] But the brilliance, candid exuberance, and the naiveté of youth pushed Galileo to set aside the problem of this obscurity for the time being, and to turn away from the bottomless abyss of groundlessness, although he returned to the brink of the abyss in his late years once more, in an emotionally charged (as he was then totally blind) and last-ditch attempt to take responsibility for the darkness that envelops nature and protects the secrecy of her first principles. Nature, however, resisted this self-inflicted assault at her virginity.

In 1604 Galileo wrote a letter to Paolo Sarpi, in which he put forward an 'erroneous' principle (according to Galileo himself who later on corrected his error) from which he claimed that he could derive the times-squared law of fall. The principle is as follows: *the speed of fall is proportional to the space fallen through.* In referring to that 'erroneous' principle Alexandre Koyré argues that Galileo already knew all the details concerning the phenomenon of fall (such as the proportionality of the space traversed to the square of the elapsed time). What Galileo had long wanted to discover, in Koyré's opinion, was a general principle from which he could deduce the law geometrically. In other words, Koyré continues, Galileo sought to find the essence, i.e., the definition, or law, of the phenomenon *fall of bodies.* Why did Galileo adopt the 'erroneous' principle, asks Koyré? The answer, for him, is clear. The key to classical physics is the geometrization of nature, which implies the application of mathematical laws to the phenomena of motion. But, in Koyré's words, it is much easier to "imagine in space rather than think in time [*imaginer dans l' espace que de penser dans le temps*]"; hence Galileo's error in 1604. Koyré's explanation of that 'error' is intriguing. It hints at a deep connection between consciousness, visual-spatial perception, and the successes and failures of mathematical imagination at reducing the foundational principles to clarity, whose ramifications might be further explored by looking at Galileo's production after he became totally blind.[6]

---

[5] For a more technical exposition, I take the liberty of referring the reader to Palmieri, 2001.

[6] As is well known, the correct principle which Galileo eventually adopted is the proportionality speed and time. In uniformly accelerated fall from rest, both along a vertical path and an inclined plane, a body's degree of speed is proportional to the time elapsed from the beginning of the fall. Galileo would not have accepted a proportionality between non-homogeneous quantities, and

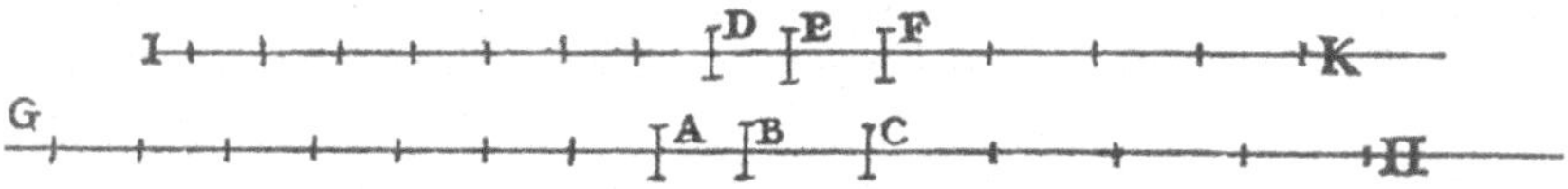

Fig. 4.1 Galileo's application of equimultiple proportionality to uniform motion. (Galilei 1890–1909, VIII, p. 192)

Galileo's mathematical natural philosophy was based on the Euclidean technique of proportional reasoning. This technique hinges on the definition of proportionality, or to be more precise of sameness of ratios. Thus for Galileo mathematizing means finding proportional relations between magnitudes. For Galileo a mathematical proof is generally a proof which leads to a proportionality, i.e., a sameness of ratios. What does proportionality mean for Galileo? He borrows the meaning of proportionality from Euclid's *Elements*. The book on proportionality is the fifth of the Elements. In it Euclid begins by furnishing the following definition of sameness of ratios.

> Magnitudes are said to be in the same ratio, the first to the second and third to the fourth, when, if any equimultiples whatever be taken of the first and the third, and any equimultiples whatever of the second and fourth, the former equimultiples alike exceed, are alike equal to, or alike fall short of, the latter equimultiples respectively taken in corresponding order.[7]

We may attempt to clarify the meaning of this definition within Galileo's mathematical physics by looking at how, in *Two new sciences*, he adapted it to the study of equable (i.e., uniform) motion as a foundation for the study of accelerated motion.

Referring to Fig. 4.1, let line IK represent time and let a movable move along line GH. Let AB be the space traversed in time DE and BC the space traversed in time EF. Galileo wishes to prove the following

> Proposition I. If a moveable equably carried with the same speed passes through two spaces, the times of motion will be to one another as the spaces passed through.[8]

would have said more precisely that the ratios of the speeds are the same as the ratios of times. In effect, the document in which Galileo derived the times-squared law from the 'erroneous' principle communicated to Sarpi, though traditionally associated with the 1604 letter, cannot be dated with certainty. It was first published in *Le opere di Galileo Galilei, Edizione Nazionale,* edited by Antonio Favaro (1890–1909). I have quoted this edition as: Galilei, 1890–1909, followed by the Roman numeral of the volume and the page numbers in Arabic numerals. Cf. Galilei, 1890–1909, VIII, pp. 373–74. See Koyré, 1966, pp. 86ff., quotation from p. 96.

[7]By *equimultiple*, or equal multiple, Euclid means multiples of magnitudes according to the same multiplication factor. Cf. Heath's comments on the long-winded philosophical debates among mathematicians concerning the obscure meaning of this definition, in Euclid, 1956, II, pp. 120–129.

[8]Galilei, 1974, p. 149.

His proof proceeds as follows. Let a number m of spaces, equal to space AB, and a number n of spaces, equal to space BC, be taken, respectively, on the left- and right-hand side of AB. Let the same number m of times, equal to time DE, and the same number n of times, equal to time EF, be taken, respectively, on the left- and right-hand side of DE. By recalling Galileo's definition of equable motion, one can assert that: (a) there being in EI as many equal times as there are equal spaces in BG, the whole EI is the time necessary to travel the whole distance BG, and the same goes for the whole time KE and the whole distance BH; (b) if space GB were equal to space BH, then time IE would be equal to time EK. In addition, by remembering Galileo's Axiom I,[9] we can assert that if space GB were greater/smaller than space BH, then time IE would be longer/shorter than time EK. It is therefore true that if $\mathbf{m}AB \gtreqless \mathbf{n}BC$ then $\mathbf{m}DE \gtreqless \mathbf{n}EF$. Thus the sought proportionality follows, i.e., AB is to BC as DE is to EF. (Note that the symbols are not in Galileo's original, which is cast in natural language; I have introduced them arbitrarily to help the reader who may be unfamiliar with this kind of linguistic mathematics.)

Galileo, however, believed that Euclid's definition of sameness of ratios was profoundly obscure and eventually set out to replace it with a new definition. In 1641, he began to dictate a special tract on proportions to the young mathematician, Evangelista Torricelli (1608–1647), who had joined him in Arcetri. The task Galileo had set himself was arduous. At the beginning of his tract, he acknowledged that he would either have to demonstrate the entire fifth book of Euclid's *Elements* with his new definition or prove that he could deduce Euclid's definition of sameness of ratios from his new definition of proportionality. He chose the second strategy. Galileo dictated to Torricelli what he intended to be a new dialogue to be added to *Two new sciences*. Once again, Galileo summoned his three literary characters, made famous in previous books, Salviati, Sagredo, and Simplicio, to discuss the difficulties of Euclid's equal multiples definition. After a brief introduction, in which Sagredo notes that the theory of motion in *Two new sciences* is founded on the definition of equable motion and this in turn is founded on Euclid's obscure definition, Salviati confesses that he too was "shrouded in the same fogs" as Sagredo for some time after studying the fifth book of Euclid's *Elements*. In Galileo's view,

> [...] in order to give a definition of the assumed proportional magnitudes suitable to produce in the mind of a reader some concept of the nature of these proportional magnitudes, we must select one of their properties. Now, the simplest [property] of all is precisely that which is deemed most intelligible even by the average man who has not been introduced to mathematics; Euclid himself has proceeded thus in many places. Remember that he does not say [for example] that the circle is a plane figure within which two intersecting straight lines will produce rectangles such that that which is made of the parts of one line will equal that which is made with the parts of the other, or [that it is a plane figure] within which all quadrilaterals have their opposite angles equal to two right angles. These would have been good definitions, had he spoken thus; but since he knew another property of the circle more intelligible than the preceding, and easier to form a concept of,

[9] 'During the same equable motion, the space completed in a longer time is greater than the space completed in shorter time'. Galilei, 1974, p. 148.

> he did much better to set forth that clearer and more evident property [equidistance from a point] as a definition [. . .].[10]

Instead of the Euclidean definition of proportionality, Galileo then proposes a new definition, which reads as follows:

> When, in order to have the same ratio to the second that the third has to the fourth, the first is neither greater nor less than it need be, then the first is understood to have to the second the same ratio as the third has to the fourth.[11]

Here I cannot articulate the entire demonstrative structure that Galileo elaborates in order to show the equivalence of his new definition with Euclid's. I will present the simplest section, which, in Galileo's view, proves that the Euclidean accord of the equimultiples does indeed follow from his new definition. Note that Euclid's and Galileo's definitions are not restricted to a particular kind of magnitude. They are intended to be valid for all kinds of magnitudes, the only restriction being on the formation of ratios between couples of homogenous magnitudes. Thus we may have a ratio between two degrees of speeds, for instance, but not one between an interval of time and a distance. The definition's independence of the kind of magnitude suggests that in the language of Euclid and Galileo "magnitude" is a placeholder for any quantity whatever. So one does not need to picture a particular type of quantity, such as, for example, a line or a surface. This was obviously very important to the then blind Galileo.

Galileo's proof is articulated in dialogue form. Salviati asks Simplicio whether he thinks that given four proportional magnitudes, say, A, B, C, D, so that the first has to the second the same ratio as the third has to the fourth, then twice the first will have to the second the same ratio as twice the third to the fourth. Simplicio agrees. Thus Salviati argues that the same will occur if we think of taking four, ten, or hundred of the first and third magnitudes. These multiples will have the same ratios to the second and fourth, respectively. This of course will be the case for any multiple of the first and third. Simplicio has no difficulty in conceding this extension ad infinitum. Now Salviati repeats the mathematical thought experiment, by asking Simplicio to consider the consequents, i.e., the second and fourth magnitudes. Whatever the multiples, providing they are equal multiples, the ratios of the first magnitude to the second one multiplied by a number of times will be the same as the third to the fourth multiplied by the same number of times. Then Salviati stretches Simplicio's imagination. He asks him to consider the equal multiple of the first magnitude, the second magnitude, the equal multiple of the third magnitude, and the fourth magnitude. Given that they are in the same ratio, and given that by multiplying the second and the fourth by the same number of times, the ratio will remain the same, as already ascertained, then the equal multiple of the first magnitude and the equal multiple of the second magnitude will be in the same ratio as the equal multiple of the third magnitude and the equal multiple of the fourth

[10] Drake, 1995, p. 424.

[11] Drake, 1995, p. 426.

magnitude. But this, Salviati claims, is exactly Euclid's definition of sameness of ratios.[12]

Note how Galileo simply goes through the sequence of the proof taking care of mentioning magnitudes always according to their ordinal sequence, the first, the second, etc. Even the letters A, B, C, D, follow the alphabetical order, but have no special referential purpose, in the sense that they do not designate either geometrical or physical entities. It is astonishing that Galileo could still pursue this kind of sophisticated analysis of the foundations of Euclidean proportionality, by simply imagining in time his mathematics of samenesses of ratios.

Galileo's concerns with the obscurity of the equimultiple definition of proportionality were probably not only the result of a purely intellectual dissatisfaction with the lack of simplicity that he attributed to this definition, but also the awareness of the difficulties he had encountered in applying it to the physical realm.

It was physics in her youth that guarded the principles in her deepest recesses from the analytical scrutiny of the first modern attempt at the mathematical investigation of nature. It is of course open to debate whether Galileo's new definition succeeds in clarifying the intuition of proportionality. But the fact is that the fundamental principle on which Galilean mathematical physics rests is that of proportionality, and the classical mathematical imagination seems to have a special attachment to, or fondness for, the idea of proportionality.

We might say that through Galileo's entire project of mathematical physics, it is physics in her exuberant youth that launches herself into an audacious exploration of the possibilities of self-scrutiny based on the aesthetic canon of the *imitation of nature*, by means of which an ideal clarity is hopefully afforded by geometrical proportionalities. Galileo theorized the aesthetic canon of imitation by comparing painting and sculpture and claiming the superiority of the former over the latter.[13]

> The contention that sculpture is more admirable than painting because the former has relief while the latter has not is so wrong that, by virtue of this very argument, painting turns out to surpass sculpture in excellence. [. . .] To the sculptors' contention that Nature herself makes the human being by means of sculpture and not by means of painting, I answer that she makes them [the bodies of the sculptors] painted as well as sculpted inasmuch as she both sculpts and colors them; and this redounds to their [the sculptors'] imperfection and is a thing which greatly diminishes the merit of sculpture. *For, the farther removed the means by which one imitates are from the thing to be imitated, the more worthy of wonder the imitation will be*. [. . .] Will we not admire a musician who moves us to sympathy with a lover by representing his sorrows and passions in song much more than if he were to do it by sobs? And this we do because song is a medium not only different from but opposite to the [natural] expression of pain while tears and sobs are similar to it.[14]

Mathematical physics, dressed up in the sensuous garb of a paintress in her splendid youth, like Artemisia Gentileschi (Fig. 4.2), imitates nature with means

[12]Drake, 1995, pp. 427–428.

[13]Cf. Galilei, 1890–1909, XI, pp. 340–343, a famous letter where Galileo expands on art, esthetics and the imitation of nature.

[14]Panofsky, 1954, pp. 35–36, my emphases.

Fig. 4.2 Artemisia Gentileschi. Self-Portrait as the Allegory of Painting (Artemisia Gentileschi (1593 – c. 1656). *Self-Portrait as the Allegory of Painting*. Picture in the public domain. Retrieved from: https://commons.wikimedia.org/wiki/File:Self-portrait_as_the_Allegory_of_Painting_%28La_Pittura%29-Artemisia_Gentileschi.jpg on 4/18/2020). I look at the painting as a symbol of mathematical physics' juvenile dilemma: her staring at nature and yet away from the surface of the painting on which she is working. She looks fixedly with the eyes wide open, spellbound by the magic of curiosity, surprise, horror, bewilderment, admiration, arousal, in ecstatic astonishment, at the light which floods her face and thus makes her gaze dazzled if not totally blind to the things that she is imitating on the canvass. Analogously, in her early stages, physics portrays nature through a mathematical means of representation that is farthest removed from natural things, namely, Euclidean geometry, and which is the most abstract form of human vision. Western philosophy and modern science have been erected on this foundation of abstract vision lending form and figure, namely, the articulation of a world, to a fluid yet earthy matter that secludes within herself the organizational principles that separate order from chaos. That geometry and hence modern mathematical physics is the offspring of this originary form of Life was already grasped by physics in her youth at the time of Galileo's terminal blindness. He left us exacting testimonies of the impossibility of a *geometry of blindness*. On 2nd January 1638, he wrote his most tormented letter on blindness to his friend Elia Diodati. He told him that he had become completely blind about a month earlier. He projected himself forward into darkness with these words: "Now, Sir, think in what a slough of despondency I have fallen, when I realize that…

of imitation that are the furthest possible removed from nature that she has at her disposal at the time, namely, abstract geometrical proportionalities, which are couched in the obscure medium of the Euclidean or Galilean languages of proportionality.

4.4 Senescence

When physics abandons the aesthetic exuberance of youth, intended towards imitating nature in her measurable manifestations, anharmonicity emerges in mathematical-physical Life. The tension intrinsic in human conscious imitation of nature through means of description that are further and further removed from the sensory modes of presentation of nature into human consciousness — that is, through mathematical means that have progressively yet quite decisively left behind the possibilities of visualization afforded by classical geometrical methods — is sharply profiled in the contrast between the ideals of Western musical aesthetics, which remained dominant until the end of the nineteenth century, and physics' attempt at resolving the mystery of the perceptual phenomena of combination tones.

Hermann Helmholtz believed in the possibility of laying the foundations for a mathematical natural philosophy of sound perception that could explain the causal connection between the neurophysiologic basis of hearing in humans and the theory of harmony as the foundation of Western musical aesthetics. In his magnificent treatise on the sensations of tone he asserted: "I was unwilling to separate the physiological investigation from its musical consequences, because the correctness of these consequences must be to the physiologist a verification of the correctness of the physical and physiological views advanced".[15]But his attempt foundered on

Fig. 4.2 (continued) *that* heaven, *that* world, *that* universe, which, with my marvelous observations and geometrical demonstrations, I had enlarged more than a thousand times beyond what had been seen by the sages of all times past, have now shrunk to the space occupied by my own body. The novelty of this fatal event has so far prevented me from accepting the fact with due patience. Assuefaction will only come with time. This transmutation burst its banks in my mind, thus causing a marvelous metamorphosis of thought, purposes, and concepts, about which I can only say little at the moment…" (Galilei 1890–1909, XVII, p. 247.). In response to an inquiry by one of his pupils, he wrote again: "… and if I could regain a less troubled condition, I would explain to you my concept; but since it is a very complex excogitation, or structure, difficult to elucidate, especially with naked words, it being impossible for a blind person to draw a diagram, I am unable to say anything more specific, except that my strategy depends on a proposition by Euclid"(Galilei 1890–1909, XVII, p. 360–1909, XVII, p. 360)

[15]Helmholtz, 1885, p. 5.

the unforeseen difficulties of explicating the mysterious emergence of a diffused anharmonicity in nature.

Senescence brings pathological states of Life. "Störungen des Zusammenklanges" is the heading under which the physics of anharmonicity announces herself. It might be rendered poignantly as "pathologies of harmony".[16] Helmholtz elaborated a physical-mathematical analysis of sound perception in humans based on the intuition that complex sounds that are periodical, that is, in his view, musical sounds par excellence, can be thought of as summations of simple tones, i.e., sinusoidal oscillations in the air pressure. Helmholtz succeeded in combining two strands of research that until then had remained separate, namely, mathematical acoustics and the physiology of hearing. But the impetus to integrate the two came from the anatomical-physiological research of Alfonso Corti. In 1851, Corti's seminal paper announced the discovery of a structure in the mammalian inner ear that would be called in his honor organ of Corti.[17] About a decade later, Helmholtz published the first edition of his celebrated treatise *On the sensations of tone*, in which the theory of the ear as frequency analyzer was fully developed.

Corti's paper was not only a milestone in neurophysiology, it suggested the crucial analogy that Helmholtz would masterfully transform into a full-blown, biomechanical theory of human hearing. To illustrate his findings, Corti drew three floating hammers above the basilar membrane that, he thought, could strike it in response to an external stimulus, thus exciting the nervous terminations which are the expansion of the acoustic nerve within the cochlea. He compared the structure of the underlying portion of the basilar membrane, known as pectinate zone, to piano strings lying very close to one another. Thus, the hammer mechanism of the piano keyboard, the most popular instrument of the Romantic epoch, suggested to Corti the perfect analogy for the mechanics of hearing. Helmholtz refined Corti's explanation by combining it with the theory that sensations of pure tones are elicited by sinusoidal pressure waves. In effect, Helmholtz postulated that the mechanism for frequency analysis in the organ of Corti must be resonance. In other words, the ear performs a Fourier analysis of the incoming external stimulus and resolves its constituent pure tones. However, as he conceded in the fourth and final revision of his book, the parts of the ear that resonate in response to the pure tones that make up the external stimulus could not for the time being be identified with certainty.[18] We hear musical sounds when we succeed in resolving the complex sounds in their harmonic constituents, i.e., the pure tones whose frequencies orderly increase following the series of the integer multiples of the frequency of the fundamental tone that corresponds to the periodicity of the sound.

There was one mind-boggling phenomenon that had long captured the imagination of physics and music. It was the appearance in perception of combination tones whenever two powerfully enough tones were sounded together such as two violin

[16] Helmholtz, 1877, pp. 251 ff.

[17] Corti, 1851.

[18] Helmholtz, 1885, p. 145.

or cello notes. For instance, if two notes are sounded at, say, 400 Hz and 300 Hz, then a third sound is easily perceived with a frequency corresponding to 100 Hz. (Nowadays, this simple experiment can be carried out with any laptop computer capable of generating sinusoidal sounds, which give rise to more clearly discernible combination tones, and headphones). When these two sounds are sounded together their summation sound has periodicity equal to 100 Hz (difference of the two frequencies) even though there is no sinusoidal sound being sounded at 100 Hz. In fact a periodic signal can be periodic with period T without having a corresponding sinusoidal component with periodicity T (as Fourier analysis teaches us).

The phenomenon of combination tones reveals the irony of physics in her senescence. Theory, according to Helmholtz, has it that we can only hear musical sounds whose perceptions correspond to aggregates of pure, sinusoidal tones. How come, then, that we hear combination tones to which no real simple tone corresponds, that is, we hear tones that do not have a correlate sinusoidal pressure wave? Physics built a monument to Western music by finally showing that the fundamental consonances recognized since ancient time such as the octave and the fifth are a manifestation of the simple progression of integers that regulates the succession of harmonics, whose structure determines salient characteristics of music perception such as pitch and the great variety in the timbre of instruments. Corti's discoveries demonstrated the existence of the biological instrument by which such variety could be appreciated in humans and other mammals. Yet, in the last analysis, combination tones emerge in the processes of perception of harmony, *Störungen des Zusammenklanges*, that not only violate the exact correlation that senescent physics requires between perception of musical sounds and the existence of sinusoidal pressure waves, but, as we shall presently see, necessitate postulating a source of anharmonic sounds in the anatomic structure of the human ear.

Since there is no corresponding pressure wave for combination tones of the type described in the above example, their emergence in consciousness must be caused, according to Helmholtz, by a non-linear mechanism in the anatomical structures of the ear that produces such tones. The anharmonic oscillator (see below for its mathematical description) that eventually became prominent in the posthumous maturity of physics, spanning the decades at the turn of the twentieth century to the present day, actually originated in the irony of physics in her senescence.

The non-linear mechanism modeled by the anharmonic oscillator could predict the emergence of the combination tone equal to 100 Hz in above example, but also obligatorily predicted the emergence of a second tone at 700 Hz (the sum of the two frequencies) whose first discovery and theoretical explanation Helmholtz proudly claimed to his credit. He also insisted that he had on many occasions satisfied himself of its existence in perception in certain experiments in which powerful tones were generated by means of large tuning forks. As irony requires these combinational summation tones have since remained a figment of the senescent imagination playing havoc in the mind of physicists and acousticians. But there is more to this beautiful story.

The postulation of the non-linear mechanism in the ear that was responsible for producing the physical correlates of combination tones that were missing in the external environment ironically placed at the centre of the hearing process in humans (and perhaps in other mammals as well) an ineliminable source of anharmonicity. When sounds are loud enough to excite vibrations with sufficiently large amplitudes, combination tones are generated. But then this implies that the normal operating regime of the ear is such that its structural non-linearities will always be effective. Contrary to Helmholtz's intuition, complex mechanical structures such as tuning forks and the chain of ossicles linked to the eardrum in the human ear respond to external forces with a large array of natural frequencies that in general are not ordered in the progression of the law of harmonics governed by the series of the integers. If you strike them powerfully, they will oscillate but not necessarily following the law of a periodic motion. In other words, they may produce sounds that Helmholtz called 'noises' because they are not strictly periodic and hence are not reducible to series of pure tones. They are not musical.

But the late appearance of the anharmonic oscillator in the Life of physics posed another fascinating riddle. The question became: How to read and how to solve the non-linear equation describing this strange object of the senescent metabolism of physics? Physics has played a game of irony with Western harmony, and still resists revealing her totality, she resists self-scrutiny through human consciousness.[19]

Helmholtz wrote the following equation for the anharmonic oscillator modeling the asymmetric structures in the human ear.

$$-m \cdot \frac{d^2x}{dt^2} = a \cdot x + b \cdot x^2 + f \cdot \sin(p \cdot t) + g \cdot \sin(q \cdot t + c) \tag{4.1}$$

To the restorative force proportional to displacement x (the linear component), he added the quadratic term with coefficient b (the non-linear component), assuming that he could justify its insertion in the modeling equation because of the lack of symmetry in the attachments of the chain of ossicles to the eardrum. Equation (4.1) also contains on the right hand side two external forces with frequencies p and q which represent the pure tones exciting the ear. Helmholtz's strategy invites speculation. Why did he not approach it by writing a Fourier expansion of the periodic solution and looking for conditions to determine the coefficients? He was well versed in Fourier analysis and, as we have seen, the results of Fourier analysis were instrumental in affording him the analogy for theorizing the whole hearing process in the organ of Corti. But he denied himself this approach. Instead he pursued another strategy, which does not require positing certain known functions as components of the solution. He assumes that the solution x(t) could be represented by another type of expansion in terms of a parameter ε as follows.

[19]A 'general' solution to a special case of the anharmonic oscillator that Helmholtz had investigated was found recently. See Rand, 1990, where a discussion of the special case is presented which, however, does not clarify what is meant by 'general'.

$$\begin{aligned} x &= \varepsilon \cdot x_1 + \varepsilon^2 \cdot x_2 + \varepsilon^3 \cdot x_3 + \ldots \\ f &= \varepsilon \cdot f_1 \\ g &= \varepsilon \cdot g_1 \end{aligned} \tag{4.2}$$

Equations (4.2) are mysterious. Functions x_1, x_2, x_3, and so on, are unknown. But also the parameter ε is unknown. The strategy resembles perturbation methods in which a parameter is introduced small enough not to disrupt the assumed solution and change its nature dramatically.[20]But none of the functions in the above expansion represents a solution to which a perturbation might be superimposed. Then Helmholtz plugged Eqs. (4.2) into Eq. (4.1) and determined the following conditions by forcing the terms in equal powers of the parameter ε to be independently equal to zero. Up the to the cube of the parameter ε, we have:

$$\begin{aligned} a \cdot x_1 + m \cdot \frac{d^2 x_1}{dt^2} &= -f_1 \cdot \sin(p \cdot t) - g_1 \cdot \sin(q \cdot t + c) \\ a \cdot x_2 + m \cdot \frac{d^2 x_2}{dt^2} &= -b \cdot x_1^2 \\ a \cdot x_3 + m \cdot \frac{d^2 x_3}{dt^2} &= -2 \cdot b \cdot x_1 \cdot x_2 \end{aligned} \tag{4.3}$$

and so on. Equation (4.3) in effect represent an infinite series of harmonic oscillators that cascade into each other. The first one is excited by the external pure tones. The second is excited by the square of the output of the first one. The third by the product of the outputs of the first two. If, for instance, one goes on to calculate the terms up to the fourth and fifth power of the parameter ε, a pattern appears in which the input to the oscillator of N power is given by a summation of products of previous outputs, each term of which has indexes whose sum is equal to N (for example, the third term in (4.3) receives as input the product $x_1 x_2$ where the indexes add up to 3). However, once the solutions to Eqs. (4.3) have been determined, one has still to perform the addition required in (4.2) by multiplying each component by the corresponding power of the unknown parameter ε. In fact Helmholtz does not at all try to determine the parameter, and by a leap of faith draws his conclusions based on an observation of the form of the solution to Eqs. (4.3). By carrying out the remaining trivial manipulations (introducing trigonometric identities), he finds that the second oscillator in (4.3) responds to the external pure tones (exciting the first oscillator) by producing two pure tones having frequencies (p + q) and (p−q), namely, the expected frequencies for combination tones whose exciting pure tones have frequencies p and q. It was, then, by assuming that the solution to the

[20]Perturbation methods came of age in the nineteenth century, rising to prominence after the publication of Lagrange's second edition of his *Mècanique Aanlitique* (1814), and particularly in an effort to determine the moon's exact orbit. "Lagrange was imagining the planet or satellite as moving at each instant in an ellipse characterized by its six orbital elements, with the elements changing from instant to instant due to perturbation. [...] Two simultaneous processes had to be taken into account: the continuous change in shape and orientation of the instantaneous elliptical orbit in which the perturbed body was conceived to be traveling, and the body's motion along this protean orbit" (Wilson, 2010, p. 17).

anharmonic oscillator has the peculiar form (4.2) that Helmholtz determined the existence of a phantom combination tone of frequency (p + q), and hence proceeded to investigate it experimentally in order to satisfy himself of its perceptual reality.

However, the solution (4.2) to the original eq. (4.1) can be read and interpreted in multiple interesting ways. Suppose, for the sake of argument, that the external pure tones are set to zero, namely, both f(t) and g(t) are zero. Then, the evolution of the dynamic system is determined by its natural frequencies of oscillation. The problem still remains of fixing the parameter ε, though. Each solution in (4.3) introduces two arbitrary constants. Energetic considerations may be brought to bear on this problem in order to help us think about how to determine the arbitrary constants and the parameter. The polynomial expansion in the parameter ε, up to any order N, may in principle have couples of conjugate complex solutions for the parameter. So, in summary, the approach followed by Helmholtz may be interpreted as an attempt to reduce the anharmonic oscillator to a constellation of virtual harmonic oscillators some of which may not be real.

In regard to the perception of sound physics undertook a fantastic excursion into her unfathomable interiority. The emergence of combination tones in perceptual consciousness seems to awaken awareness of the hidden presence in the world of virtual oscillators whose form of existence may not be described by quantities symbolized by real numbers but must be articulated in mathematical language by factoring in imaginary numbers.

The poetic imagination of physics, which creates the marshaling image of the chain of harmonic oscillators cascading into each other, clashes with the more powerful yet obscure virtue of the mathematical imagination. The latter conjures into existence imaginary numbers that resist being pictured in the mind. Girolamo Cardano, the magician, mathematician, and astrologer who invented imaginary numbers in the sixteenth century, resorted to the language of imagination to adumbrate the nature of the unknown *number* (if such it is) obtained by the square root of a negative number.[21] He spoke of the sophistical nature of the imaginary number he had excogitated to manipulate square roots of negative numbers. However, he also cautioned the reader that it would be vain to pursue the nature of this being, or research the operations to perform with it by stretching the human imagination. Yet modern mathematics and mathematical physics are unthinkable without the imaginary numbers, that is, without that monstrous stretching of the imagination feared by Cardano.

4.5 Posthumous Maturity

Senescence metamorphoses into posthumous maturity, a stage of the Life of physics during which works of classical mechanics are translated into novel physical-

[21] Cardano, 1570, p. 131.

mathematical languages. More precisely, I refer to the "posthumous maturity" of physics with the word *Nachreife*. I borrow this precise item of linguistic description from an intuition by Walter Benjamin. He theorized the translation process between two natural languages insightfully. His text is important for my purposes. It is worth quoting the original in which he speaks of *Nachreife* extensively.[22]However, I avoid translating Benjamin's difficult passage literally, and I prefer to interpret his thoughts somewhat more freely as follows.

There is a posthumous maturity of the words that appear to have been fixed such as in works of art or creative linguistic enterprises. A certain tendency that was the driving force of the poetic language of an author at a given point in time might exhaust itself at a later time, while novel tendencies may be born from the text during its after life. What was new and original may become old and worn out, or sound archaic. It is a mistake to look for the essence of these processes in the psychological attitudes of posterity, in other words, in human subjectivity. What must be scrutinized is the Life of language and its works. Only thus is the error avoided of mistaking the essence of the historical process for the occasional motivations and circumstances of its temporal declension. Even if one assumes that the author's signature on their work should be taken as their final word about its meaning, the received theory of translation (according to which, translation must strive for the literal transmission of the meaning as accurately as possible) cannot be accepted. For, as the tonality and meaning of great poetic works radically change over the centuries, so the target language in which a linguistic work will be rendered, namely, the mother tongue of the translator undergoes dramatic changes over the course of time. Poetic words endure in their original language while even the best translations are destined to be metabolized in the Life of the language of the translator, and thus finally to perish. Far from being the plain equation of two dead languages, the translation process carries the ultimate and delicate responsibility of remaining sensitive to the needs of the posthumous maturity [*Nachreife*] of the

[22]"Es gibt eine Nachreife auch der festgelegten Worte. Was zur Zeit eines Autors Tendenz seiner dichterischen Sprache gewesen sein mag, kann später erledigt sein, immanente Tendenzen vermögen neu aus dem Geformten sich zu erheben. Was damals jung, kann später abgebraucht, was damals gebräuchlich, später archaisch klingen. Das Wesentliche solcher Wandlungen wie auch der ebenso ständigen des Sinnes in der Subjektivität der Nachgeborenen statt im eigensten Leben der Sprache und ihrer Werke zu suchen, hieße — zugestanden selbst den krudesten Psychologismus — Grund und Wesen einer Sache verwechseln, strenger gesagt aber, einen der gewaltigsten und fruchtbarsten historischen Prozesse aus Unkraft des Denkens leugnen. Und wollte man auch des Autors letzten Federstrich zum Gnadenstoß des Werkes machen, es würde jene tote Theorie der Übersetzung doch nicht retten. Denn wie Ton und Bedeutung der großen Dichtungen mit den Jahrhunderten sich völlig wandeln, so wandelt sich auch die Muttersprache des Übersetzers. Ja, während das Dichterwort in der seinigen überdauert, ist auch die größte Übersetzung bestimmt in das Wachstum ihrer Sprache ein-, in der erneuten unterzugehen. So weit ist sie entfernt, von zwei erstorbenen Sprachen die taube Gleichung zu sein, daß gerade unter allen Formen ihr als Eigenstes es zufällt, auf jene Nachreife des fremden Wortes, auf die Wehen des eigenen zu merken." (Benjamin, 1972, pp. 12–13.)

foreign language, while at the same time caring for the suffering of its own mother tongue.

I already noticed the uncannily emotional connotation that Husserl's question evokes about the obscurity of the fundamental principles. There is a tormented suffering about this obscurity, the presence of a dark zone that prevents insight, a suffering that could be qualified as non-human in so far as it is a kind of suffering that manifests itself in the pain of the mother tongue. This pain is due to a process of non-destructive antagonism between two creatures whose Life is symbiotically related for a certain period of time, until the final moment of separation, when the drama of birth — or, to avoid the biological metaphor, shall we say, 'disarticulation'— becomes inevitable, and the final event that terminates the antagonism must occur correctly, according to natural norms that we cannot entirely envisage, for the sake of the healthy continuation of both Lives.

In Heisenberg's ground-breaking translation of the classical linguistic framework that profiled the identity of the anharmonic oscillator into the younger mother tongue of quantum mechanics, the juvenile language of quanta and spectral lines, the Life of physics enters the stage of posthumous maturity. There is no better description of the emergence of this epoch-making transformation in the language of physicists at the turn of the twentieth century than the summary of the *status quaestionis*, concerning physic's understanding of the structure of the atom, given by Arnold Sommerfeld in his fundamental textbook, *Atombau und Spektrallinien*.

> After the discovery of spectral-analysis no physicist could doubt that the problem of the atom would be solved when we had learned to understand the language of spectra. So manifold was the enormous amount of material that had been accumulated in sixty years of spectroscopic research that it seemed at first beyond the possibility of disentanglement. [...] What we are listening to nowadays, when the language of spectra is spoken, is the *music of the spheres* of the atom, the consonances of simple ratios of integers, in a crescendo of order and harmony in diversity. [...] All the laws of spectral lines bear the signatures of the whole numbers, and atomic theorizing flows from the theory of quanta. This fledgling theory is the hidden *organon* on which nature plays the music of the spectra, and on whose rhythmic patterns she designed the architecture of the atom and the nucleus.[23]

In classical theory a given magnitude x(t) can be represented by a totality of magnitudes,

$$A_{\alpha}(n) \cdot e^{i\omega(n)\alpha t} \tag{4.4}$$

whose meaning, according as the motion is periodic or not, is explicated by the following sum or integral,

[23] Freely adapted from Sommerfeld 1923, p. viii. Cf. the original text, in Sommerfeld, 1921, pp. vii–viii.

$$x(n,t) = \sum_{-\infty}^{+\infty} A_\alpha(n) \cdot e^{i\omega(n)\alpha t}$$
$$\text{or} \tag{4.5}$$
$$x(n,t) = \int_{-\infty}^{+\infty} A_\alpha(n) \cdot e^{i\omega(n)\alpha t} d\alpha,$$

but cannot be translated into the language of quanta and spectral lines in a straightforward and unique manner.[24]Indeed, a unification of the corresponding quantum theoretical magnitudes appears to be arbitrary, in view of the equal importance of the magnitudes n, n-α. However, the totality of the following magnitudes,

$$X(n, n-\alpha, t) = A_\alpha(n, n-\alpha) \cdot e^{i\omega(n,n-\alpha)t} \tag{4.6}$$

can be regarded as representing the magnitude x(t). The meaning of the square of x(t) will be explicated, based on the combination rules for frequencies, which are stipulated in the grammar of the language of spectral lines, as follows:

$$Y(n, n-\beta) \cdot e^{i\omega(n,n-\beta)t} = \sum_{-\infty}^{+\infty} A(n, n-\alpha) A(n-\alpha, n-\beta) e^{i\omega(n,n-\beta)t}$$
$$\text{or}$$
$$Y(n, n-\beta) \cdot e^{i\omega(n,n-\beta)t} = \int_{-\infty}^{+\infty} A(n, n-\alpha) A(n-\alpha, n-\beta) e^{i\omega(n,n-\beta)t} d\alpha. \tag{4.7}$$

In Heisenberg's 1925 paper, we notice quite readily that the translation of classical kinematics and mechanics (more precisely, of classical kinematical and mechanical relationships) into the new language of quanta and spectral lines proceeds literally by importing the exact syntactical structures by which statements about the composition of magnitudes are meaningfully formulated in the classical language of Fourier analysis. What the translator is concerned with is the words that must be chosen properly as adequate representatives of the words spoken in the classical language. In this case, we see that the freedom of the translator is immense in that he moves in his own proper element when he finally makes the choice of words. It is not the syntax that can be negotiated as this would violate the received wisdom about the value of literalness in translations. The latter, which is not disputed, can only be achieved in terms of an obsessive accuracy as to the form, namely, as to the syntax of language. Meaning is not preserved in translations of great works or art or science, though this luminous insight worked out by Benjamin runs counter to the widespread delusion that translation always operates by preserving meaning as accurately as possible. The great freedom enjoyed by the translator in determining the words to be used in the target language of the translation, in this case the young language of quanta and spectral lines

[24]I will loosely follow the exposition in Heisenberg, 1925.

indicated by Sommerfeld as the secret *organon* of nature, reflects the requirement that fidelity of translation must be a guiding value in the work of the translator, but a value that aims at the harmonization of the languages in which the life of the works of science, in our case physics, endures. The crucial stage in the entire, 'magical' paper that Heisenberg published in 1925[25] must be regarded as being the choice of words expressed in (4.6). The novel, indeed 'magical' word that in Heisenberg's translation into the language of quanta and spectral lines corresponds to the classical word $x(n,t) = \sum_{-\infty}^{+\infty} A_\alpha(n) \cdot e^{i\omega(n)\alpha t}$ is spoken as $X(n, n-\alpha, t) = A_\alpha(n, n-\alpha) \cdot e^{i\omega(n,n-\alpha)t}$, and with this new word felicitously chosen, Heisenberg proceeds to translate literally the classical Fourier statement, $\left\{x(n,t) = \sum_{-\infty}^{+\infty} A_\alpha(n) \cdot e^{i\omega(n)\alpha t}\right\}^2$, into the enigmatic quantum theoretical statement, $Y(n, n-\beta) \cdot e^{i\omega(n,n-\beta)t} = \sum_{-\infty}^{+\infty} A(n, n-\alpha)\, A(n-\alpha, n-\beta)\, e^{i\omega(n,n-\beta)t}$, where a mysterious, double index transformation is indicated in the beautiful symmetry of the same index, or phonetic structure, (n-α), that is nested twice in between the two extremes.

In Heisenberg's glorious translation there is no question of preserving the meaning of the original sentence. The epoch of glory of the classical framework, which is this posthumous maturity, justifies and indeed calls for the translation, for a departure towards the uncharted territory of unforeseen meaning. The originary word, x(t), represents a magnitude flowing in the course of time (a survival of Newton's fluxional magnitudes cast in the language of the Leibnizian calculus), an isolated individual manifesting itself at any point in time, a temporal being whose evolution follows the flow of time. The novel word that translates the originary x(t) in the target language represents a totality of magnitudes, not in the sense of an assemblage of individual beings, i.e., a community of individuals, which can be referred to by naming the individual magnitudes, but as a novel being whose relation to time and ontological status remain undecided because entrapped (at least for the time being?) in the articulation of the letters and syllables that form the sonorous architecture of the 'magical' formula uttered by physics playing her *organon* in (4.6). Indeed, we may invoke a musical analogy to scrutinize the essence and the form of meaning of the enigmatic phrase $X(n, n-\alpha, t) = A_\alpha(n, n-\alpha) \cdot e^{i\omega(n,n-\alpha)t}$. When in the classical language we assemble the individual words that we call Fourier components into the sentence, $x(n,t) = \sum_{-\infty}^{+\infty} A_\alpha(n) \cdot e^{i\omega(n)\alpha t}$, we obtain a periodic form that, once it has been interpreted in terms of sonorous energy and actually performed, will elicit in human consciousness the perception of a tone whose pitch and timbre and other salient qualities are adumbrated in, though not

[25] Aitchison, MacManus, Snyder 2004. The word 'magical' emphasized by Aitchison, MacManus, and Snyder is quite appropriate here, I take it literally.

straightforwardly correlated with, the structure of the harmonics and their temporal patterns. When physics in her posthumous maturity makes a statement such as $X(n, n - \alpha, t) = A_{\alpha}(n, n - \alpha) \cdot e^{i\omega(n, n - \alpha)t}$, we do not find in the new language of the spectra and the series of spectral lines an immediate correlate for the classical $x(n, t) = \sum_{-\infty}^{+\infty} A_{\alpha}(n) \cdot e^{i\omega(n)\alpha t}$ because we do not have a new rule (or a set of rules) of composition of the multiplicity into the unity of an individual, by the interpretation of which we might generate the individual utterance that once performed will elicit in our consciousness a corresponding perception in any of the sensory modalities whatever. Such rules of composition remain obscure, absentee landlords of a Kafkian opera house whose musicians and singers have all turned into dumbstruck mimes.

What the novel language of quanta and line spectra seems to be lacking is a principle of individuation. Classical scholastic philosophy produced a motley array of solutions to the problem of individuation, a veritable philosophical riddle, but in essence failed to satisfactorily determine the rules for individuation.[26]This is, to my mind, one profound reason, perhaps the most profound, for the depletion of the scholastic tradition in early modern philosophy, and for its eclipse during the seventeenth century. The decline into obscurity was brought about not so much by the emergence of the scientific revolution and its philosophical side effects, but by an internal dynamics that enervated and eventually exhausted scholasticism itself, which in the end failed to survive the extreme tension between being and individuation. The same situation seems to affect the epoch of posthumous maturity in the Life of physics, where an analogous tension is all too evident between the claim to being of novel entities, or modes of being-in-the-world, such as are spoken of in $X(n, n - \alpha, t) = A_{\alpha}(n, n - \alpha) \cdot e^{i\omega(n, n - \alpha)t}$, and physics' dramatic failure to produce within the current framework of the quantum language a cogent grammar for forming individuated sentences, i.e., sentences that can be spoken of individual existences and their diffracted references in the quantum language.

To extenuate the musical analogy, I might argue that in the new language of quanta and spectral lines, a music can be played but not heard by the human neurophysiologic apparatus, a silent music, though not necessarily silent for non human hearing, or perhaps for an altered state of human consciousness in which not only hearing but the other sensory modalities and the language centers are conditioned in the presence of opportune neurochemicals. There is, however, one more possibility that the hearing analogy suggests. The neurophysiology of mammal hearing is evolutionarily more recent than its visual counterpart. The sense of hearing in humans is a young system on the evolutionary scale of biological life. The entire history of Western thought is dominated by categories and concepts that reflect the predominance of vision and ocular demonstration. The eye grazes the surface of things like grass. But the internals of things cannot be reached by vision except under certain instrumental circumstances. Hearing has not enjoyed

---

[26]Gracia, 1984.

sufficient time, compared to vision, to integrate itself into the conceptual system of Western philosophy and science. It is conceivable that a novel way of thinking, indeed, the birth of a novel form of scientific thinking and hence of physics, will have to be accompanied by novel transformations in thought induced by a stronger influence of the other senses and particularly hearing. Mathematics itself will have to respond to the emergence of hearing into the conceptual system of scientific knowledge. We still have no choice but describe sound events in terms of 'phenomena', the very word that has been inherited into European vernaculars from the Greek philosophical vocabulary, and which in essence means 'the things that manifest themselves to vision'. Helmholtz's dilemma of combination tones appears as a particular case of the phenomenon known to psychoacousticians as the paradox of the 'missing fundamental'. Indeed, modern technology makes it quite possible to surgically remove from sounds their fundamental frequency and/or any of the harmonics, and thus produce artificial sounds that manifest themselves to consciousness as if they had lost their 'ground', or possibly their 'identity'. This depersonalization is due to the fact the fundamental and the lowest harmonics play a decisive role in human recognition and categorization of sounds. Yet these depersonalized sounds, whose identity, or individuation, has been weakened, remain nevertheless perfectly perceivable by the sense of hearing. They are not produced by traditional musical sounds or other natural sounds. They offer an unprecedented experience made possible by electronic technologies. It is only a crude analogy, and it is obvious that it can only be discussed in the context of traditional conceptual, namely, visual language. But it invites original speculation as to how a future language of science might sound like. For the time being we still face the dilemma of the obscurity of nature, this time transposed from the modality of vision to the modality of hearing, in so far as a straightforward sonorous interpretation of $X(n, n - \alpha, t) = A_{\alpha}(n, n - \alpha) \cdot e^{i\omega(n, n - \alpha)t}$ is not possible, whence the silence that nature still imposes on her and our self-scrutiny. This pessimistic view is consistent with Sommerfeld's imagery of a music of the spheres within the atom, a classical form of the human imagination that in the Renaissance and later on became ever more popular and was named *musica mundana*, or *universalis*, or *divina*, and whose essence consisted in knowing the order of the whole cosmos, its divine structure, or ratio, or proportion, for, according to this doctrine, the order of all singularities existing in the cosmos is nothing but a divine concert, a sacred melody, yet a music that is not perceptible to the mortal ears of fallen humans.

Rather than by preserving the clarity of meaning, then, we must realize that translation seems to operate like a weird metabolism of unknown origin, or "a hermeneutical transformation",[27] by transforming the unfathomable, the obscure —

[27] "He [Heisenberg] proposed to replace the kinematical framework LN of classical mechanics by a new quantum theoretic framework, let us call it LQ, which would fulfill the five conditions of the relativistic model. Condition Hi is satisfied by that part of ordinary and scientific language LP which is neutral to the transposition from LN to LQ and includes, therefore, the language of electromagnetic theory as well as the language of the manifest image of the world. Conditions Hii and Hiii represent different aspects of Bohr's Correspondence Principle. Condition Hiv states that

the secret, hidden score by reading from which nature plays the *organon* of the atomic spectral lines —, and thus by jealously keeping the secret that guards the fundamental principles from our indiscreet, direct gaze.

The translation work of physics in the phase of posthumous maturity included also classical mechanics (i.e., dynamics), as announced in the title of Heisenberg's paper, and we find in it an analogous process of hermeneutical transformation regarding classical dynamical relations. Particularly, Heisenberg focused on the challenge of translating into the new language the classical equation of the anharmonic oscillator, which he stated as follows.

$$\ddot{x} + \omega_0^2 x + \lambda x^2 = 0 \tag{4.8}$$

It is the same archetypal object investigated by Helmholtz in the context of classical human psychoacoustics and combination tones. It brought Helmholtz into a dramatic confrontation with the uncanny in nature. Combination tones afforded him the puzzle of perfectly perceivable sounds that are deprived of energy at the fundamental frequency (no harmonic in the spectrum is present at the fundamental frequency), and hence appear to violate his dictum that human hearing works by perceiving the individual harmonic components of musical sounds and by unconsciously compounding them into salient qualities such as pitch (directly correlated to the fundament frequency, according to Helmholtz) and timbre, and that when hearing fails to do so it is because it finds itself in the presence of those weird, non-musical (non periodic) sounds that Helmholtz calls noises. The archetypal anharmonic oscillator performed the same service to the quantum physicists, bringing them into confrontation with the same uncanniness, that is, the obliging persistence of the syntactical classical Newtonian form, expressed in the equation of motion (4.8), and yet its being at the same time devoid of individuated existence once translated into the new language. Painful awareness of the dilemma is symptomatically signaled in the Heisenberg paper by the puzzling event that the statement of a quantum correlate for (4.8) remains wonderfully conspicuous for its absence throughout the entire paper. The absence has been so conspicuous that eventually the need was felt for reinstating its glorious presence posthumously. This was courageously done in a recent paper, almost a century after Heisenberg, as if to heed a call and heal a menacing warning in physics' collective unconscious to the effect that the pain due to psychic tension and unresolved conflict should not be repressed.[28]The translation is the following, but note that the array

LQ will contain 'only relations between quantities which are observable in principle' [Heisenberg 1925, p. 879]. Condition Hv implies that a semantical re-interpretation of the variables — a hermeneutical transformation — accompanies the transposition from LN to LQ." (Heelan, 2016, p. 30).

[28]Aitchison, MacManus, Snyder 2004. Puzzled by the mystery of the recursive formulae for a quantum correlate of (4.8), which had been put forward by Heisenberg without any hint of the processes by which they became present to his consciousness, the authors reinvented the calculation by which, according to them, Heisenberg must have discovered them.

$X(n, n-\alpha, t)$ is not plugged into (4.8) in its general form and the equation simply rewritten. An assumption is also made about its specific periodic form that must be $X(n, n-\alpha, t) = A_\alpha(n, n-\alpha) \cdot e^{i\omega(n, n-\alpha)t}$. Then we have:

$$\left[-\omega^2(n, n-\alpha) + \omega_0^2\right] A(n, n-\alpha) \cdot e^{i\omega(n,n-\alpha)t} + \lambda \sum_\beta A(n, n-\beta)\, A(n-\beta, n-\alpha) \cdot e^{i\omega(n,n-\alpha)t} = 0 \tag{4.9}$$

An equation like (4.9) was never published by Heisenberg, as already noted, but was given birth recently by physicists Aitchison, MacManus, and Snyder in order to demonstrate that the recursive formulas given by Heisenberg for the quantum case of the anharmonic oscillator could in principle be obtained by a perturbative approach very analogous to that which Heisenberg himself had proposed to solve the classical case (4.8).[29]Clearly the translators were still motivated by a strong interest in the correct choice of the quantum words that needed to be employed in the translation while respecting the absolute literalness imposed by the Newtonian syntax of Eq. (4.8).

The array layout of (4.9), especially when written down in more explicit matrix symbolism, is part and parcel with the extended, two-dimensional surface on which the ocular thinking that has presided over the originary formation of Western philosophical and scientific categories grazes in search for the individuation criterion that appears to have gone into hiding during the process of rendering the classical point-like existence adumbrated in (4.8) into the language of the music of the atomic spheres. The persuasiveness by which the periodic form of the single components of the array $X(n, n-\alpha, t) = A_\alpha(n, n-\alpha) \cdot e^{i\omega(n, n-\alpha)t}$ imposes itself to the attention of the physicists speaks volumes about physics' irony. She obscures the fundamental principles of individuation that beguile the imagination of human consciousness and yet lures the human being into ghost images of individuated existence twinkling from the inside of the array of multiple frequencies that emerge into the Life of physics in the epoch of posthumous maturity.

[29]Aitchison, MacManus, Snyder 2004, pp. 1372–1372. Already Patrick Heelan, in 1970, had written the same form for the quantum anharmonic oscillator by plugging the array $X(n, n-\alpha, t)$ into (4.8), but he did not pursue the question of how Heisenberg had arrived at the recursive formulas derived from the application of perturbation theory to the quantum case. See Heelan, 2016, pp. 15, 29–32.

4.6 Conclusion

"... the usual feeling that to consider the universe as a whole is at least immodest, if not blasphemous" is a religious sentiment expressed by physicists J. S. Bell and M. Nauenberg, subservient to the dogma that speaking of the wholeness of the universe is uttering a blasphemy.[30]They concluded that what "is much more likely is that the new way of seeing things will involve an imaginative leap that astonishes us. In any case it seems that the quantum mechanical description will be superseded. In this it is like all theories made by man. But to an unusual extent its ultimate fate is apparent in its internal structure. It carries in itself the seeds of its own destruction".[31]

I ask the question why physics carries in herself the seeds of her own destruction, thus bringing to fruition the resistance to clarity that Husserl had exposed in the question of the obscurity of the fundamental principles of mathematical physics.

More pressingly, I feel the urgency of the question of the blasphemous nature of physics. Blasphemy is an utterance or the harboring of a thought that diminishes God's absolute goodness. It is both conceptual and emotional. When it is manifested in thought, it is a cognitive blasphemy, otherwise it is linguistic. Physics must have recognized the morally evil character of her own existence. She invokes her own self-destruction. Indeed blasphemy is a mortal sin as it removes the blasphemous agent at an infinite distance from the principle of Life, namely, divine goodness.

Husserl's question seems to be exalted in this dénouement, and to metamorphose, like Kafka's hero, inexplicably, into the question of the hidden blasphemous nature of physics. Physics cannot survive her own self-reflective scrutiny, and most likely the imagination of physics, insofar as form of Life, will destroy her while giving birth to an unprecedented and yet unimaginable form of Life.

Acknowledgments I am grateful to my colleague John Norton for many casual conversations, failed jokes, and insightful platitudes on the parasitic nature of quantum mechanics.

References

Aitchison, I. J. R., MacManus, D. A., & Snyder, T. M. (2004). Understanding Heisenberg's "magical" paper of July 1925: A new look at the calculational details. *American Journal of Physics, 72*, 1370–1379.

Bell, J. S. (1987). *Speakable and unspeakable in quantum mechanics*. Cambridge, UK: Cambridge University Press.

Bell, J. S. & Nauenberg, M. (1987). The moral aspect of quantum mechanics. In Bell 1987 (pp. 22–28).

Benjamin, W. (1972). Die Aufgabe des Übersetzers. In: *Gesammelte Schriften* (pp. 9–21). Band. IV/1, Frankfurt am Main, Germany: Suhrkamp.

[30]Bell & Nauenberg, 1987, p. 27.

[31]Bell & Nauenberg, 1987, p. 27.

Cardano, G. (1570). *Opus novum de proportionibus numerorum, motuum, ponderum, sonorum ... praeterea Artis magnae, sive de regulis algebraicis liber unus* ... Basel, Switzerland: Ex officina Henricpetrina.

Corti, A. (1851). Recherches sur l'organe de l'ouïe. *Zeitschrift für Wissenschaftliche Zoologie, 3*, 109–169.

Croce, B. (1937). *La storia come pensiero e come azione* (pp. 1–35). XXXV: La critica.

Croce, B. (1941). *History as the story of liberty* (S. Sprigge, Trans.). London: George Allen and Unwin.

Drake, S. (1995). *Galileo at work. His scientific biography*. New York: Dover. (1st ed. Chicago: The University of Chicago Press, 1978).

Euclid. (1956). *The thirteen books of the elements*. Translated by Sir Thomas Heath (2nd ed). 3 vols. New York: Dover Publications.

Galilei, G. (1890–1909). *Le Opere di Galileo Galilei, Edizione Nazionale*, edited by Antonio Favaro. 20 vols. Firenze, Italy: Barbèra.

Galilei, G. (1974). Two new sciences. *Including centres of gravity and force of percussion* (S. Drake, Ed & Trans.). Madison, WI: The University of Wisconsin Press.

Gracia, J. J. E. (1984). *Individuation in scholasticism: The later middle ages and the counter-reformation, 1150–1650*. New York: SUNY Press.

Heelan, P. A. (2016). The observable. In *Heisenberg's philosophy of quantum mechanics*. New York: Peter Lang.

Heisenberg, W. (1925, December). Über quantentheoretische Umdeutung kinemathischer und mechanischer Beziehungen. *Zeitschrift für Physik, 33*(1), 879–893.

Helmholtz, H. (1877). *Die Lehre von den Tonempfindungen, als physiologische grundlage für die Theorie der Musik. Vierte Ausgabe*. Braunschweig, Germany: Friedrich Vieweg und Sohn.

Helmholtz, H. (1885). *On the sensations of tone as a physiological basis for the theory of music* (A. J. Ellis, Ed & Trans.). London: Longmans & Co.

Husserl, E. (1954). *Die Krisis der europäischen Wissenschaften und die transzendentale Phänomenologie* (W. Biemel, Eds.). Haag, The Netherlands: Martinus Nijhoff.

Koyré, A. (1966). *Études galiléennes*. Paris: Hermann. (1st ed., 3 vols, Paris: Hermann, 1939).

Kuhn, T., Heilbron, J., Forman, P., & Allen, L. (1967). Sources for the history of quantum physics. In *An inventory and report*. Philadelphia: The American Philosophical Society.

Palmieri, P. (2001). The obscurity of the equimultiples Clavius' and Galileo's foundational studies of Euclid's theory of proportions. *Archive for History of Exact Sciences, 55*, 555–597.

Panofsky, E. (1954). *Galileo as a critic of the arts*. The Hague, The Netherlands: Martinus Nijhoff.

Rand, R. (1990). Using computer algebra to handle elliptic functions in the method of averaging. In A. K. Noor, I. Elishakoff, & G. Hulbert (Eds.), *Symbolic computations and their impact on mechanics* (pp. 311–326). New York: The American Society of Mechanical Engineers.

Sommerfeld, A. (1921). *Atombau und Spektrallinien* (Zweite Auflage ed.). Braunschweig, Germany: Friedrich Vieweg und Sohn.

Sommerfeld, A. (1923). *Atomic structure and spectral lines*. Translated from the third German edition by H. L. Brose. London: Methuen & Co.

Wilson, C. (2010). *The Hill-Brown theory of the moon's motion. Its coming-to-be and short-lived ascendancy (1877–1984)*. New York/Dordrecht, The Netherlands/Heidelberg, Germany/London: Springer.

Chapter 5
Unities of Knowledge and Being – Weyl's Late "Existentialism" and Heideggerian Phenomenology

Norman Sieroka

Abstract Most of the secondary literature on Hermann Weyl's philosophical writings and on his interest in phenomenology focuses on the 1910s and 1920s and on the relation to the work of Edmund Husserl. In contrast, little attention has been paid to Weyl's later writings and to how they relate to later phenomenology. The present paper aims to fill part of this gap by considering Weyl's work of around 1950 in which he critically evaluates several phenomenologically inspired notions from Martin Heidegger's *Being and Time*. As it turns out, Weyl here aims for a third way in between Heideggerian phenomenology and Cassirer's neo-Kantianism.

5.1 Introduction

A lot has been written on the influence of Husserl on Weyl and on the parallels and differences between Weyl's work in philosophy of mathematics and of physics and Husserlian phenomenology (see, e.g., Tieszen, 2005 and Ryckman, 2005). Comparatively little, however, has been written on the question of the extent to which Weyl might have been influenced by Heideggerian phenomenology and by "existential philosophy" more generally. Among the few papers which touch upon this issue are Scholz (2005, 2006). This lack of a broader scholarly interest is surprising, given that we find Weyl, especially around 1950, discussing the work of Heidegger in considerable detail. Weyl acknowledges the huge influence authors such as Heidegger and Jaspers have had on a broader philosophically interested public, and he notes that their influence has contributed to a shift away from the type of transcendentalist viewpoint previously prominent in the German speaking realm and to which Husserlian phenomenology belonged.

N. Sieroka (✉)
Universität Bremen, Institut für Philosophie, Bremen, Germany
e-mail: sieroka@uni-bremen.de

H. A. Wiltsche, P. Berghofer (eds.), *Phenomenological Approaches to Physics*, Synthese Library 429, https://doi.org/10.1007/978-3-030-46973-3_5

What is striking here is that Weyl thinks the same kind of shift has also taken place in his own primary field of research, the exact sciences. He explicitly wonders whether "the development of modern mathematics and physics points in the same direction as the movement we observe in current philosophy, away from an idealistic toward an 'existential' standpoint?" (Weyl, 1949c, p. 264). But what might that mean? What, according to Weyl, characterizes such an existential standpoint in mathematics and physics and how does it relate both to Weyl's understanding of Heidegger around 1950 and to Weyl's own earlier attempts to address philosophical issues in mathematics and physics? This is what this paper is about.[1]

By the same token, the present paper also provides an alternative and much more science-friendly perspective on some of the central concepts in *Being and Time*. That is, contra to Heidegger's own intention, Weyl's critical evaluation of some of Heidegger's concepts—such as his "tool-analysis" (*Zeuganalyse*)—reveal direct and enlightening parallels with and possible adaptation to and within the exact sciences. In fact, Weyl here mediates between Heideggerian phenomenology and the neo-Kantianism of Ernst Cassirer.

5.2 Symbolic Construction and the Concrete Use of Tokens of Symbols

According to Weyl, modern physics and mathematics are dominated by what he calls "theoretic" or "symbolic construction." This notion derives from Weyl's interest in Leibniz and Fichte (see Sieroka, 2007, 2010a, 2012), and Weyl first uses this term in the mid-1920s in his *Philosophy of Mathematics and Natural Science* (English trans., Weyl, 1949c).

"Symbolic construction," as understood by Weyl, consists in positing a coherent system of scientific concepts (see Sieroka, 2018). It is a rational or rationally guided enterprise and not a naïve trial-and-error procedure, and the formation of an axiomatic system in mathematics might count as a paradigm case. This is not meant to deny the tentative character of symbolic constructions in physics, which always are to be compared to empirical findings. However, what is denied in these comparisons is a simple one-to-one correspondence between single concepts and elementary experiences or events. Instead, symbolic construction implies holism.

This much is true for Weyl's notion of theoretical or symbolic construction during the 1920s as well as around 1950. What may shift a little is his emphasis on the concreteness and the actual usage of the symbols within such a construction. It is not that that aspect is absent during the 1920s—Weyl's concept always had a pragmatist or operationalist ring to it—but it certainly comes to the fore more strongly around 1950. And this emphasis might be viewed as a kind of

[1] The following discussion is partially based on earlier works of mine in Sieroka, 2010a, pp. 333–349.

"existentialist" move—that is, as a move away from conceptual issues and issues of transcendental constitution and toward a stronger appreciation of the concrete presence of the symbols used (see, e.g., Weyl, 1947).

Symbolic construction, to the Weyl of around 1950, is instantiated by the concrete actions performed by scientists in their daily work (see Weyl, 1949b). For Weyl, symbolic construction is intersubjective, based on daily practice and communication (see also Scholz, 2006, pp. 305–307). Scientists use pens, ink, pieces of chalk (and today computer keyboards) to do their calculations and to pin down their conceptual investigations within an intersubjective space created by their peers and the wider public.

For Weyl, concrete objects and ordinary daily practices provide phenomenological grounding for symbolic construction and for science in general. Even mathematics and theoretical physics, like other practices, have pre-theoretical elements to them (see Weyl, 1949a).[2] They are parts of our real life, not at least because they are done in the real world and bear real consequences:

> Hence logical thinking and logical inferring is not the core of theoretical procedure as performed in mathematics and the sciences, but rather [it is] the practical management of symbols in accordance with certain rules. (Weyl, 1934, p. 55)

There is one way, however, in which symbolic construction fails to connect to daily life, and that is in its inability to evoke the experiences we have as individual persons. Let me borrow a simple example from Weyl here, namely the propagation of electromagnetic waves of a certain wavelength which I see and which appear to me as yellow light. What science aims at is a deepening of the understanding of the waves, of how they are received and processed on and by my retina, my nervous system, etc. Thus, in this case, symbolic construction aims at narrowing the gap between scientific description and sense experience. However, according to Weyl there will always remain a difference between scientific description and lived experience. Following Fichte, Weyl uses the term "living eye" in order to name this kind of embodied and unaccountable residuum:

> If the electric field strength depends on the space-time coordinates in a certain mathematically defined way, and if an eye that is awake and sees and which *I am* is present in the field, then yellow appears. The existence of the eye must, of course, be described here in the same objective symbolical way as that of the light wave (a rather complicated affair, that demands the reduction of the whole of physiology to physics). On the other hand, as we change from the transcendental sphere of objects to immanent consciousness, the further assumption that I am the living eye, is not less essential; for an apparition can only be an apparition for me. (Weyl, 1934, p. 33)

---

[2]For the sake of completeness, it should be added that indeed similar views can be found in Husserl's later philosophy (as, for instance, in his "Origin of Geometry"). In the present paper, however, the focus is on Weyl's self-assessment—that is, the focus is on how Weyl himself located the claims of other thinkers in relation to his own view.

Apart from this necessary failure in accounting for first-person experiences there is another limiting factor in symbolic construction. Consider writing symbols on a blackboard or piece of paper:

> As scientists we may be tempted to argue: As we know, the chalk on the blackboard consists of molecules, and these are swarms of charged and uncharged electrons, neutrons, etc., which ultimately dissolve again into mere symbols and formulas, which in their turn are written with chalk on a blackboard ... You see the ridiculous circle. (Weyl, 1949a, p. 187)

Similar passages can also be found in several of Weyl's other writings (such as in Weyl, 1947, 1949b, p. 342). He claims that this ridiculous circle could only be overcome "if we understand the way, in which we engage with objects and other people in everyday life, as being an irreducible foundation" (Weyl, 1949b, p. 342; my trans.).[3]

Thus, it is in fact the actual use of concrete sign tokens, consisting of chalk or ink, which is fundamental and which, in turn, prevents the symbolic construction from becoming a mere and even ridiculous game. Signs on paper, on a blackboard (or screen), are always concrete objects in the real world relating to other concrete objects and actions. Or as Weyl illustrates it:

> In this view the number symbols, e.g., $4 = ||||$, consist of strokes following each other, and the term "follows" is here to be understood in that concrete, inexact and "pragmatic" spatial sense on which we rely when we move around in our study or cross a street. (Weyl, 1947)

Thus, such signs, as used by science, are not abstract. They do not have a superior mode of existence by belonging to some realm of ideas (Weyl, 1953, p. 529; see also Weyl, 1949b, p. 342). Nor is science meant to ground or fully account for human existence:

> Science is not engaged in erecting a sublime, truly objective world ... It simply endeavors to prolong a certain important line already laid out in the structure of our practical world. By no means does it pretend to exhaust concrete existence. (Weyl, 1949a, p. 188)

Another important point about such daily practices is that they are usually intersubjective; no scientist works completely on his or her own. Such common efforts are only successful when organized in a suitable fashion, and such organization in turn relies on communication. For Weyl, communication is a fundamental ingredient in any human enterprise—and, hence, also in science (see Beisswanger, 1966). Of course, science is characterized by an "engagement in isolating facts"; at the same time, however, one must not forget "the engagement to find an adequate language in which to communicate these facts"; both are "creative acts of human beings which must go hand in hand" (Weyl, 1949b, p. 345; my trans.). For instance, in the context of mathematics this has direct implications for Weyl's understanding of Hilbert's metamathematics as a level of communicative discourse (on which, in this case, consistency is the central topic). In this regard, both mathematical formalism

[3]Here one can also understand Weyl's late appreciation of, if not enthusiasm for, Paul Lorenzen's work on an operationalist foundation of mathematics (see ETH-Bibliothek, Archive, Hs91: 365–369; see also Sieroka, 2010a, pp. 101–102).

and existential philosophy reach some common ground, according to Weyl, because "their existential origin and challenge lies first in *communication*. On this, thinkers do agree who otherwise show huge intellectual differences, such as Jaspers and Hilbert" (Weyl, 1953, p. 527; my trans.).

5.3 Being-in-the-World, Insight, and Reflection

What about Heidegger? Should we see in Weyl's emphasis on pens, ink, and pieces of chalk the influence of Heidegger? Indeed, Weyl himself claims that the relationship between his own work and that of Heidegger is an obvious one:

> From here, I think, the relation to existential philosophy, e.g. in its Heideggerian form, becomes evident. Since it is Heidegger who puts a certain emphasis on the fact that all insight (*Erkenntnis*) presupposes this natural understanding and open mindedness by which I encounter the world and my fellow human beings. (Weyl, 1949b, p. 343; my trans.)

This passage stems from Weyl's 1949 paper on "Science as symbolic construction of man." This paper surely marks Weyl's most intensive and explicit written engagement with the work of Heidegger. However, this does not mean that Weyl was not aware of the work of Heidegger before 1949. In fact, Weyl seems to have read and talked about the work of Heidegger at least a couple of years earlier. This is suggested by a short article Weyl wrote in June 1948 about his first wife, who had died shortly before. In that article Weyl claims that he and his wife "engaged together in existential philosophy, in Heidegger and Jaspers" (Weyl, 1996, p. 381; my trans.).

Thinking in terms of phenomenology, one might describe Weyl's intellectual development from about 1920 to about 1950 as a shift away from focusing on a more abstract notion of pure consciousness and toward emphasising individual and historically contextualised human beings—and such a shift may be labelled "a move away from the early Husserl and toward Heidegger." By the way, a similar shift can be found also in Husserl's own earlier master pupil when it comes to the philosophy of the exact sciences, namely Oskar Becker. In his 1927 work on "Mathematical Existence," Becker describes his own "transition from the 'formal' phenomenology (Husserl) to the 'hermeneutic' phenomenology (Heidegger)" and that it meant "a tightening up of 'pure consciousness' to 'historical being'," which meant both a philosophical "narrowing, but also a concretisation" (Becker, 1927, p. 755).

For Heidegger, engaging in a formal or abstract theoretical investigation marks a deficient mode of doing philosophy (see Heidegger, 1962, 497 [§64, fn]); and the shift of Weyl and Becker shows that they agree with this to some extent. However, Weyl and Becker do not subscribe to Heidegger's anti-scientific attitude. They rather view this attitude as an individual and idiosyncratic contingency and not as something inherent in Heidegger's philosophical position (see Becker, 1927, p. 636). Hence, both Weyl and Becker understand that exact scientists also have pre-theoretic relationships to the world, and in the case of Weyl, as already mentioned, such relationships are based to the fact that exact scientists know *how to handle* symbols and symbolic constructions.

Notably, for Weyl the need for such a shift in focus had already become apparent during the second half of the 1920s. During that period, he first acknowledged a tension, to be found within each human being, between being a rational being in a transcendent sense, on the one hand, and being a living person having a concrete existence in the world, on the other. He called this tension the "riddle" or the "double position of the ego" (Weyl, 1949c [first published 1927], p. 215; for further details and more extended references see Sieroka, 2010a, pp. 166–174). From the timeless perspective of the rational subject, the concrete details of who I am, as *this* individual person with *this* history, are irrelevant. This, of course, is again the issue of symbolic construction not being able to account for a first-person perspective, and Weyl vividly illustrates it by the example of Judas: even if one assumes the complete determinism of the fate of each person, there is still the possibility and intelligibility of a "desperate outcry of Judas 'Why did I have to be Judas?' ... Knowledge is incapable of harmonizing the luminous ego with the dark erring human being that is cast out into an individual fate" (Weyl, 1949c, pp. 124–125).

Heidegger famously introduces the term "being-in-the-world" (Heidegger, 1962, pp. 78–90 [§§12–13]) in order to cover the concrete individual existence of a human being within a social and natural environment. And later on, in his 1949 paper "Science as symbolic construction of man," Weyl explicitly confirms the close relationship between his own considerations regarding the "double position of the ego" and Heidegger's understanding of being and of its being-in-the-world:

> When one denies the fundamental phenomenon of being-in-the-world, one must reweave the subject, an isolated remnant, into a world itself in tatters; but this remains patchwork. By understanding myself as being-with, I also understand other being (other human beings). This, however, is not knowledge gained and developed by insight (*Erkenntnis*); instead, it is a primary existential way of being, which itself marks the *conditio sine qua non* for insight and knowledge. (Weyl, 1949b, p. 344; my trans.)[4]

In Weyl's thinking, *Erkenntnis*, "insight," pairs with *Besinnung*, "reflection," and the two words can be fit into a Heideggerian context as well. These mark the central and complementary activities of every thinking human being (see again Becker, 1927, p. 543). We can think of the first as the "labour of learning," the second as the "process of reflection." The first involves the work done within a sphere of symbolic construction, whether in the sciences or arts. "Reflection," in contrast, is the work of gaining an overview, of broadening one's perspective. When the perspective is on oneself, "reflection" fosters the peace that comes through accepting painful or worrying experiences rather than trying to explain them away or shove them aside. Accordingly, when speaking about the exact sciences, Weyl emphasises the importance of *Erkenntnis*, insight, whereas when speaking about philosophy, he

[4]Weyl mixes up the Heideggerian terminology a little: Heidegger's term is "In-der-Welt-sein," whereas Weyl writes "Sein-in-der-Welt." This, however, does not have any serious consequences (and the English translation makes use of the changed word order anyway).

emphasises the need for *Besinnung*, reflection—also and especially about the exact sciences (see, e.g., already Weyl, 1932, p. 347).

> I do not share the scorn of many creative scientists and artists toward the reflecting philosopher. Good craftsmanship and efficiency are great virtues, but they are not everything. In all intellectual endeavors both things are essential: the deed, the actual construction on the one side; the reflection on what it means, on the other. (Weyl, 1946a, p. 163)

Ideally, when balanced in the right way, insight and reflection allow for a convergence between objectivity as created by or in symbolic construction, on the one hand, and individual and historical humanity, on the other (see also Weyl, 1954b, 1985 [a piece written presumably after 1953]). Weyl describes insight and reflection as being in "an essential and healthy tension," and adds that he "deliberately seek[s] to swing back and forth" between them, like the bob on a pendulum (ETH-Bibliothek, Archive, Hs91: 258, letter to Erich Hecke).[5]

In 1946, Weyl characterised philosophical reflection as "intellectual mediation between the luminous ether of mathematics and the dark depths of human existence" (Weyl, 1946a, p. 168). This phrasing obviously relates to the double position of the ego with its (transcendent) rational abilities and its individual being-in-the-world. More specifically, given the date when this phrase was written, it must also be viewed as a response to the building and first use of atomic weapons. This use had a huge impact on Weyl (see Sieroka, 2010a, pp. 160–161) and it made the need for reflection—which for Weyl has a moral dimension—even more obvious and pressing. Here are two passages to illustrate this claim:

> [T]he physicist's contemplation is not a purely passive attitude—it is creative construction in symbols, resembling the creative work of the musician. ... To what extent shall and can the theorist take responsibility for the practical consequences of his discoveries? ... [T]here is great danger indeed that in the fight for the basic values of our existence we may lose these values themselves; that the relentless pursuit of science—strange antinomy!—may imperil its very foundation in man's life. (Weyl, 1946b, p. 267)

> For us today the idea that the Gods from which we wrestled the secret of knowledge by symbolic construction will revenge our *hybris* has taken on a quite concrete form. For who can close his eyes against the menace of our self-destruction by science; the alarming fact is that the rapid progress of scientific knowledge is unparalleled by a congruous growth of man's moral strength and responsibility, which has hardly chance in historical time. (ETH-Bibliothek, Archive, Hs91a: 72, p.7; English manuscript, dated 1949 and bearing the German title "Entwicklungslinien der Mathematik seit 1900")

According to Weyl, the world was in a dangerous state during the second half of the 1940s. Symbolic construction and hence insight were not balanced by reflection as they should have been in order to appreciate and safeguard the lives of individual human beings. Symbolic construction lacked its necessary attenuation, as it were.

---

[5]See also Fichte's notion of a "wavering of the imagination" (*Schweben der Einbildungskraft*) which marks the same kind of see-saw mechanism (see Sieroka, 2007, 2010a, 2010b for details). Moreover, in the letter to Hecke just quoted, Weyl uses the German term *Schöpfung* ("creation") instead of *Erkenntnis* ("insight"). This nicely emphasises the common and active character of insight and symbolic construction.

And even without considering atomic weapons, there was his general concern about an imbalance between insight and reflection, with insight turning into a seemingly self-sufficient but in fact deceptive and futile enterprise. Weyl had already addressed this general concern in 1921 and, strikingly enough, he described it using the term “business” (*Betrieb*). That is, he used the term Heidegger was to use a few years later to describe the same kind of phenomenon—though Weyl, once more, was rather unique in applying this terminology in the context of the exact sciences:

> Indeed: any serious and honest reflection must lead to the conclusion that the insalubrities in the border areas of mathematics must be classified as symptoms; what comes to the fore in those symptoms is exactly that what is hidden by the outwardly shining and frictionless business: the inner instability of the foundations on which the structure of the empire rests. (Weyl, 1921, p. 143; my trans.; see also Heidegger, 1962, pp. 221–224 [§38])

5.4 A Tool-Analysis of the Exact Sciences

According to Weyl, the way a mathematician or theoretical physicist operates with symbols on a piece of paper (or a screen) is not that different from the way people use such things as doors, telephones, and hammers in other daily contexts:

> We are left with our symbols, tokens drawn with chalk on a blackboard. With them we deal on the same footing as with other utensils of our daily life, as we open a door to enter a room, sit down in a chair, travel to a meeting, or call on a friend, and we rely on the same kind of understanding. We move in the world of our seeing, acting, caring, natural life ...; in a world so infinitely more obvious and familiar to every one of us, although the suspicious analyzing intellect finds it bewilderingly complex and muddy. (Weyl, 1949a, pp. 186–187)

On the one hand, Weyl’s position reminds one of what Heidegger says about the tool-like character of daily objects, of things being “ready-to-hand” (*zuhanden*) such as, for instance, a hammer (Heidegger, 1962, pp. 95–122 [§§15–18]). On the other hand, Heidegger would obviously object to saying symbols are the equivalent of hammers. According to Heidegger, the exact sciences are fundamentally characterised by *lacking* such a moment of being ready-to-hand. This is because, according to Heidegger, the exact sciences do not take part in the care-taking encounter (*besorgender Umgang*) of daily life. Symbols, according to Heidegger, are always “present-at-hand” (*vorhanden*) but never “ready-to-hand” (*zuhanden*); meaning that they are encountered as abstract “ob-jects,” as standing opposed to human beings and their ordinary engagement in the world. That implies that the exact sciences always act from within a deficient (that is, a purely theoretical) mode of being (see Heidegger, 1962, esp. pp. 412–414 [§69b]).

Weyl disagrees with such a view. For him, the exact sciences are a serious and honest enterprise, and an exact scientist engages in a care-taking encounter. For him or her, symbols are not just “present-at-hand.” Symbols are given by concrete strokes on a blackboard, dots of ink on paper, illuminated pixels on a screen etc., and hence these concrete objects are as “ready-to-hand” as are the numerous other tools in our daily and professional lives. Take a carpenter as an example: the carpenter modifies a piece of wood by means of a rasp or hammer in order to solve or attenuate a concern of daily practice, namely to build, say, a bed in order to gain a better place for sleeping. Similarly, a structural engineer manipulates the dots of inks on his piece of paper in order to figure out, say, the maximum resilience of a bridge; and a theoretical physicist manipulates the strokes on the blackboard in order to determine some parameters for an experiment which, again, is something concrete, with concrete material, measuring devices, etc. involved. And as little as the carpenter can build the "idea of a bed" can the theoretical physicist act within a sphere of abstract objects or of acts of pure consciousness.

The analogy between formulas and hammers goes even further. Heidegger illustrates the difference between something being ready-to-hand and being present-at-hand by the following example (see Heidegger, 1962, pp. 102–107 [§16]): If I am engaging in nailing in order to hang a picture, then the hammer is ready-to-hand. I do not notice it as such. The hammer is a tacit part of my engagement with the world. However, if the hammer breaks, then my action stops and suddenly I become aware of it. The hammer is now present-at-hand; it becomes part of some reflection or even of theoretical considerations, and I begin to ask myself questions such: Did I do something wrong? How to fix this hammer? Where do I get a new hammer? Can I get the nail into the wall without using a hammer?

Admittedly, Weyl does not discuss all this in detail. However, these analogies are at least suggested, and they seem obvious. A symbolic construction (usually a mathematical formalism or a single formula) is a tool by which a certain problem is to be solved. Instead of putting a picture up a wall, one may want to know the maximum load a certain bridge can carry. Hence, instead of using a hammer, one uses differential equations from classical mechanics. And as soon as a symbol is written down, it has a moment of resistance in the sense that now there are structural restrictions or constraints. The formalism does no longer allow for any arbitrary move—similarly to the way that there are constraints to the possible use and structural stability of a hammer.

The concrete use of formalisms by manipulating dots of inks on a piece of paper might indeed break down in a way analogous to the breaking of the hammer. A calculation that leads to a meaningless result is as useless as a broken hammer. Imagine a calculation that gives the result that the bridge might carry "minus five thousand kilograms." At the same time, as with the hammer, it is the tool itself that gains attention now. Just as I started to wonder about the hammer, I now start to wonder about the initial formula and my calculation. How can this meaningless result be fixed? Is the ansatz really correct? Is there an error in my calculation?

Thus, with the concept of "tool-analysis" (*Zeuganalyse*)—which Heidegger himself would never have applied to the exact sciences—Heidegger implicitly provides us with an interesting and important concept for better understanding symbolic construction in science and engineering as well—a better understanding of how they form parts of daily life.

Speaking on a more general level, Heidegger's focus on sometimes rather archaic examples from handicraft should not stop one from considering more modern examples, and they must not make one blind to the fact that many daily practices build upon scientific and technological insights. If putting a picture up a wall counts as daily practice, then solving a sudoku (by doing formal inferences) is a daily practice as well. If ploughing an acre is a practice, then preparing a cell culture dish or programming an algorithm is too. Note also that the latter examples are not reducible to the former—it is not that the practices of immunology and computer science are rooted in farming in any straightforward sense.

Thus, Heidegger in fact underestimates the pluralism of daily life and of the way humans are born and culturally enrooted in the world in many ways. It is not that human beings are simply "thrown" (*geworfen*) into this world, as Heidegger would have it (see Heidegger, 1962, pp. 219–224 [§38]). Of course, Weyl would agree with Heidegger that there is a lot of "business" in the above sense; and, of course, scientific and technological developments allow not only for salutary but also for disastrous applications, as in the case of atomic weapons. This, however, is no reason to engage in praise for an archaic world and to consider our role a passive one. Instead, as mentioned above, it should be an encouragement to engage in both insight and reflection, aware that we are the ones who actively shape this world through our knowledge and culture.

This critique of Heidegger's understanding of our daily life and practice was also prominently raised by Fritz Medicus. This is worth mentioning here because Medicus was Weyl's colleague at the ETH Zurich from 1913 to 1930, and the two engaged in a continuous exchange about philosophical issues in the exact sciences and beyond (see Sieroka, 2007, 2010a, 2012, pp. 31–35, for details).

Weyl was certainly familiar with Medicus's critique of the huge element of passivity, of receptivity and "thrownness," in Heidegger (see esp. Medicus, 1954, which was first published 1950, and Medicus, 1951, p. 126). In contrast, Medicus's own position was strongly influenced by an existential or pragmatist understanding of Fichte. Medicus subscribed to Fichte's concept of intersubjectivity, according to which human interaction and togetherness in a multitude of social environments (friends, colleagues, fellow-citizens, religious communities) are constitutive of human beings as individuals. In a sharp response to Heidegger, Medicus wrote that human beings are "born" (*geboren*) into the world, not "dropped" (*geworfen*) like litters of piglets (see again Medicus, 1954; see also Fichte, 1995, pp. 18–22, 172–173, 269 [= *System of Ethics* §§1, 15, 21]).

It is probably to his friendship with Medicus that we can trace Weyl's own belief in the primacy of intersubjectivity, as well as his strong interest in Fichte (see Sieroka, 2010a, pp. 122–134). Weyl then adapted the concept explicitly to the context of mathematics. When analyzing the notion of the continuum during the 1920s, Weyl claimed the concept of a point to be an abstraction based on the existence of the overlapping "surroundings" (*Umgebungen*) in which it is situated, and he took this to be analogous to the way in which individual human beings are constituted by the overlapping intersubjective "surroundings" in which they live (see Sieroka, 2010a, pp. 122–134, and Sieroka, forthcoming, for details).

Moreover, Fichte (and with him Medicus) claimed that a crucial binding element of such environments or communities is the use of a common language or, more generally speaking, of a common symbolic system (see Fichte, 1995, pp. 237–248

[= *System of Ethics,* §18]). When read in a sufficiently broad sense, this Fichtean view sounds congenial to what Weyl would say about symbolic constructions. For Weyl too, symbolic constructions are anchored intersubjectively because they are based on a common communicative practice of applying the same language—where "language" is understood in a broad sense such that it includes natural languages as well as mathematical formalisms.

5.5 Windbags of Profundity

Given that Weyl was concerned about all these issues (the "double position of the ego," insight, reflection, the concrete usage of symbols as tools, etc.) from the 1920s onward, one may well wonder about Weyl's motivation for engaging with the writings of Heidegger as late as around 1950. What, after all, would have been Weyl's aim in discussing notions advanced by Heidegger, given that Weyl (at least from his own perspective) had already discussed similar issues for several decades previously?

The short answer to this question seems to be that Weyl's engagement in Heidegger was indeed less about addressing some fundamentally new issues as it was about expressing old or constant concerns in a new and maybe more pregnant fashion. Or this is at least what Weyl seemed to realise quite quickly after a short phase of thinking that there may be some hope of finding in Heidegger something intriguingly new. But let me put this claim a bit more into context: Heidegger himself surely took his notions such as "being-in-the-world" to be highly innovative, and he even claimed that questions about the tension between reason and individual historicity simply cannot be asked within the context of earlier transcendental philosophy (see Heidegger, 1962, p. 497 [§64, fn]). This is not the place to go into a detailed discussion about such self-ascription, and it suffices to mention that such self-ascriptions have been criticised extensively in the literature (see, e.g., Rockmore, 2000, especially the contribution by Hedley) and that Weyl's own work from the 1920s obviously provides a good counterexample. In fact, Weyl's work itself stands in a strong transcendental (Fichtean and Husserlian) tradition—a tradition which, it seems, brought much more to Heidegger's *Being and Time* than Heidegger himself acknowledged.[6]

Indeed, Weyl was rather sceptical regarding Heidegger's self-ascribed originality. He wondered whether Heidegger's idiosyncratic rhetoric might indeed be nothing more than terminological airiness. Already in his "Science as symbolic construction of man," Weyl put "a big question mark" behind his own experimentation with Hei-

[6]Notably, even within Heidegger scholarship it has been claimed that it is exactly the idea of a pre-theoretic foundation of human existence that marks a specific Fichtean inheritance in Heidegger (see Denker, 2000, p. 115); and Weyl's notion of a ridiculous circle in combination with such a pre-theoretical encounter of the world fits extremely well into Weyl's own reading of Fichte in a particularly anthropological and pragmatist way (see once more Sieroka, 2010a, for details).

deggerian terminology (Weyl, 1949a, 1949b, 1949c, p. 344). In another manuscript, written in the same year, Weyl is even more explicit; he claims that he himself made similar observations as did Heidegger, and that he did so "some years before Heidegger's *Sein und Zeit* appeared" (Weyl, 1949a, p. 192). And a few years later and with obvious reference to Heidegger, Weyl then even speaks of "windbags of profundity" in philosophy (Weyl, 1954a, p. 201).

This is indeed harsh language, for Weyl was usually conciliating and appreciative in his writings. Accordingly, we might be allowed to speculate about the psychological background and causes for such harsh comments. It may be that, by 1954, Weyl had arrived at some inner discontent with the fact that, first, he himself had fallen prey to such a windbag for several years; and, second, that during the 1920s he himself failed to present his thoughts in a way which would have been as striking and influential as was the work and terminology of Heidegger.

5.6 Unities of Knowledge and Being

For Weyl, as already mentioned, symbolic constructions are anchored in the intersubjective use of a common language, usually a mathematical formalism. This, however, does not imply the inverse conclusion that any practical encounter with the world is after all a symbolic encounter. In order to round off this paper and to further clarify the aim of Weyl's late "existentialism," it is helpful to put this claim into relation to the work of another eminent German-speaking philosopher; namely to, as it were, Heidegger's great antagonist during the 1920s and 1930s: Ernst Cassirer.

Cassirer's *Philosophy of Symbolic Forms* (Cassirer, 1965; originally published 1923–29) provides what might be called an overarching philosophy of human culture. It starts by distinguishing five fundamental cultural enterprises: science, language, art, religion, and myth; and Cassirer characterizes all five in terms of the way they use symbols. Of course, symbolic constructions in modern science work differently from the way symbols work in myth. However, by distinguishing and categorizing the different sign-theoretic relations involved, one (allegedly) gains a unified interpretational framework for the whole of human culture.

On the one hand, Weyl must have been sympathetic to such an approach because he himself characterizes the exact sciences by the way they use symbols and symbolic construction. Moreover, in some of his writings, Weyl even plays around with the idea that, one day, symbolic construction might be applied much more widely. Already during the 1920s (and in between the initial appearances of the three volumes of Cassirer's *Philosophy of Symbolic Forms*), Weyl claimed:

> There is no reason to see why the theoretical symbolic construction should come to a halt before the facts of life and of psyche. It may well be that the sciences concerned have not as yet reached the required level. But that this limitation is neither fundamental nor permanent is already shown by psychoanalysis, in my opinion. The fact that in nature "all is woven into one whole," that space, matter, gravitation, the forces arising from the electromagnetic field, the animate and inanimate are all indissolubly connected, strongly supports the belief

> in the unity of nature and hence in the unity of scientific method. There are no reasons to distrust it. (Weyl, 1949c, p. 214)

On the other hand, this widening of the application range of symbolic construction is very different from what Cassirer suggests. Weyl is talking only about science and about the "unity of scientific method." That is, one day not only mathematics and physics but also chemistry, biology, and medicine (and maybe such borderline cases as psychoanalysis) might partake in symbolic construction. This, however, is different from Cassirer's much broader claim according to which all human enterprises are marked by their symbolic character.

In the passage just quoted, Weyl does not refer to Cassirer explicitly. However, he does so in a later paper (Weyl, 1954a). There he acknowledges the unity that can be gained by Cassirer's philosophy of symbolic forms. In fact, like the way in which symbolic construction might provide a "unity of scientific method," Cassirer's approach may lead to a "unity of knowledge." That is, Cassirer might well provide us with some general framework in which to understand the epistemological dimension of human culture. However, not all (in fact not very many) things that are relevant for us in daily life are based on epistemological concepts and concerns. For this reason, Weyl's own approach started from a much wider understanding of daily life as a practice. This kind of pragmatist or existential dimension allows Weyl to aim for more than a mere "unity of knowledge"—namely, for a "unity of being" (Weyl, 1954a).

Thus, taken together, Weyl seems to aim for a third way, avoiding both Heidegger's anti-scientific affections, which tend to decouple science from life and human existence, as well as Cassirer's "scientism," which shifts every relevant cultural enterprise into the realm of symbolic construction. Or, to put it in a positive fashion: while Weyl's reference to ordinary daily practices marks an important parallel to Heideggerian phenomenology, Weyl's understanding of science as symbolic construction marks an important parallel to Cassirian neo-Kantianism.[7]

5.7 Conclusion

The present paper provided insights into Weyl's later "existentialism," especially into how it relates to the work of Heidegger (and Cassirer) and to Weyl's own earlier work.

[7] Another important parallel to Cassirer are Weyl's "historical dialectics of science," as one might call it. Again and again, Weyl presents the development of mathematics and physics as a historical unfolding of human reasoning (see, e.g., Weyl, 1921, 1925, 1949c; cf. also Weyl 1968, 2009, passim). However, whereas in Cassirer this view might be largely influenced by the work of Hegel, in Weyl it is the work of Fichte; including Fichte's attempt to write a "pragmatic history of the human mind" (see Sieroka, 2007, 2010a, p. 26).

Even though Weyl strongly disliked Heidegger's self-proclaimed originality and bloated terminology, Weyl was able to relate several of Heidegger's notions to the context of the exact sciences in indeed a striking way—and one which Heidegger himself would have denied. For instance, the famous distinction between being "ready-to-hand" and being "present-at-hand" seems to apply not only to hammers in the hands of craftsmen but also to symbols used by scientists in their symbolic constructions. This is possible because, for Weyl, individual symbols are physical objects (namely blots and lines of chalk or ink) to be manipulated. The understanding of, for instance, the chemical ingredients of chalk may very well start from using chalk in order to write down the structural formula of chalk—a "ridiculous circle," as Weyl calls it, but something one cannot get rid of.

Indeed, what one cannot get rid of here is a general rooting in daily practices. Both our individual historicity and our transcendent way of being are marked by the fact that we permanently encounter options for actions and that we permanently have to decide among them. This is what Weyl described as the "double position of the ego." Human existence, one might claim, is characterized by accepting contingencies (by accepting things which might have been otherwise) and by adapting future indeterminacies to some general rational framework. Leading a human life is about, as it were, coordinating the past and ever-new current experiences into a coherent picture. This relates back to Weyl's notions of reflection and insight, and it relates back to the notion of symbolic construction, which he sometimes characterizes by the phrase "being is projected upon the background of the possible" (Weyl, 1949c, p. 37; similarly, already in Weyl, 1925, p. 511, and elsewhere). At the same time, symbolic construction is not adaptable to all areas of human existence but only to those areas in which epistemological concerns are important. Here Weyl offered a critical assessment of the place of philosophy of science in relation to other areas of philosophy and life in general. Contra a neo-Kantian framework such as that of Cassirer, Weyl's claim is that the double position of the ego and with it the unity of being cannot to be fully accounted for by sign-theoretic relations and symbolic constructions.

Acknowledgments I would like to thank Richard Allen for his very careful comments and suggestions on content, language, and grammar.

References

Becker, O. (1927). Mathematische Existenz: Untersuchungen zur Logik und Ontologie mathematischer Phänomene. *Jahrbuch für Philosophie und phänomenologische Forschung, 8*, 440–809.

Beisswanger, P. (1966). Herman Weyl and mathematical texts. *Ratio, 8*, 25–45.

Cassirer, E. (1965). *The philosophy of symbolic forms* (Vol. 3). New Haven, CT/London: Yale University Press.

Denker, A. (2000). The young Heidegger and Fichte. In Rockmore, 2000, 103–122.

Fichte, J. G. (1995). *Das System der Sittenlehre nach den Prinzipien der Wissenschaftslehre (1798)*. Hamburg, Germany: Meiner.

Heidegger, M. (1962). *Being and time*. Oxford, UK: Blackwell Publishing.
Medicus, F. (1951). *Menschlichkeit: Die Wahrheit als Erlebnis und Verwirklichung*. Zurich, Switzerland: Artemis.
Medicus, F. (1954). Philosophiegeschichtliches und Geschichtsphilosophisches: Ein Beitrag zur Würdigung des Existentialismus. In F. Medicus (Ed.), *Vom Überzeitlichen in der Zeit: Beiträge zu humanistischer Besinnung* (pp. 179–196). Zurich, Switzerland: Artemis.
Rockmore, T. (Ed.). (2000). *Heidegger, German idealism and neo-Kantianism*. Amherst, MA: Humanity Books.
Ryckman, T. A. (2005). *The reign of relativity: Philosophy in physics 1915–1925*. Oxford, UK: Oxford University Press.
Scholz, E. (2005). Philosophy as a cultural resource and medium of reflection for Hermann Weyl. *Révue de Synthèse, 126*, 331–351.
Scholz, E. (2006). Practice-related symbolic realism in H. Weyl's mature view of mathematical knowledge. In J. Ferreirós & J. J. Gray (Eds.), *The architecture of modern mathematics: Essays in history and philosophy* (pp. 291–309). Oxford, UK: Oxford University Press.
Sieroka, N. (2007). Weyl's "Agens theory" of matter and the Zurich Fichte. *Studies in History and Philosophy of Science, 38*(1), 84–107.
Sieroka, N. (2010a). *Umgebungen: Symbolischer Konstruktivismus im Anschluss an Hermann Weyl und Fritz Medicus*. Zurich, Switzerland: Chronos.
Sieroka, N. (2010b). Geometrization versus transcendent matter: A systematic historiography of theories of matter following Weyl. *British Journal for the Philosophy of Science, 61*(4), 769–802.
Sieroka, N. (2012). Hermann Weyl und Fritz Medicus: Die Zürcher Fichte-Interpretation in Mathematik und Physik um 1920. *Fichte-Studien, 36*, 129–143.
Sieroka, N. (2018). Theoretical construction in physics: The role of Leibniz for Weyl's "Philosophie der Mathematik und Naturwissenschaft.". *Studies in History and Philosophy of Modern Physics, 61*, 6–17.
Sieroka, N. (forthcoming). Neighbourhoods and intersubjectivity: Analogies between Weyl's analyses of the continuum and transcendental-phenomenological theories of subjectivity. In J. Bernard & C. Lobo (Eds.), *Weyl and the problem of space: From science to philosophy*. Dordrecht, The Netherlands: Springer.
Tieszen, R. (2005). *Phenomenology, logic, and the philosophy of mathematics*. New York: Cambridge University Press.
Weyl, H. (1921). Über die neue Grundlagenkrise der Mathematik. *Mathematische Zeitschrift, 10*, 39–79. Reprinted in and quoted from Weyl 1968, 2:143–180.
Weyl, H. (1925). Die heutige Erkenntnislage in der Mathematik. *Symposion, 1*, 1–23. Reprinted in and quoted from Weyl 1968, 2:511 542.
Weyl, H. (1932). Zu David Hilberts siebzigstem Geburtstag. *Die Naturwissenschaften, 20*, 57–58. Reprinted in and quoted from Weyl 1968, 3: 346–347.
Weyl, H. (1934). *Mind and nature*. Philadelphia, PA: University of Pennsylvania Press. Reprinted in and quoted from Weyl, 2009, 83–150.
Weyl, H. (1946a). Address at the Princeton bicentennial conference. ETH-Bibliothek, Archive, Hs91a: 17, 18. Reprinted in and quoted from Weyl 2009, 162–174.
Weyl, H. (1946b). Encomium (Wolfgang Pauli). *Science, 103*, 216–218. Reprinted in and quoted from Weyl 1968, 4: 265–267.
Weyl, H. (1947). Discussion remarks at the first symposium of the Institute International des Science Théoriques, Brussels. ETH-Bibliothek, Archive, Hs91a: 19.
Weyl, H. (1949a). Man and the foundations of science. ETH-Bibliothek, Archive, Hs91a: 28. Reprinted in and quoted from Weyl 2009, 175–193.
Weyl, H. (1949b). Wissenschaft als symbolische Konstruktion des Menschen. *Eranos-Jahrbuch 1948*, 375–431. Reprinted in and quoted from Weyl 1968, 4: 289–345.
Weyl, H. (1949c). *Philosophy of mathematics and natural science*. Princeton, CT: Princeton University Press.

Weyl, H. (1953). Über den Symbolismus der Mathematik und mathematischen Physik. *Studium Generale, 6*, 219–228. Reprinted in and quoted from Weyl 1968, 4: 527–536.
Weyl, H. (1954a). Address on the unity of knowledge delivered at the Bicentennial Conference of Columbia University. Reprinted in and quoted from Weyl 2009, 194–203.
Weyl, H. (1954b). Erkenntnis und Besinnung (Ein Lebensrückblick). *Studia Philosophica: Jahrbuch der Schweizerischen Philosophischen Gesellschaft/Annuaire de la Société Suisse de Philosophie.* Translated in and quoted from Weyl, 2009, 204–221.
Weyl, H. (1968). *Gesammelte Abhandlungen* (Vol. 4). Berlin, Germany: Springer.
Weyl, H. (1985). Axiomatic versus constructive procedures in mathematics. *Mathematical Intelligence, 7*(10–17), 38.
Weyl, H. (1996). In memoriam Helene Weyl. In I. Lang (Ed.), *Arnold Zweig, Beatrice Zweig, Helene Weyl: Komm her, wir lieben Dich. Briefe einer ungewöhnlichen Freundschaft zu dritt* (pp. 379–390). Berlin, Germany: Aufbau-Verlag.
Weyl, H. (2009). *Mind and nature: Selected writings on philosophy, mathematics, and physics*. Princeton, NJ: Princeton University Press.

Part II
Phenomenological Contributions to (Philosophy of) Physics

Chapter 6
A Revealing Parallel Between Husserl's Philosophy of Science and Today's Scientific Metaphysics

Matthias Egg

Abstract One of the central motivations for Husserl to develop his transcendental phenomenology is what he perceives as the crisis of the sciences of his time (physics in particular), which have forgotten their meaning-fundament by substituting the life-world with mathematically structured idealities and mistaking the latter for true being. It thus seems that Husserl would have had little sympathy for today's attempts to draw metaphysical conclusions from highly mathematized scientific theories within the project known as scientific metaphysics. Nevertheless, I argue in this chapter that there is an important parallel between Husserl's approach to science and the currently most influential version of scientific metaphysics. As a consequence, I will show that a certain line of criticism against Husserl's phenomenology holds important lessons for the contemporary debate on scientific metaphysics.

6.1 Introduction

In §2 of *The Crisis of European Sciences and Transcendental Phenomenology*, Edmund Husserl bemoans science's loss of its significance for life. The reason for this loss, according to Husserl, has nothing to do with any failure of the sciences themselves. Quite the contrary, it is rather the impressive success of the sciences that has led to a shift in the overarching worldview, which then resulted in a neglect of some crucial questions:

> The exclusiveness with which the total world-view of modern man, in the second half of the nineteenth century, let itself be determined by the positive sciences and be blinded by the "prosperity" they produced, meant an indifferent turning-away from the questions which are decisive for a genuine humanity ...: questions of the meaning or meaninglessness of the whole of this human existence. (Husserl 1970, pp. 5–6)

M. Egg (✉)
Institut für Philosophie, University of Bern, Bern, Switzerland
e-mail: matthias.egg@philo.unibe.ch

H. A. Wiltsche, P. Berghofer (eds.), *Phenomenological Approaches to Physics*, Synthese Library 429, https://doi.org/10.1007/978-3-030-46973-3_6

The deeper analysis of this shift and its detrimental consequences constitutes the main theme of Part II (§§8–27) of the *Crisis*. Of particular interest for the philosophy of physics is the beginning of this analysis (§9), in which Husserl locates the origin of science's neglect of our life-world in Galileo Galilei's mathematization of nature. Again, Husserl does not claim that there is anything wrong with mathematization itself, nor does he dispute the impressive successes of this new scientific method. What he criticizes is rather "the surreptitious substitution of the mathematically substructed world of idealities for the only real world, ... our everyday life-world" (pp. 48–49).

This substitution results in what Husserl views as a misguided metaphysics, which mistakes the idealities with which mathematical science is concerned for objective reality, supposedly underlying the life-world of our "subjective" experience. Such a view forgets that the life-world is actually the "meaning-fundament of natural science" (p. 48) and is in that sense prior to the objects that science describes:

> Mathematics and mathematical science, as a garb of ideas, ... *represents* the life-world, *dresses it up* as "objectively actual and true" nature. It is through the garb of ideas that we take for *true being* what is actually a *method* ... (p. 51)

Husserl's conception of science thus seems diametrically opposed to an influential trend in contemporary analytic philosophy, which goes by the name of *scientific metaphysics* and pursues "the articulation of a unified world-view derived from the details of scientific research" (Ladyman & Ross 2007, p. 65). This is a hotly debated topic at the intersection of metaphysics and philosophy of science (see Ross, Ladyman, and Kincaid (2013) and Slater and Yudell (2017) for some important contributions) and there are different versions of scientific metaphysics on the market. In the following, I will mainly focus on the version defended by James Ladyman and Don Ross, which is arguably one of the most influential ones. These authors claim that the scientific world-view (where, as we will see, "scientific" refers primarily to highly mathematized sciences) is the only legitimate description of objective reality, and their view thereby precisely subscribes to the objectivism with respect to natural science that Husserl criticizes.

What I want to show in this chapter is that in spite of this apparent opposition, there is an interesting parallel between Husserl's philosophy of science and today's scientific metaphysics. As a consequence, thinking about Husserl's approach holds important lessons for the contemporary debate in metaphysics. To substantiate these points, I will first provide a general comparison of the two approaches (Sect. 6.2), and then discuss how each of the two views seeks to recover a life-orienting role for science (Sects. 6.3 and 6.4). The conclusion from these considerations will be drawn in Sect. 6.5.

6.2 Transcendental Phenomenology and Scientific Metaphysics

As we have seen, the impressive success of the empirical (positive) sciences plays a key role in what Husserl perceives as a crisis in the philosophy of his time. Similarly, today's scientific metaphysicians are impressed by the success of science and the contrasting fruitlessness of philosophy, but their account of the connection between these two elements is rather different. On their view, the problem is not that philosophers have been blinded by the success of science (as Husserl supposes), but that they have not paid sufficient attention to science and are instead holding on to dubious methods of armchair reasoning, without taking notice of the increasing relevance of scientific findings for what used to be regarded as purely philosophical questions (Ladyman & Ross 2007, Ch. 1). This expresses a thoroughly naturalistic approach to knowledge, according to which the only acceptable methods of inquiry are the ones used in the sciences, as opposed to distinctively philosophical methods relying on human intuition. The contrast to Husserl's approach is obvious: by suggesting (in Part III of the *Crisis*) that transcendental phenomenology can overcome the present crisis, he is advocating the use of philosophical methods that differ radically from the methods used in the empirical sciences.

Apart from this difference in methodology, there is also a difference in the ontological commitments involved in the two approaches. While the project of scientific metaphysics is usually taken to presuppose at least a modest kind of scientific realism, according to which scientific claims refer to an objectively existing reality, we already saw that Husserl warns us against mistaking the supposed referents of scientific notions for "true being". However, the discussion in the following sections will show that the two positions may actually be closer to each other (in terms of ontology) than it initially seems.

Another interesting point of comparison concerns the way in which the two approaches view the role of mathematization. Here as well, there is an obvious contrast between the two views, but also an underlying commonality which will turn out to be important for what follows. The contrast is that today's scientific metaphysicians do not share Husserl's worry about "the surreptitious substitution of the mathematically substructed world of idealities for the only real world". For them, the mathematically structured world described by our scientific theories *is* the only real world, and mathematization is what enables us to acquire objective knowledge about it:

> Fortunately, people learned to represent the world and reason mathematically—that is, in a manner that enables us to abstract away from our familiar environment, to a degree that has increased over time as mathematics has developed—and this has allowed us to achieve scientific knowledge. (Ladyman & Ross 2007, p. 2)

This quote, however, also contains a thought of a somewhat Husserlian spirit, namely the idea that the ability to abstract away from our familiar environment is constitutive of knowledge. It is just that Husserl would not view this ability as fully realized in mathematical science, but in a kind of philosophy that grows out of what he calls "the theoretical attitude", which he describes (in the *Vienna Lecture*) as follows:

> Man becomes gripped by the passion of a world-view and world-knowledge that turns away from all practical interests and, within the closed sphere of its cognitive activity, in the times devoted to it, strives for and achieves nothing but pure *theoria*. In other words, man becomes a nonparticipating spectator, surveyor of the world; he becomes a philosopher. (Husserl 1970, p. 285)

Husserl thus shares with Ladyman and Ross the appreciation of a theoretical attitude that frees those who take it from the bias of everyday interests and parochial concerns. In the next two sections, we will see how this appreciation is connected to science's significance for life.

6.3 An Unacknowledged Cosmology in Pure *theoria*?

After his diagnosis of the nature and the origin of the present crisis (in Parts I and II of the *Crisis*), Husserl turns (in Part III) to an exposition of how it might be overcome. The basic idea is that, since science's loss of significance for life originated in its neglect of the life-world as its meaning-fundament, restoration of such significance requires a new kind of science, a "science of the life-world" (§34). In the *Vienna Lecture*, Husserl (1970, p. 298) describes this as a "form of universal, responsible science, in which a completely new mode of scientific discipline is set in motion where all conceivable questions—questions of being and questions of norm, questions of what is called 'existence'—find their place".

In this context, Husserl expresses his confidence that transcendental phenomenology has the resources to generate this new kind of science, but he does not seem to have explicitly addressed the question how exactly this is to be achieved. It is therefore not surprising that, 30 years after Husserl, Jürgen Habermas suspected that Husserl's hope of a phenomenologically renewed science regaining significance for life was unfounded. Whether Habermas was right in his suspicion is not my issue here. For the purpose of the present chapter, Habermas's critique is important because it further reveals the connection between Husserl's thought and today's project of scientific metaphysics.

Habermas (2005, Sect. III) offers the following three-step reconstruction of Husserl's attempt to renew the scientific world-view by means of phenomenology and thereby to restore to the sciences their significance for life:

1. By pointing to the life-world as the (forgotten) meaning-fundament of natural science, phenomenology undermines the objectivism of the sciences.
2. Instead of such a naïve objectivism, phenomenology brings with it a truly theoretical attitude, which no longer suffers from an unacknowledged dependence on practical interests.

3. This attitude engenders a novel sort of practice, rendering humanity "capable of an absolute self-responsibility on the basis of absolute theoretical insights" (Husserl 1970, p. 283, quoted in Habermas 2005, p. 313).

Now Habermas largely accepts the first two steps in Husserl's reasoning, but criticizes the transition encapsulated in the third step as unwarranted. Indeed, Husserl does not explain how the theoretical insights afforded by phenomenology lead to a transformed praxis, but simply states that they must unavoidably do so:

> If we reflect a little more closely on the manner of this transformation, we immediately understand the unavoidable result: if the general idea of truth-in-itself becomes the universal norm of all the relative truths that arise in human life, the actual and supposed situational truths, then this will also affect all traditional norms, those of right, of beauty, of usefulness, dominant personal values, values connected with personal characteristics, etc. (Husserl 1970, p. 287)

Husserl's idea seems to be that, by its very character of being an absolute norm, theoretical truth automatically transforms practical norms as well, but this neglects the difference between theoretical and practical reasoning, as Habermas (2005, p. 313) points out:

> At best, phenomenology grasps transcendental norms in accordance with which consciousness necessarily operates. It describes (in Kantian terms) laws of pure reason, but not norms of a universal legislation derived from practical reason, which a free will could obey.

In the classical understanding of *theoria*, there was a way to bridge the gap between theory and practice, which Habermas (2005, Sect. I) locates in the platonic notion of *mimesis*: having grasped the cosmic order through theorizing, the philosopher brings himself into accord with it, whereby theory enters the conduct of life. This, however, seems to presuppose an ontological assumption (namely, that this cosmic order preexists the subject that comes to grasp it) which is at odds with Husserl's attack on objectivism. Insofar as Husserl expects pure theory to have an impact on practice, Habermas (2005, p. 313) thus takes him to be committed to a kind of unacknowledged cosmology:

> Theory in the sense of the classical tradition only had an impact on life because it was thought to have discovered in the cosmic order an ideal world structure, including the prototype for the order of the human world. Only as cosmology was *theoria* also capable of orienting human action. ... While criticizing the objectivist self-understanding of the sciences, Husserl succumbs to another objectivism, which was always attached to the traditional concept of theory.

One might object that this is an unfair reading of Husserl and that phenomenology can very well claim to have an impact on life without depending on some hidden cosmology. For example, a thoroughly phenomenological understanding of physics would do away with the metaphysical hypostatisation of mathematical idealities that has been prevalent since Galilei, and this would presumably make a substantial difference to the practice of physics.[1] On behalf of Habermas, I would

[1] I owe this suggestion to Harald Wiltsche.

reply that even in this case, the practical impact of phenomenology depends on the assumption that (in some sense at least) it gets the cosmic order (which now includes the transcendental subject and its role in constituting the objects of physics) right. Although naturalistic cosmology has successfully been cast out, Habermas might still object to the cosmological (or more accurately: metaphysical) character of the system that has been put in its place. Again, I do not insist on taking Habermas's side in this dispute. Maybe the notion of metaphysics employed in the response just given is so weak that it is no longer problematic for Husserl. What is important for the following is that the demand for practical significance of *theoria* inevitably brings with it at least a certain amount of metaphysics.

6.4 Scientific Metaphysics As a Guide to Life?

Husserl's demand that scientific theorizing (even in domains without direct practical application, such as fundamental physics) ought to have significance for our practical lives is shared by some representatives of scientific metaphysics, as the following quote shows:

> The best motivation for trying to synthesize our scientific knowledge into a unified picture—that is, for building naturalistic metaphysics—is the crucial service this activity potentially performs in extending the Enlightenment project. If science is not seen to provide the basis for a general worldview, then people will continue to collectively confabulate alternative general pictures. This in turn matters because the confabulated pictures inspire groundless and usually wasteful and destructive politics and policy. We see no reason to be coy about the fact that, like the logical positivists, our philosophizing is inspired by a normative commitment: while acknowledging the importance of conserving what is valuable, we abhor conservatism, which we view as a sad refusal to explore the magnificent range of possibilities that our ability to do mathematics allows us, and thus betrays the best reason for caring passionately about objective truth. (Ladyman & Ross 2013, p. 113)

This passage explicitly draws a kind of connection between metaphysics and life-orientation that was, according to Habermas, already implicit in Husserl. Of course, Ladyman and Ross do not invoke platonic *mimesis*, but the idea that a true theoretical account of the world (as opposed to "collectively confabulated pictures") inspires a better (that is, less "wasteful and destructive") way of life is clearly present. This is not to say that their approach would appeal to Habermas, who does not conceal his critical attitude towards metaphysics: "The insight that the truth of statements is linked in the last analysis to the intention of the good and true life can be preserved today only on the ruins of ontology" (Habermas 2005, p. 320).

In fact, Ladyman and Ross themselves acknowledge the limits of metaphysics in their (2007, Sect. 1.7) response to the anti-metaphysical critique of Bas van Fraassen (2002). They accept van Fraassen's criticism of what they call *strong metaphysics*, understood as a commitment to philosophical *doctrines* involving claims about reality that go beyond what the sciences imply. By contrast, they pursue a kind of *weak metaphysics* that treats philosophical positions as *stances* (in van Fraassen's sense) and consists in articulating a unified world-view on the basis of a given stance.[2] In the case of Ladyman and Ross, the basis is what they call the *scientistic stance* (cf. Ladyman 2011). This kind of metaphysics, they admit, is of limited use in engaging with adherents of other stances. With respect to people who resist adopting the scientistic stance, they write: "we would not try to convert them with metaphysics, for van Fraassen is right that that would require strong metaphysics, and strong metaphysics can't get off the ground" (Ladyman & Ross 2007, p. 64).

In view of the above quotation, one might ask whether building naturalistic metaphysics in order to dissuade people from "confabulated alternative general pictures" is not really a way of trying to "convert people with metaphysics", but it is fair to say that this is a weaker kind of metaphysics than one that would be required for changing one's way of life by means of platonic *mimesis*. Instead of metaphysics, Ladyman and Ross view questions of direct practical interest as driving the discussion with people who reject the scientistic stance: "Their resistance to science, which must be quite thoroughgoing if it is not to be unprincipled, will confront them with serious policy problems in the management of social affairs, and we will want to press them as hard as possible on these" (p. 64). The upshot is that the role of metaphysics in such debates is rather limited, and this calls into question what was above described as "the best motivation ... for building naturalistic metaphysics".

6.5 A Surprising Convergence and a Way Forward in Metaphysical Debates

The foregoing investigation of the way in which science may provide us with orientation in life has unearthed two somewhat contrary movements within the approaches of Husserl and of Ladyman/Ross, respectively: On the one hand, Husserl's expectations towards a life-orienting role of the theoretical attitude were shown (in Sect. 6.3) to presuppose more metaphysics than he might be prepared to admit. On the other hand, it turned out (in Sect. 6.4) that metaphysics is less relevant to practical decision-making than the rhetoric of scientific metaphysicians suggests.

These two movements converge on a kind of weak metaphysics that is present in both approaches. Being of the same *kind* does not, however, mean that it is the *same* metaphysics. As explained in Sect. 6.4, weak metaphysics is the articulation of a world-view based on a certain stance, and the stances underlying the two approaches discussed in this chapter are decisively different: phenomenological in one case, scientistic in the other.

[2] Interestingly, van Fraassen's notion of *stance* is in turn inspired by Husserl's notion of *attitude*. See Ratcliffe (2011) for a further exploration of this connection.

The shift from doctrines to stances within this weak understanding of metaphysics highlights a problem that is often encountered in metaphysical debates: Insofar as each stance comes with its own set of values, adherents of different stances tend to talk past each other, lacking a common measure against which different stances could be compared. And even if there is some shared notion of rationality, it often turns out not to be sufficiently discriminating, since many different stances can count as equally rational. Consequently, Anjan Chakravartty (2017, p. 243) has recently argued that there can be no objective ranking of rational stances:

> Since rationality is the only stance-neutral criterion for the acceptability of a stance, there are no further grounds on which to prosecute a non-question-begging case for the epistemic superiority of one over another; they are, *qua* rationality, the only relevant measure, "equally strong".

Now it seems to me that the lessons learnt in the previous sections point to a way forward in such situations. The convergence just described is rooted in a tension that Husserl's approach shares with the one championed by Ladyman and Ross: Both approaches subscribe to an ideal of disinterested theorizing (cf. Sect. 6.2), but at the same time demand that such theorizing should matter to our lives (Sect. 6.4). This tension is the focus of Habermas's critique, which culminates in his postulate of a "knowledge-constitutive interest"[3] (Habermas 2005, 314) that guides even the kind of theorizing that presents itself as completely disinterested. The two stances discussed in this chapter thus share such a knowledge-constitutive interest, which is to say that one may find a more substantial basis for comparing them than the slim notion of rationality that Chakravartty found insufficient for a non-question-begging assessment of epistemic superiority.

It is not the aim of this chapter to speculate about the result of such an assessment. What I hope to have shown is that two philosophical approaches to science, despite radical differences on the doctrinal and methodological level, have much in common on the motivational (and, to some extent, axiological) level. Acknowledging these common elements gives further substance to the recent trend of regarding philosophical positions as stances rather than mere doctrines, and thereby facilitates a new way of thinking about philosophical approaches to scientific theorizing.

References

Chakravartty, A. (2017). *Scientific ontology: Integrating naturalized metaphysics and voluntarist epistemology*. Oxford: Oxford University Press.

Habermas, J. ([1965] 2005). Knowledge and human interests. In G. Gutting (Ed.), *Continental philosophy of science* (pp. 310–321, trans: Shapiro, J.J.). Malden: Blackwell.

[3] Habermas uses the term "erkenntnisleitendes Interesse", a more accurate translation of which would be "knowledge-guiding interest".

Husserl, E. ([1936] 1970). *The crisis of European sciences and transcendental phenomenology* (trans: Carr, D.). Evanston: Northwestern University Press.

Ladyman, J. (2011). The scientistic stance: The empirical and materialist stances reconciled. *Synthese, 178*, 87–98.

Ladyman, J., & Ross, D. (2007). *Every thing must go: Metaphysics naturalized*. Oxford: Oxford University Press.

Ladyman, J., & Ross, D. (2013). The world in the data. In D. Ross, J. Ladyman, & H. Kincaid (Eds.), *Scientific metaphysics* (pp. 108–150). Oxford: Oxford University Press.

Ratcliffe, M. (2011). Stance, feeling and phenomenology. *Synthese, 178*, 121–130.

Ross, D., Ladyman, J., & Kincaid, H. (Eds.). (2013). *Scientific metaphysics*. Oxford, Oxford University Press.

Slater, M. H., & Yudell, Z. (Eds.) (2017). *Metaphysics and the philosophy of science: New essays*. Oxford: Oxford University Press.

van Fraassen, B. C. (2002). *The empirical stance*. New Haven: Yale University Press.

Chapter 7
Physical Things, Ideal Objects, and Theoretical Entities: The Prospects of a Husserlian Phenomenology of Physics

Lee Hardy

Abstract Husserl's phenomenological philosophy of the physical sciences is commonly understood to be committed to some form of scientific anti-realism and to an instrumentalist interpretation of scientific theories. It denies that the unobservable entities posited by the sciences exist; and it takes scientific theories concerning those entities to be little more than tools used in the prediction of empirical states of affairs. For that reason, many commentators have taken Husserl's phenomenology to be of limited value in illuminating the rationality of the physical sciences as they exist today given their rapid expansion in the theoretical domain since the beginning of the last century. In this contribution I argue that Husserl's phenomenology is compatible with a realist interpretation of scientific theories. I begin with the generally accepted distinction between scientific laws and scientific theories, and proceed to argue that Husserl's phenomenology offers an instrumentalist account of scientific laws, not scientific theories. I then suggest that a phenomenology of the theoretical dimension of the physical sciences could be carried out as a description and analysis of the constitution of indicative sign consciousness, where the givenness of an observed entity or event comes to count as a sign of an unobserved (and in some cases unobservable) entity or event.

7.1 Introduction

In section 20 of *Ideas Pertaining to a Pure Phenomenology and to a Phenomenological Philosophy* (1913) Edmund Husserl claimed that if "Positivism" means a commitment to the "absolutely prejudice-free grounding of all the sciences on the 'positive,' that is, on what is directly apprehended, then we are the true Positivists" (Husserl, 1913/1982, 39, translation modified). But what Husserl claimed as a badge

L. Hardy (✉)
Calvin University, Grand Rapids, MI, USA
e-mail: Lhardy@calvin.edu

H. A. Wiltsche, P. Berghofer (eds.), *Phenomenological Approaches to Physics*, Synthese Library 429, https://doi.org/10.1007/978-3-030-46973-3_7

of honor in 1913 became a major reason for turning him away at the gates of the philosophy of science in the latter part of the twentieth century. Positivism, hobbled by an empiricist epistemology, a verificationist theory of meaning, an ontology restricted to perceivable physical objects—or, in some cases, to sense data—and an instrumentalist interpretation of scientific theories, seemed incapable of doing justice to the rapid and impressive advance of the physical sciences into the hidden regions of the unobservable. And it seemed to many that Husserl had irrevocably hitched his phenomenological wagon to that fading positivist star. Aurelio Rizzacasa, for instance, claims that "Husserl's approach to the science of nature with its idea of the objectivity of principles and its insistence on induction and verification, is very close to positivism in spite of the fact that he has rejected its philosophical consequences" (Rizzacasa, 1979, 78). Theodore Kisiel expands on the point: "The favorite theses of logical positivism in its nadir of instrumentalism and operationalism still seem to lurk behind Husserl's formulations of his own phenomenological positivism: the empty language of mathematics is applied to the invariant mass of the lifeworld merely in order to acquire a measure of predictive control over it. Physical theories are thus reduced to merely an abstract interlude and useful complication in our practical concerns, and therefore can be suppressed at any time without the loss of any real knowledge" (Kisiel, 1973, 222–23). Ernan McMullin, in comments on a paper by John Compton in the *Review of Metaphysics*, asserts that Husserl's approach to the philosophy of science is "broadly instrumentalist" because "Husserl shared in the generally positivist understanding of natural science in the middle Europe of his day" (McMullin, 1979, 31, 34). Joseph Rouse aptly sums up the consensus opinion on the prospects of a Husserlian phenomenology of science in his article "Husserl's Phenomenology and Scientific Realism": "those philosophers of science at all familiar with Husserl tend to associate him with views akin to instrumentalism, which has been largely discredited today; he is therefore thought to be of historical interest at best" (Rouse, 1987, 222).

The consensus opinion articulated by Rouse is not without some basis in Husserl's own writings. One of the main theses of the positivist philosophy of science is that scientific theories do not aim at providing us with a straightforwardly true account of the unobservable deep structure of nature, but rather with sophisticated symbolic machinery for generating useful predictions about observable phenomena within nature. As Moritz Schlick, founder of the Vienna Circle, put it, "what every science seeks, and seeks alone, are … the rules which govern the connections of experiences, and by which alone they can be predicted" (Schlick, 1932–1933/1991, 44). It would seem that Husserl is in agreement on that point. In the Prolegomena to the *Logical Investigations* (1900), Husserl claims that the practitioners of the empirical sciences are "more concerned with practical results and mastery than with essential insight" (Husserl, 1900/1970, I 245). In *Ideas I*, he maintains that the utility of physics consists in the fact that "any cognition in physics serves as an index to the course of possible experiences with the things pertaining to the senses and their occurrences found in those experiences. It serves, therefore, to orient us in the world of current experience in which we all live and act" (Husserl, 1913/1982, 85). Later, in the *Crisis of European Sciences and*

Transcendental Phenomenology (1936), he writes that in the laws of the physical sciences, "the functional co-variations of empirical phenomena are generalized and fixed with exact, mathematical precision.... Thus one can outline the empirical regularities of the practical life-world which are to be expected. In other words, if one has the formula, one already possesses, in advance, the practically desired prediction of what is to be expected with empirical certainty in the intuitively given world of concretely actual life, in which mathematics is merely a special praxis" (Husserl, 1936/1970, 43). Thus far, all appears to be in line with Schlick, generally recognized as the founder of Logical Positivism as promulgated by the Vienna Circle.

Husserl not only locates the value of the physical sciences in the expansion of predictive control over the world of sense experience, he also appears to embrace a strong form of scientific antirealism. In the *Crisis* he argues that the objective correlates of the mathematical laws of the physical sciences simply do not exist in the physical sense. They are ideal mathematical objects, not real physical things. In the realistic construal of the physical sciences we witness the "surreptitious substitution of the mathematically substructed world of idealities for the only real world, the one that is actually given through perception, that is ever experienced and experienceable—our everyday life-world" (Husserl, 1936/1970, 48–49). Here we have been misled "into taking the formulae and their formula meaning for the true being of nature itself" (Husserl, 1936/1970, 44–44). We fail to realize that "the 'objective' world is a mere ideal construct, developed for the sake of making exact laws possible. Mathematics and mathematical science, as a garb of ideas ... encompass everything which, for scientists and the educated generally, represents the life-world, dresses it up as 'objectively actual and true' nature. It is through the garb of ideas that we take for true being what is actually a method"—a method for making more precise predictions within the life-world of perception and practical life (Husserl, 1936/1970, 51–52).

So it would seem that Husserl's philosophy of science is committed to some form of instrumentalism. Instrumentalists claim that the acceptance of a theory for scientific purposes does not commit one to believing that the theory is true in any straightforward sense. On the instrumentalist account, a scientific theory is itself not the kind of thing that is true or false; it is, rather, a more or less reliable tool for generating conditional statements about empirical phenomena. As Peter Godfrey-Smith puts it in *Theory and Reality*, instrumentalism "holds that we should think of theories as predictive tools rather than as attempts to describe the hidden structure of nature" (Godfrey-Smith, 2003, 15). The virtue of a theory, then, is not its truth, but its empirical adequacy, its track record in producing reliable predictions about observable states of affairs. Acceptance of a theory does not at the same time commit one to the view that its theoretical terms refer, or that the entities it seems to postulate actually exist. Theories may be admirable syntactical machines, but they have no real semantic value. To assign a semantics to scientific theories would represent a serious misreading of their nature and intent.

In this chapter on the prospects of a Husserlian phenomenology of the physical sciences, I will argue that Husserl's phenomenological critique of modern mathematical physics is entirely compatible with a realistic construal of scientific theories. On my reading, Husserl was indeed an instrumentalist of some sort, but his instrumentalism is restricted to an interpretation of scientific laws, not scientific theories. In support of this contrarian thesis, I will argue that the construction of ideal objects in mathematical physics is not a falsification of the real physical world, but rather a way of achieving objective and exact knowledge of the real physical world by way of approximation; that Husserl, despite his positivist sympathies, did not deny the possibility of the existence of physical entities beyond the reach of human sense perception; that Husserl did not deny the possibility that experience could properly motivate the positing of the existence of physical entities beyond the reach of human sense perception; moreover, I argue that Husserl possessed the tools for a phenomenological analysis of the rationality of positing the existence of physical entities beyond the reach of human sense perception—although he did not make use of them in his own approach to science. Those tools, briefly alluded to in the first of the *Logical Investigations*, outline how a phenomenology of the of theoretical dimension of the physical sciences would be conducted—that is, how a phenomenology of the non-phenomenal would proceed.

The plan of this chapter is as follows: after clarifying the distinction between laws and theories, I will move on to preliminary and general issues regarding the relation between sense perception and existence. I then turn to the ontology of physical things and ideal objects. Finally, I will outline a phenomenological approach to the scientific rationality of positing the existence of theoretical entities given Husserl's reflections on indicative sign consciousness in the *Logical Investigations*.

7.2 Laws and Theories

My primary thesis hangs on a distinction between scientific laws and scientific theories. Scientific laws specify the functional interdependence of quantified physical variables. Their intent is to capture lawlike regularities in the behavior of empirical phenomena. Familiar examples of such laws are Galileo's law of free-falling bodies, which determines the instantaneous velocity of a free-falling body as a function of lapse time; and Boyle's law, which specifies the pressure of gas in a given container as a function of volume. Such laws state how empirical objects behave. Free-falling bodies accelerate at a rate proportional to the square of the lapse time of their descent; the pressure of a gas within a container is inversely proportional to the volume of that container.

But scientific laws do not explain why empirical objects behave the way they do. That's the job of scientific theories. Scientific theories typically explain the empirical behavior of things by postulating unobservable entities that causally interact in such a way as to produce the behavior of empirical objects captured in scientific laws. For the law governing the rate of acceleration of free-falling bodies, we have the theory of gravitational force; for the gas laws covering the relation of pressure, temperature, and volume, we have kinetic theory. The "generally accepted" account of theories, as Bas van Fraassen puts it, is that "theories account for the phenomena (which means, the observable processes and structures) by postulating other processes and structures not directly accessible to observation" (Van Fraassen, 1980, 29).

Van Fraassen's statement fairly represents the received view of theories in Anglo-American philosophy of science. The causal interactions of the entities posited by a scientific theory, writes Richard Boyd, are to "explain the predicted regularities in the behavior of observable phenomena" (Boyd, 1973, 1). Peter Kosso counts a theory as "any description of the unexperienced world that is part of what accounts for and helps us understand the experienced world" (Kosso, 1992, 15). In his work on scientific realism, Stathis Psillos states that theories "explain and predict observable phenomena by reference to unobserved phenomena" (Psillos, 1999, 40).

The distinction between laws and theories, then, depends upon a bi-level analysis of science. Typically, laws state in the exact language of mathematics the regular functional interdependence between observable physical phenomena; theories seek to explain why such regularities hold by postulating unobservable entities and specifying their causal capacities. Laws make predictions possible; theories provide explanations. The realist/instrumentalist dispute within contemporary philosophy of science is predicated upon this bi-level analysis of science. At root, it is a dispute over the semantic value of theories, which go beyond what is observable in order to explain it. The instrumentalist insists that science ultimately refers only to that which is observable. Science does so in its laws, which are empirical generalizations. Theories are no more than instruments by which the predictive power of empirical science is unified, enhanced, extended, and made more efficient. Theories may appear to refer to unobservable entities, forces, and processes. But we should understand these, says the instrumentalist, as no more than convenient fictions. The realist, on the other hand, holds that science refers not only to observable phenomena, but also, in its theories, to unobservable entities, forces, and processes. Furthermore, the empirical adequacy of a theory provides good grounds for believing that the theory is true, that its terms refer, and that the entities it postulates exist. Theories are more than just useful tools in the making of empirical predictions. They give us scientific knowledge of an unseen, but real, world.

My interpretation of Husserl trades on this generally accepted distinction between scientific laws and scientific theories. On my view, even though Husserl's phenomenology advances an instrumentalist view of scientific laws, it gives us no reason to deny the existence of entities postulated by scientific theories. I will not claim that the bi-level analysis of science is without its own problems. The line between the observable and the unobservable is notoriously difficult to draw. Moreover, the lawlike behavior of theoretical entities often calls for additional levels of explanation and, with them, the positing of a new round of theoretical entities. In the case of the ideal gas law (which enriches Boyle's law, formulated in the seventeenth century, with the universal gas constant and variables for temperature and moles) we make use of the explanatory powers of kinetic theory. Positing the existence of molecules as the unobservable constituents of a gas, kinetic theory proceeds, under the standard Maxwell-Boltzmann interpretation, to understand temperature in terms of the average kinetic energy of the molecules and pressure in terms of the change of momentum as the molecules strike the inner surface of a container. An increase in temperature raises the average kinetic energy of the molecules, resulting in an increase in their momentum. Assuming the volume of the container remains relatively constant, the pressure of the gas—now understood in terms of molecular momentum—rises as a direct result. Hence the theoretical explanation of why pressure is directly proportional to temperature. But the molecules that do the work in kinetic theory exhibit regular chemical behavior that in turn calls for theoretical explanation. Molecules are resolved into atoms, the smallest units of any given element; and atoms are in turn resolved into protons, neutrons, and electrons whose relations are governed by electromagnetic forces that explain the chemical behavior of the molecules they constitute. At a still deeper level, the behavior and transformations of protons and neutrons are explained by the postulation of up and down quarks—the entities that make them up—and weak nuclear forces. The law/theory distinction, then, is not limited to the observable/unobservable distinction, even if it was initially built on that distinction. But the initial build is what frames the realist/instrumentalist controversy, and it is the initial build that I want to employ in examining the question of Husserl's alleged instrumentalism.

7.3 Truth, Perception, and Existence

There are a number of passages where Husserl seems to tie existence to perception, and, in doing so, to limit what can exist to what can be perceived. Consider the following statements drawn from Husserl's works: "the sphere of real objects ... is in fact no other than the sphere of possible sense perception." For "we define a real object as the possible object of a straightforward percept" (Husserl, 1900/1970, II 791). Hence, "nothing exists that cannot be perceived" (Husserl, 1900/1970, II 822). In *Ideas I* Husserl writes, "it must always be borne in mind that whatever physical things are ... they are as experienceable physical things" (Husserl, 1913/1982, 106). It would be a "completely groundless assumption" on my part to think that there exists anything that cannot be connected, at least potentially, to my perceptual experience (Husserl, 1913/1982, 100). If this is indeed the case, then theoretical entities are in deep existential trouble, since they cannot be perceived. Edward Ballard takes it that these and like statements commit Husserl to "a radical empiricism. Any object that is not originally given in perception or is not derived in a determined manner from something originally given in my experience must be suspect" (Ballard, 1971, 179).

The chain of reasoning that ties existence to perception in Husserl's thought can be traced in more detail. There is, Husserl states in the *Logical Investigations*, an "*a priori* togetherness" of truth and objective states of affairs, of *Wahrheiten* and *Sachverhalten* (Husserl, 1900/1970, I 225–26). A proposition is true just in case the corresponding state of affairs obtains. The proposition expressed the English sentence "Snow is white" is true just in case the state of affairs—snow's being white—obtains. "Nothing can be without being thus or thus determined, and that it is, and that it is thus and thus determined, is the self-subsistent truth which is the necessary correlate of the self-subsistent being" (Husserl, 1900/1970, I 225–26). In a claim that anticipates Wittgenstein's *Tractatus*, Husserl asserts that the world "is merely the unified objective totality corresponding to, and inseparable from, the ideal system of all factual truth" (Husserl, 1900/1970, I 143). The relation between truth and states of affairs is, logically speaking, biconditional: if a proposition is true, the corresponding state of affairs obtains; and if a state of affairs obtains, the corresponding proposition is true.

Truth, then, is necessarily connected to the way the world is. But Husserl also connects truth to evidence, where evidence is defined as the intuitive givenness of a state of affairs in an act of consciousness. On some readings of this connection, Husserl holds that it is also the case that a proposition is true just in case it is evident, that is, just in case the corresponding state of affairs is intuitively given in some act of perceptual consciousness. It seems natural to give this relation a bi-conditional reading as well: if a state of affairs is given, the corresponding proposition is true; if a proposition is true, then the corresponding state of affairs is (or has been) given.

If a biconditional captures a relation that is more than merely logical, if causality is also involved, then one can bring up a separate question concerning the ground of that relation. If we examine the relation between a flagpole and its shadow, following the example provided by Wesley Salmon (Salmon, 1989, 47), we could establish that a necessary correlation obtains between the length of the flagpole and the length of the shadow, given the position of the sun. We could say that if the flagpole is 30 feet high, then the shadow is 20 feet long; and we could just as easily say that if the shadow is 20 feet long, then the flagpole is 30 feet high. The relation is biconditional; it runs in either direction. But we would not say that the flagpole is 30 feet high *because*—in any causal sense of "because"—the shadow is 20 feet long. Rather, we would say that the shadow is 20 feet long because the flagpole is 30 high. The height of the flagpole explains why its shadow is 20 long; its shadow does not explain why the flagpole is 30 high, even if the height of the flagpole logically follows from the length of the shadow. The height of the flagpole makes the shadow 20 feet long; the length of the shadow does not make the flagpole 30 feet high. Causal relations are not equivalent to logical relations.

The evidence theory of truth often ascribed to Husserl holds that there is a biconditional logical relationship between the truth of a proposition and the givenness of a corresponding state of affairs. But it also asserts that what makes the proposition true is not actuality of the corresponding state of affairs but rather the givenness of the corresponding state of affairs in an act of consciousness. "What makes a statement (about the world at hand) true is not a magical union between two masses—a meaningful sentence and a thing—but a more or less adequate fit between the meaning intended in and through speech-acts [a proposition] and what is given in and through perceptual acts," claims Donn Welton in his exposition of Husserl's phenomenological account of truth (Welton, 1983, 139). The givenness of a state of affairs not only counts as evidence for the truth of the corresponding proposition, it makes the corresponding proposition true.

If we line up the connections thus understood between truth, evidence, and the way the world is, we can now see why many commentators have taken Husserl to be an idealist in his metaphysics and an instrumentalist in his philosophy of science. Let "A" stand for a state of affairs, and p for some proposition. According to Husserl, there is a necessary correlation between truth and the way the world is:

A obtains if and only if p is true.

But he also maintains a necessary relationship between truth and evidence:

P is true if and only if p is evident.

Given the phenomenological gloss on "evident," we get:

P is true if and only if A is intuitively given in an act of consciousness.

Since biconditional relations are not only symmetric but transitive, we can bring together the left hand of the first biconditional with the right hand of the last, yielding:

A obtains if and only if A is intuitively given in an act of consciousness.

If the ground of this biconditional relation is indeed intuitive givenness, then we can see how phenomenological ontology is defined by perceptual experience. This is why Husserl commentators such as Henry Pietersma will claim that the relation Husserl posits between truth and evidence will inevitably "involve the idealism of the Husserlian position: the scope of the mind defines reality" (Pietersma, 1977, 43). Likewise, Günther Patzig claims that Husserl's evidence theory of truth accounts for Husserl's "turn to idealism" (Patzig, 1977, 179), where idealism is understood as the claim that what exists somehow depends on the mind, that the way the world is somehow depends on acts of consciousness. This view will also entail an instrumentalist/anti-realist interpretation of scientific theories, since the range of objects that can be given in an act of consciousness is limited to what can be intuitively given. Theoretical entities cannot be intuitively given. They cannot be observed. That's what makes them theoretical. On this interpretation of Husserl's theory of truth, then, theoretical entities cannot exist. Given his position on the necessary correlation between truth and existence, Husserl holds that theoretical entities exist if and only if it is true that they exist. But it can be true that they exist if and only it is evident that they exist. And it can be evident that they exist if and only if they can be given in an intuitive act of consciousness. But as long as theoretical entities are by definition unobservable, it could never become evident that they exist. Hence it could never be true that they exist. So they do not exist. On this interpretation of Husserl, phenomenology is incompatible with a realist interpretation of scientific theories.

I think this interpretation of Husserl is deeply mistaken. A closer reading of Husserl will reveal that the relation between truth and evidence should be recast in terms of a theory of justification. On this view, evidence does not make a proposition true. Rather, it makes those who have the relevant evidence justified in believing that it is true. What hangs on evidence is justification, not truth. And what evidence justifies is not a proposition in the abstract, but a person in particular. Introducing S as a variable ranging over persons, the way to represent the relation of evidence to truth and justification is this:

S is justified in believing p if and only if the correlative state of affairs A is given to S in an intuitive act of consciousness.

Husserl expressed a view like this in the *Logical Investigations*, where he writes that evidence is "the experience in which the correctness [the truth] of his judgment is brought home to a judging subject" (Husserl, 1900/1970, I 191). In the *Cartesian Meditations* Husserl's theory of evidence is not providing truth conditions for a proposition, but conditions under which a proposition comes to acquire a sense of truth for us: "it is clear that truth or the true actuality of objects is to be obtained only from evidence, and that it is evidence alone by virtue of which an 'actually' existing, true, rightly accepted object of whatever form or kind has sense *for us* ... Every rightness [in the sense of being justified, every *Recht*] comes from evidence" (Husserl CM 1931/1969, 60, my emphasis). Evidence is the original source of justification, the original *Rechtsquelle* (Husserl, 1913/1982, 338).

This interpretation makes truth and existence independent of the acts of consciousness. What depends on the acts of consciousness are evidence and justification. A state of affairs may obtain whether or not we are aware of it, or I am aware of it, or someone is aware of it. So the lack of intuitive givenness of theoretical entities does not put us in a position to deny the existence of theoretical entities. It simply gives us no reason to affirm their existence. Nonetheless, Husserl's formulation of the relation between evidence and justification appears to limit the justification of belief to perceptual experience. If this is the case, then even if theoretical entities do exist, we would never be justified in believing that they exist, since they are imperceptible. We may be able to give up the idealism—but not the instrumentalism.

The last point to clear up, then, in making way for the compatibility of Husserlian phenomenology and a realistic interpretation of scientific theories pertains to the modal force of the perceivability condition on real existence. When Husserl spoke of the perceivability condition on existence—the notion that if a physical object exists it must be possible to perceive that object—he makes it clear that he is speaking of an *ideal* possibility, not a real possibility. "Perceivable" does not mean perceivable by human beings. In *Ideas I*, Husserl writes: "Obviously, there are physical things and worlds of physical things which do not admit of being definitely demonstrated in any human experience; but that has purely factual grounds which lie within the limits of such experience" (Husserl, 1913/1982, 109). Real entities, hidden to us, may be perceived by "other egos who see better and further" (Husserl, 1913/1982, 119). The perceivability condition that Husserl lays on real existence in general, then, speaks to an ideal possibility, not a real possibility tied to the sensory capacities of the human species. It follows that the category of the unperceivable—in the sense of a real possibility for humans—may well include that part of real existence which cannot be perceived by human beings because it is too small, too distant, too fast, or occupies the wrong part of the wave spectrum, as well as those objects that are unperceivable because, as ideal objects, they occupy the wrong ontological category. The fact that gas molecules, or the atoms that make them up, cannot be perceived by human beings does not, for Husserl, automatically put them in the category of useful fictions. It is logically possible they exist. To be justified in believing that they actually exist will depend on epistemic motivations to be found in the course of scientific research, to which I turn in the next section.

7.4 Justification in Science

In *Ideas I*, Husserl proclaimed that "we" are the true positivists insofar as we resolve to ground the sciences in what is directly and adequately given in immediate intuition. The methodological ideal is clear enough, perhaps. But who are "we"? Not all of us, as Husserl will point out, but rather phenomenologists on those occasions when they engage in the infinite task of rigorously grounding scientific knowledge. The rule of phenomenological rationality—basing all beliefs on adequate intuitive givenness—is not meant as a general rule for all persons at all times. Husserl's concept of rationality is in fact differentiated with respect to epistemic project. Different standards of rationality hold for everyday life, and for the "positive" (unreflective) sciences.

In "Philosophy as a Rigorous Science" (1910) Husserl writes that "in the urgency of life that in practice necessitates adopting a position, man could not wait until—say, after a thousand years—science would be there, even supposing that he already knew the idea of rigorous science" (Husserl, 1910/1965, 138, translation modified). What has been scientifically established can have an impact on practical life. But the practical exigencies of life require that decisions be made and actions carried out on the basis of the beliefs that often fall outside the circle of what has been established with certainty in the sciences. At any point in their development, the sciences cannot speak to the full range of present human concerns. The idea of science is an "infinite idea" in the Kantian sense, regulating theoretical activity of successive generations, forever approximating the idea but never arriving at it (Husserl, 1910/1965, 136). In the meantime, practical life requires answers to vital questions. It cannot wait for science. Here guidance is provided by *Weltanschauung* philosophy, a form of achievable, finite wisdom that is available in any well-formed culture. And this form of philosophy has it own legitimacy: "The value of *Weltanschauung* stands with utmost firmness on its own foundation" (Husserl, 1910/1965, 143).

Similar considerations hold for the positive sciences. Here too, in the practice of the sciences, we are justified in going beyond what is, strictly speaking, evident to us, beyond what is adequately given in direct intuition. Practicing positive scientists are not required, as the phenomenologist is, to make a regressive inquiry in into the ultimate foundations of their discipline. Were that the case, the discipline would never move forward. "Simple positivity," writes Husserl in *Formal and*

Transcendental Logic (1929), "as a naïve devotedness, not only of the practical but also of cognitive living, to the world that is indeed given beforehand, has in it a legitimacy—unclarified, to be sure, and therefore still undelimited, but nonetheless a legitimacy (Husserl, 1929/1978, 226). In addition, practicing scientists are not only under no obligation to go to the evidential foundations of their own disciplines, but are also permitted to go beyond intuitively given evidence by way of hypotheses and inference. For, "the *de facto* course of our human experience is such that it constrains our reason to go beyond intuitionally given physical things ... and base them upon the 'truth of physics'" (Husserl, 1913/1982, 105). "When engaged in natural science we effect experientially and logically ordered acts of thinking in which these actualities [experienced physical things], being accepted as they are given, become conceptually determined and in which likewise, on the basis of such directly experienced and determined transcendencies [external objects], new transcendencies are inferred" (Husserl, 1913/1982, 114). Although immediate givenness is the "primal source" of all justification, it is not the exclusive source. One can draw upon the primal source "indirectly" in order to posit with justification entities that are not immediately experienced (Husserl, 1913/1982, 338). Through the use of symbolic means, science, unlike phenomenology (Husserl, 1913/1982, 114), can and does go far beyond the narrow confines of the intuitable (Husserl, 1900/1970, I 201), given that "objects are certainly possible, that in fact lie beyond the phenomena accessible to any human consciousness" (Husserl, 1900/1970, I 428). The constraints of phenomenological rationality, as the foundational science of consciousness, do not hold for everyday life of straightforward, positive engagement with the world. Nor do they hold for the positive sciences. Positive scientists may be entirely justified in positing the existence of entities that are not intuitively given—entities the existence of which the practicing phenomenologist must adopt a position of neutrality (Husserl, 1913/1982, 114).

7.5 Physical Things

If the exact sciences of nature are, as Husserl claims, about the world of ideal objects, it follows that they are not about the real world. But this analysis does not mean that Husserl's work in the *Crisis* "negates" the truth-claims of the natural sciences, as Hans Wagner suggests (Wagner, 1974, 175). Rather, it means that the claims of the exact sciences of nature will always be indirect and only approximately true of the real world. In order to achieve a mathematically exact and objective representation of the real physical world, the physical sciences first resort to ideal cases and then proceed to approach real cases by successive approximation. One begins with the ideal case of a free-falling body (which exists nowhere in nature), and then adds in the resistance of the medium. One begins with frictionless planes (which again exist nowhere in nature), and then adds in the coefficient of friction. Through acts of abstraction and idealization there is an initial movement away from real objects and processes, and then a return to real objects by way of application

(Husserl, 1936/1970, 32). Exact, objective knowledge is possible only by way of a passage through the ideal; and for that very reason it will never be more than approximate knowledge of the real.

Husserl's main concern in the *Crisis* is not with the methods of the exact sciences of nature, but rather with a certain ontological interpretation of their results. Here he worries that the idealized version of the natural world as represented in modern mathematical physics has been effectively substituted for the real world, the world given in perceptual experience (Husserl 1936, 48–49). Following the Pythagorean/Platonic paradigm, this interpretation of the physical sciences degrades the physical world we perceive to the status of mere appearance, while the idealized version of the world as conceived in the sciences is elevated to the position of objective reality, the world that "really" exists. The phenomenological critique of modern mathematical physics is to expose the objective physical world as projected in the physical sciences for the construct it is by recourse to the mental processes of abstraction and idealization by which it is constituted.

Husserl's initial response to this dualist ontology, which posits a thing behind the thing, an objective physical thing behind the subjectively perceived object, begins in his 1910 essay, "Philosophy as a Rigorous Science" and reaches a provisional conclusion in *Ideas I*. In the section of "Philosophy as a Rigorous Science" devoted to naturalism, Husserl address himself to the ontological distinction between physical being and mental being with respect to the way they are given in experience. Unlike mental processes, physical things are always given in and through a series of appearances. In any case of such givenness we can distinguish between the thing that appears and its appearances. The appearances of a thing can change as the spatial orientation of the thing and the percipient vary with respect to each other, as the sensory apparatus of the percipient varies, or as the real conditions of the perceptual situation change. Yet, in and through the multiplicity of its changing appearances, the physical thing is experienced as one and the same. It is given as a unity throughout the changes in its "diverse sensible appearances" (Husserl, 1910/1965, 104). What the physical thing is "in itself," then, must be distinguished from its appearances. It is not to be identified with any one of its appearances, or even with the totality of its appearances. It retains its self-same identity though shifting phenomenal properties, and it is intersubjectively accessible as the same because, Husserl claims, it is a unity of real, causal properties in one all-embracing space and time (Husserl, 1910/1965, 104). Its appearances are not objective, if we mean by "objective" belonging to the object itself. Rather, they are subjective in the sense that they are relative to the orientation and sensory constitution of the perceiving subject.

Following up on this distinction, the physical sciences seek to penetrate the "vague medium of appearances" in order to make an exact determination of what the physical thing is "in itself" (Husserl, 1910/1965, 105). Such determination will always be carried out on the basis of the way the physical object appears, but in such a way as to leave the appearances behind (Husserl, 1910/1965, 105). "Physics … eliminates the phenomenal in order to seek for the nature that presents itself therein" (Husserl, 1910/1965, 101, translation modified). Every scientific inquiry into the nature of the physical thing "necessarily leads over into causal connections and terminates in the determination of corresponding objective properties subject to law" (Husserl, 1910/1965, 105). The physical sciences, then, seek to determine the object with respect to those properties it has "in itself," not with respect to properties it has only in relation to perceiving subjects. If an object appears red to us, it is not because it is red, but because it is in possession of a property in itself (here, a reflectance property) that appears red to the likes of us. To say that the object is red is to express what Husserl calls a "subjectively relative truth," that is, the truth that the object typically appears red to the standard perceiving subject of a certain species under conditions that have been intersubjectively established as normal. Physical science, on the other hand, wants to ascertain the objective truth about the physical thing, the truth that holds irrespective of the way the thing appears to a particular perceiving subject of a particular natural kind. In the physical sciences, the so-called secondary properties are taken to be mere appearances. They are not determined by the object itself, but in its relation to a perceiving subject whose sensory apparatus just happens to be constituted in a certain way (Husserl, 1930/1980, 53). In seeking to determine the object "objectively," the physical sciences countenance only those determinations of the objects that would hold for all subjects at all times. Such objective determinations will be "intersubjectively valid."

But if the intersubjective community for which these determinations are to hold is construed as the community of human beings, the physical sciences have yet to achieve the goal of objectivity to which they aspire. While the initial form of intersubjective validity will overcome the subjectivity of human individuals, it does not transcend the subjectivity of the human species as a whole. As Husserl indicates in *Ideas III*, the scope of the intersubjective community that constitutes the sense of objectivity in the physical sciences must be expanded to include not only all human subjects, but all possible subjects of whatever sensory constitution. Although the objective properties of the physical thing will always be determined on the basis of some specific form of sense experience, the properties themselves must be independent of all sense experience (Husserl, 1930/1980, 54). Thus the objectivity of nature, as conceived and pursued in the natural sciences, is essentially correlated to the ideal plurality of all possible perceiving subjects (Husserl, 1930/1980, 55). What belongs to the physical thing is what "any possible subject of the pre-delineated ideal community can bring out and determine in rational experiential thought on the ground of his 'appearances' and the communications of others concerning their 'appearances'" (Husserl, 1930/1980, 55). The world as perceived by human beings, then, receives the status of an appearance of "objective nature exclusively determined by 'exact' mathematical-physical predicates, absolutely not intuitable, not experienceable" (Husserl, 1930/1980, 56). On the horizon of the physical sciences there arises a "unique physical nature, with the one objective space and the one objective time, consisting of nothing but physical things that are characterized purely by concepts having the exactness ascribed in physics" (Husserl, 1930/1980, 56).

At this point, however, an ambiguity enters the scene. The physical thing as conceived in the physical sciences is to be distinguished from its appearances.

Should it also be distinguished from the perceived object? Should we think of the perceived object itself as a "mere appearance" of the thing of physics? The thing of physics is, in some important sense, never perceived (Husserl, 1930/1980, 58). Should we then think of the physical thing as an entity hidden behind perceived objects? Of the mathematically determined thing as the reality behind a veil of appearances? Husserl addresses himself this question section 52 of *Ideas I*. There he rejects the idea that the physical thing is the hidden cause of its own appearances, which, taken together, constitute a second and distinct object. The physical thing is in fact identical to the perceived object. The physical thing is itself given in "sensuous modes of appearance" (Husserl, 1913/1982, 120). Moreover, "the physical things which he [the physicist] observes, with which he experiments, which he continually sees, takes in his hand, puts on the scales or in the melting furnace; that physical thing, and no other, becomes the subject of the predicates ascribed in physics, such as weight, temperature, electrical resistance, and so forth" (Husserl, 1913/1982, 120–21). According to the Husserl of *Ideas I*, then, the mid-sized physical thing as determined with mathematical exactitude by the physical sciences is identical to the physical thing as perceived in everyday life. The reflectance property of a thing, as determined by the physical sciences, which appears red to us, is a property of the thing that appears red to us. It does not belong to a thing distinct from the object we perceive. "The perceived physical thing itself is always and necessarily precisely the thing which the physicist explores and scientifically determines following the method of physics" (Husserl, 1913/1982, 119). The sensuous intuition of a thing delivers a "mere this" an "empty X," which then "becomes the bearer of the exact determinations ascribed in physics which do not themselves fall within experience proper" (Husserl, 1913/1982, 119). While exact determinations ascribed to a physical thing cannot themselves be perceived, they are nonetheless determinations of a perceivable thing, a thing that has, in addition, determinations that can be perceived. Although the thing is determined quite differently in the physical sciences than in straightforward perception, the determinations are of the same thing and are "quite compatible." (Husserl, 1913/1982, 120).

7.6 Ideal Objects

In Part I of *Ideas I*, Husserl treats the relation between the mathematical discipline of geometry and the physical sciences as a special case of the relation between eidetic science and empirical science in general. It is a matter of principle, Husserl claims, that every empirical science is founded upon those eidetic disciplines that pertain to the essence of the abstract genera to which the objects in its research domain belong. In fact, the realization of the ideal of a completely rational empirical discipline, where "every particular included in it has been traced back to the particular's most universal and essential grounds" (Husserl, 1913/1982, 19), depends on the development of the relevant eidetic disciplines. Thus, the completion of the empirical sciences of nature depends upon the development of the regional ontology of nature. The more "rational" a science becomes, the more it approximates the ideal of an exact nomological science, the greater its scope and power will be (Husserl, 1913/1982, 19). The rationalization of a science is directly proportional to the degree to which the foundational eidetic disciplines have been developed and utilized. In the case of physics, "people made clear to themselves that the material thing is essentially *res extensa* and that geometry is therefore the ontological discipline relating to an essential moment of material thinghood, namely the spatial form" (Husserl, 1913/1982, 19). Here Husserl is thinking of spatial forms as universals that can be instantiated by material individuals. Thus the exact determination of physical things gained through the employment of geometry and allied mathematical disciplines provides access to the determinations of real physical things.

In his later work, however, Husserl gives a markedly different analysis. The objects of the exact science of geometry are not universals capable of instantiation on the part of real material things, but rather idealized objects wholly distinct from the realm of material things. While in *Ideas I* Husserl wanted to underscore the identity of the thing as determined in the physical sciences and the thing perceived, in the *Crisis* he emphasizes the difference between the two and assigns them to ontologically distinct camps. "The bodies familiar to us in the life-world are actual bodies, but not the bodies in the sense of physics" (Husserl, 1936/1970, 139).

The basis of the shift in Husserl's ontology of the referents of the modern physical sciences lies in his analysis of the mental processes and correlative objectivities that form the basis of their founding mathematical disciplines—especially geometry. In the *Ideas I*, geometry seizes upon the essential spatial structures and relations of the physical world (Husserl, 1913/1982, 36–37). But in the *Crisis*, Husserl maintains that the science of geometry is not based upon the process of ideating abstraction, an eidetic reduction from spatial fact to spatial essence, but rather on the process of idealization. The two processes in question yield categorically different objectivities: one, a universal; the other, an ideal object. Idealization produces an ideal object or an "Idea" in the Kantian sense. As a matter of principle, the latter cannot be instantiated by real material things. It can only be approximated. Thus ideas, or ideal objects, will never be given in sensuous intuition. Husserl makes this point early in the *Logical Investigations* in a statement that foreshadows his later analysis, "For a figure understood geometrically is known to be an ideal limit incapable in principle of intuitive exhibition in the concrete" (Husserl, 1900/1970, II 777). On the basis of the sensuous intuition of round objects, ideating abstraction delivers the concept of the inexact essence "roundness," which is, in turn, instantiated by round objects. But real round objects are not perfect circles. And they never will be. They are always more or less circular. The extension of the concept of roundness is inexact and open to differences in subjective interpretation. Inexact predicates, then, are not "objective" in the strict sense. They cannot produce or compel intersubjective agreement with respect to their precise extension. Idealization, by contrast, delivers an exact concept. Its extension is not a cluster of real objects bound together by rough resemblances, but the unique ideal object constituted in the process of idealization. There is only one perfect circle in the geometrical sense (or infinitely many, if one wants to produce a perfect circle for every possible radius—but again there would be only one perfect circle for each possible radius). In idealization, empirically given and imaginable real objects are aligned in a series converging on an ideal limit where real variation in one respect is reduced to zero. The idealized object then emerges at the limit of the series. As a limit, it differs qualitatively from the series of objects that converge upon it. In the example of round objects, the process of idealization aligns more or less round objects in a series converging on the ideal limit where the variation in distance between the center and all points on the circumference is reduced to zero. In this way there arises consciousness of a circle in the geometric sense. But, again, the circle is not the spatial essence of round objects. It is an ideal object. In the world of actual perceptual experience, "we find nothing of geometrical idealities, no geometrical space or mathematical time with all their shapes" (Husserl, 1936/1970, 50).

If the object of the exact, mathematically expressed laws of physics is not the real world, or the essential structure of the real world, but a distinct realm of ideal objects, one might conclude that the physical sciences do not give us knowledge of the real world, that there is indeed something fraudulent about them. Yet Husserl says that the method of the physical sciences "has the sense of achieving knowledge about the world" (Husserl, 1936/1970, 47). It does so indirectly and by way of approximation. It approaches the real by way of the ideal. The ideal gas laws hold for ideal, not real gases. But such laws can be modified to approximate the behavior of real gases through the incorporation of empirical constants as well as adjustments for differences in molecular size and inter-molecular attraction. In this way the physical sciences retain their exact character while at the same time counting as knowledge of the real world. Such knowledge, as long as it remains exact, will only be approximate. In mathematical physics, the real world, to use Husserl's metaphor, is measured for a "well-fitting garb of ideas" (Husserl, 1936/1970, 51). The garb of ideas is not the real world itself. To think so would be to commit the platonic error of misplaced concreteness. But the garb of ideas fits the real world to a greater or lesser degree. For that reason, it is informative. One can tell the height and proportions of a man by examining his suit.

Pure geometry affords objective (and exact) knowledge of ideal objects; applied geometry affords objective (but approximate) knowledge of real objects. In its employment of the mathematical disciplines, physical science follows a basic two-step process in preparing the natural world for representation through the exact formulation of empirical laws, a process that moves between the pure and the applied. First: the step of idealization, which produces an ideal case. Unlike real cases, the ideal cases will admit of exact and uniform determination. Second: the ideal case is applied to real cases by way of approximation. As an example of step one of this method as practiced in the early period of modern physics, take Galileo's paradigmatic treatment of free-falling bodies in the *Two New Sciences* (1638). In our common, everyday experience of falling bodies, we see that the rate of fall varies with respect to a number of real factors—the weight and shape of the falling body, the density of the medium through which it falls, and the like. A bowling ball falls faster through the air than a feather. A stone falls faster through the air than through water. With respect to real, observed cases of falling bodies, Galileo notes that "the inequality of speeds is always greater in the more resistant mediums of different resistances" (Galileo 1638/1974, 72). This observed fact suggests a convergent series: "Movables of different weight differ less and less in speed as they are situated in more and more yielding mediums; and that finally, despite extreme difference in weight, their diversity of speed in the most tenuous medium of all ... is found to be very small and almost unobservable" (Galileo, 1638/1974, 76). On the horizon of such a series there then arises the ideal limit, where the resistance of the medium is reduced to zero: "In the void all speeds would be entirely equal" (Galileo 1638/1974, 76). But note the shift to the subjunctive counterfactual in the statement of the ideal case: as a void exists nowhere in nature, there are, strictly speaking, no free-falling bodies. The free-fall is the ideal case. But precisely because it is an ideal case, where the rate of acceleration is absolutely uniform, it can be expressed in terms of an exact empirical law that holds irrespective of the variations that inevitably crop up in real cases. This law can, in turn, be applied to real cases through successive complication (i.e. by adding in the resistance of the medium and other relevant factors).

7.7 Theoretical Entities

In *Reading the Book of Nature*, Peter Kosso expressed a widely shared view in contemporary circles of the philosophy of science when he wrote that instrumentalism "gives up what is most worth doing in science, namely, understanding what is happening behind the scenes in the realm of unobservables" (Kosso, 1992, 95). If Husserl's phenomenological philosophy of science is indeed committed to instrumentalism in the standard sense, it is unlikely that it has much to offer to an understanding of the physical sciences as they are practiced and understood today, sciences that have ever since the turn of the last century been operating in a robust, impressive, and ever-expanding theoretical dimension. Taking Husserl's critique of science in the *Crisis* as fundamentally hostile to the theoretical dimension of science, Patrick Heelan claimed that "contemporary phenomenology cannot contribute much of value to the philosophy of science unless it is enlarged in some way" (Heelan, 1967, 375). Heelan proposed an enlargement by way of a hermeneutical addition to Husserl's purely descriptive methods.

I maintain that Husserl's phenomenology is wholly consistent with a realistic interpretation of scientific theories, and that his "instrumentalism" is limited to an account of scientific laws. What Husserl denies exist as real physical things are such entities as frictionless planes, perfectly elastic bodies, extensionless mass points, incompressible fluids, and the like. Such idealized objects are the proper objects of the exact and uniform laws of the physical sciences. Expressing functional correlations between abstract and simplified physical variables in the exact language of mathematics, these laws make the prediction of events in the lifeworld possible on a scale unknown to prescientific practical life. Husserl's phenomenological analysis of the meaning of such laws and the ontological status of their objects does not, however, touch upon the question of the existence of the entities postulated by scientific theories designed to explain the functional correlations captured in those laws. Although the behavior of the entities postulated by such theories may in turn be represented in exact mathematical laws, the theories themselves are not purely mathematical, as Ernan McMullin and Nancy Cartwright have correctly pointed out (McMullin 1979, 32; Cartwright, 1983, 6). They possess real physical content, and the entities they postulate—atoms, electrons, quarks, strong and weak nuclear forces, and the like—if they exist, exist as real physical entities with causal capacities.

Husserl himself may have been disinclined to develop a phenomenology of theoretical science. An ambitious phenomenology of theoretical science would certainly count as an enlargement of the scope of phenomenological research as Husserl himself practiced it. But it need not represent a foreign addition to its native conceptual resources, as Heelan and others have suggested.

To identify those resources, it may help to consider one way in which Husserl commentators have sought to enlarge phenomenology in order to do justice to theoretical dimension of the physical sciences. The problem, as they identify it, is that Husserl unduly limited his phenomenology to unaided perception, to what could be given to the naked human eye. If he had considered the augmentation of our perceptual powers through instrumentally mediated perception, he would have opened phenomenological inquiry to experience beyond immediate sense perception. The central claim in this approach is that advances in technology and instrumentation have made it possible to bring previously postulated theoretical entities into the realm of the observable and thus render them open to phenomenological description on Husserl's own terms (in which case, we should note, they are no longer theoretical entities, since they are now observable). In *Husserl's Missing Technologies*, Don Idhe argues such entities as gas molecules are real, material entities that we can now perceive by way of scientific instruments. "Put simply, CFGs, CO_2, and ozone are not pure shapes [i.e., ideal mathematical objects] but materially presentable material entities instrumentally mediated" (Idhe, 2016, 82). Again: "If science is technoscience and instrumentally embodied, such particles and gases become perceivable in visualizable form through imaging technologies. This, however, calls for a phenomenology of instrument use which recognizes seeing through instruments" (Idhe, 2016, 81). The drawback of Husserl's phenomenology is that it limits the real to unaided reach of human perception and in so doing ignores the role scientific instrumentation can play in the extension of human perceptual powers into previously unexplored regions of physical reality.

In a similar line of criticism, Charles Harvey claims that Husserl "was victim to the historical contingencies surrounding the vision of egos" (Harvey, 1986, 304). The invention of scientific instruments such as the cloud chamber now makes it possible to perceive what were once posited as mere theoretical entities. Invented by the Scottish physicist Charles Thomson Rees Wilson in 1911, cloud chambers, Harvey claims, now make it possible for us to see ions. Alpha or beta particles are shot through a chamber containing a supersaturated atmosphere. The particles, so the accompanying theory goes, create tracks of condensation as they grab or knock off electrons from the ambient gas molecules, thus creating negatively charged ions that serve as momentary centers of small water or alcohol droplets. Some, like Harvey, have claimed that the development of cloud chambers has now made it possible for us to see ions. They are not just hidden theoretical entities to be described in mathematical terms alone; they are now to be admitted into membership of our real perceivable universe thanks to the advance of technology (Harvey, 1986, 304).

But is this the case? What do we actually see in the cloud chamber? To the untutored eye, nothing more than wispy lines of droplets in a sealed chamber. To the informed eye, condensation or ionization tracks. But the presence of ions—the alpha or beta particles—is clearly inferred, not seen. At most we "see that" ions were present on the basis of seeing the condensation track, where the "see that" locution is just another way of saying "apprehend that." More formally, we see that p is the case on the basis of seeing X, where p is some proposition and X is some object. Seeing that p is the case is the result of an inference from what is seen, an inference undergirded by a relevant causal theory—in this case a theory having to do with electrostatic forces. Philosopher of science Bas van Fraassen was entirely right, it seems to me, when pointing out that, "while the particle is detected by means of the cloud chamber, and the detection is based on observation, it is clearly not a case of the particle's being observed" (Van Fraassen, 1980, 17).

If we reject this way of making theoretical entities into observable objects and thus available for phenomenological description, what avenues remain for a reconciliation between phenomenology and theoretical science? Does Husserlian phenomenology possess the resources to illuminate, rather than reject or ignore, theoretical science? I believe the required resources lie in Husserl's early and brief description of indicative sign consciousness, which establishes the conditions under which a perceived physical thing can serve as an indicative sign of the existence of another physical thing. In the first chapter of the First Logical Investigation, Husserl describes the way in which the perception of one physical thing can properly "motivate" the belief in the existence of another physical thing, where the first physical thing is perceived and the second is not (Husserl, 1900/1970, I 270). To a person well acquainted with the relevant causal regularities, the perception of smoke motivates the belief in the existence of a fire even when the fire itself is not seen. To that person, the smoke perceived serves as an "indicative sign" of the presence of fire.

In this prosaic example, however, we have not moved beyond the domain of Humean association between the "constant conjunction" of observed events, where the perception of one motivates the belief in the other. Both events are perceivable; and if, in the past, both have been repeatedly perceived in succession, the conjunction of the two gives rise to the sub-rational habit of associating the one with the other. In the experimental contexts of theoretical science, however, the effect is observable, but the cause is not. The motivation, although based in the perceptual experience, depends on the imagination and theoretical construction of the causal relation where the presence and features of the observable effects serve as indicators of the presence and features of the hidden cause. Consider Earnest Rutherford's alpha particle scattering experiments at the beginning of the last century. In those experiments, the spatial distribution of scintillations on a zinc sulfide screen served to indicate the mass structure of the gold atom. Against the background of a theory pertaining to the causal interaction of alpha particles, gold atoms, and the zinc sulfide screen, the observation and measurement of the angles of deflection represented by the scintillations on the screen motivated beliefs about the existence and physical properties of a theoretical entity. The scintillations became indicative signs of a hidden reality.

The motivation for belief in theoretical contexts cannot be based on a simple induction over observables, the "constant conjunction" of which Hume fondly speaks, but rather on controlled experimental conditions in which theoretical hypotheses are tested. The testing will not only be based on the "empirical adequacy" of a theory, its ability to generate predictions that are borne out in observation, but also on other methodological criteria such as simplicity, internal plausibility, and consistency with what we already think we know. These factors, involving complex founded judgments and inferences to the best among competing explanations, would call for a sophisticated and nuanced phenomenology–a phenomenology that is undeveloped at this point largely because of long-standing prejudices against the possibility of a robust phenomenology of theoretical science of the Husserlian sort. If I have not advanced that phenomenological project here, I hope to have at least eliminated many of the prejudices against it and indicated how it might proceed.

References[1]

Ballard, E. (1971). *Philosophy at the crossroads*. Baton Rouge, LA: Louisiana State University Press.

Boyd, R. N. (1973). Realism, underdetermination, and a causal theory of evidence. *Noûs, 7*(1), 1–12.

Cartwright, N. (1983). *How the laws of physics lie*. Oxford, UK: Oxford University Press.

Galilei, G. (1638/1974). *Two new sciences* (S. Drake, Trans.). Madison, WI: University of Wisconsin Press.

Godfrey-Smith, P. (2003). *Theory and reality*. Chicago: University of Chicago Press.

Harvey, C. (1986). Husserl and the problem of theoretical entities. *Synthese, 66*, 309–291.

Heelan, P. (1967). Horizon, objectivity and reality in the physical sciences. *International Philosophical Quarterly, 7*, 375–412.

Husserl, E. (1900/1970). *The logical investigations* (Vols. 1&2, J. N. Findlay, Trans.). London: Routledge & Kegan Paul.

Husserl, E. (1910/1965). Philosophy as a rigorous science. In *Phenomenology and the crisis of philosophy* (Q. Lauer, Trans.) (pp. 71–147). New York: Harper Row.

[1]Note: In the citations for Galileo, Husserl, and Schlick I give the date of original publication—or composition in the case of Husserl's Crisis and Ideas *III*—as well as the date of publication for the English translation. This to help the reader keep track of the chronological order of the texts.

Husserl, E. (1913/1982). *Ideas pertaining to a pure phenomenology and to a phenomenological philosophy*, first book (F. Kersten, Trans.). The Hague, The Netherlands: Martinus Nijhoff Publishers.

Husserl, E. (1929/1978). *Formal and transcendental logic* (D. Cairns, Trans.). The Hague, The Netherlands: Martinus Nijhoff Publishers.

Husserl, E. (1930/1980). *Ideas pertaining to a pure phenomenology and to a phenomenological philosophy*, third book (T. E. Klein, & W. E. Pohl, Trans.). The Hague, The Netherlands: Martinus Nijhoff Publishers.

Husserl, E. (1931/1969). *Cartesian meditations* (D. Cairns, Trans.). The Hague, The Netherlands: Martinus Nijhoff Publishers.

Husserl, E. (1936/1970). *The crisis of European sciences and transcendental phenomenology* (D. Carr, Trans.). Evanston, IL: Northwestern University Press.

Idhe, D. (2016). *Husserl's missing technologies*. New York: Fordham University Press.

Kisiel, T. (1973). On the dimensions of a phenomenology of science in Husserl and the young Dr. Heidegger. *Journal of the British Society for Phenomenology, 4*, 217–234.

Kosso, P. (1992). *Reading the book of nature*. Cambridge, MA: Cambridge University Press.

McMullin, E. (1979). Compton on the philosophy of nature. *Review of Metaphysics, 33*(1), 29–38.

Patzig, G. (1977). Husserl on truth and evidence. In J. N. Mohanty (Ed.), *Readings on Edmund Husserl's "Logical Investigations"* (pp. 179–196). The Hague, The Netherlands: Martinus Nijhoff.

Pietersma, H. (1977). Husserl's views on the evident and the true. In F. A. Elliston & P. McCormick (Eds.), *Husserl: Expositions and appraisals* (pp. 38–53). Notre Dame, IN: University of Notre Dame Press.

Psillos, S. (1999). *Scientific realism: How science tracks the truth*. London: Routledge.

Rizzacasa, A. (1979). The epistemology of the sciences of nature in relation to the teleology of research in the thought of the later Husserl. *Analecta Husserliana, 9*, 73–84.

Rouse, J. (1987). Husserlian phenomenology and scientific realism. *Philosophy of Science, 54*(2), 222–232.

Salmon, W. (1989). Four Decades of Scientific Explanantion. In W. Salmon & P. Kitcher (Eds.), *Minnesota studies in the philosophy of science* (Vol. 13) (pp. 3–219). Minneapolis, MN: University of Minnesota Press.

Schlick, M. (1932–1933/1991). Positivism and realism. In R. Boyd, P. Gasper, & J. D. Trout (Eds.), *The philosophy of science* (pp. 37–55). Cambridge, MA: MIT Press

Van Fraassen, B. (1980). *The scientific image*. Oxford, UK: Oxford University Press.

Wagner, H. (1974). Husserl's ambiguous philosophy of science. *The Southwestern Journal of Philosophy, 5*(2), 169–185.

Welton, D. (1983). *The origins of meaning: A critical study of the thresholds of Husserlian phenomenology*. The Hague, The Netherlands: Martinus Nijhoff Publishers.

Chapter 8
A Match Made on Earth: On the Applicability of Mathematics in Physics

Arezoo Islami and Harald A. Wiltsche

Abstract Anyone interested in understanding the nature of modern physics will at some point encounter a problem that was popularized in the 1960s by the physicist Eugene Wigner: Why is it that mathematics is so effective and useful for describing, explaining and predicting the kinds of phenomena we are concerned with in the sciences? In this chapter, we will propose a phenomenological solution for this "problem" of the seemingly unreasonable effectiveness of mathematics in the physical sciences. In our view, the "problem" can only be solved—or made to evaporate—if we shift our attention away from the *why-question*—Why can mathematics play the role it does in physics?—, and focus on the *how-question* instead. Our question, then, is this: How is mathematics actually used in the practice of modern physics?

8.1 Introduction

Mathematics is everywhere. For some to their pleasure, for some to their agony and perhaps for some to their bafflement. We use it in every-day life as well as in the sciences. We use it as a tool of calculation and inference, and it also gives us a "deeper", more quantitative, more exact understanding of "how things really are". We use it to make predictions about the future of our universe or to trace things back to the past, as in the big bang model. Theoretical physicists sit at their desks and make quantitative predictions that later, sometimes decades later, experimentalists are able to verify. You open any textbook in engineering and science, from physics to economics, and you will encounter a plethora of mathematical symbols. The mathematics can be arithmetic, geometry, algebra,

A. Islami (✉)
Department of Philosophy, San Francisco State University, San Francisco, CA, USA
e-mail: arezooi@sfsu.edu

H. A. Wiltsche
Department of Culture and Society (IKOS), Linköping University, Linköping, Sweden
e-mail: harald.wiltsche@liu.se

H. A. Wiltsche, P. Berghofer (eds.), *Phenomenological Approaches to Physics*, Synthese Library 429, https://doi.org/10.1007/978-3-030-46973-3_8

calculus, abstract algebra, linear algebra, topology, algebraic geometry and so on. In public discourse some theologians consider the applicability of mathematics even as strong evidence for the existence of God.

The philosophical problem surrounding the relation between mathematics and the empirical sciences is rather obvious: Why is it that mathematics is so effective and useful for describing, explaining and predicting the kinds of phenomena we are concerned with in the sciences? Philosophers, in their attempt to make sense of the enormous success of science, thus face what is commonly called the *applicability problem*, the problem of explaining the intimate tie between mathematics and science.

This problem is revived and reformulated by the physicist Eugene Wigner under the striking title of the *Unreasonable Effectiveness of Mathematics in the Natural Sciences* (Wigner 1960). While Wigner mostly focuses on the case of modern theoretical physics, and its relationship with mathematics, he still fails to find a satisfactory solution for the applicability problem (Islami 2017). In a nutshell, Wigner's "solution" is that there is no solution: all we can say about the applicability problem is that "the enormous usefulness of mathematics in the natural sciences is something bordering on the mysterious" (Wigner 1960, 223). On Wigner's view,

> [t]he miracle of the appropriateness of the language of mathematics for the formulation of the laws of physics is a wonderful gift which we neither understand nor deserve. We should be grateful for it and hope that it will remain valid in future research and that it will extend, for better or for worse, to our pleasure, even though perhaps also to our bafflement, to wide branches of learning. (Wigner 1960, 237)

Wigner is not the only one who thinks that the applicability problem cannot be solved. The physicist Paul Dirac echoes Wigner's remarks:

> [T]he mathematician plays a game in which he himself invents the rules while the physicist plays a game in which the rules are provided by Nature, but as time goes on it becomes increasingly evident that the rules which the mathematician finds interesting are the same as those which Nature has chosen. (Dirac 1939, 124)

And David Hilbert in a course lecture in (1919) says:

> We are confronted with the peculiar fact that matter seems to comply entirely to the formalism of mathematics. There arises an unforeseen harmony of being and thinking, which for now we have to accept like a miracle. (Hilbert 1992, 69; our translation)

The physicist David Gross takes the same line when he writes that it is "something of a miracle that we are able to devise theories that allow us to make incredibly precise predictions regarding physical phenomena" (Gross 1988, 8372).

8.2 The Applicability Problem(s)

The applicability problem, commonly viewed, concerns the relationship between mathematics and sciences, social as well as natural. Thus commentators on Wigner's applicability problem have argued for a range of positions from the *Unreasonable*

Ineffectiveness of Mathematics in Economics (Velupillai 2005) to the *Unreasonable Ineffectiveness of Mathematics in Biology* (Lesk 2000, 29). The underlying assumption in treating applicability as a unified problem is a view of science as having a universal experimental method (as well as assigning a single method to mathematics itself). For us, however, modern physics presents a peculiar and especially interesting case of the applicability problem where mathematics plays a more fundamental role than being a mere language for the description of the physical phenomena. Following Wigner, then, we focus on the case of modern theoretical physics.

At the heart of the applicability problem lies what we call the *distinctness thesis*, i.e. the thesis that mathematics and the physical sciences are categorically distinct. According to a widespread view, the challenge presented by the applicability problem is then to explain why, despite their distinctness, mathematics and physics are so closely intertwined. Yet, it is important to realize at this point that what the distinctness in question precisely amounts to can vary greatly depending on one's philosophical background assumptions.

Consider, for instance, a realist both with regard to mathematics and physics. A realist of this kind might be happy to follow Gödel in viewing mathematics as analogous to the physical sciences concerning their basic methodological outlook: All disciplinary differences notwithstanding, mathematics and physics are fundamentally similar in that both seek to describe a mind-independent reality that determines the truth-values of propositions in the respective areas (Gödel 1983, 456). However, even if mathematics and physics are viewed as analogous in this way, there is still an important sense in which they are categorically distinct: While the subject matter of physics is typically said to consist of concrete physical objects, mathematical realism is usually associated with the view that mathematical objects such as primes or polynomials exist outside of space and time, and independently of the causal relations in which concrete physical objects stand. And it is this *ontological distinctness* that gives rise to one specific version of the applicability problem: Why is it that knowledge about the *abstract* realm proves to be so enormously effective in generating knowledge about world of *concrete* physical phenomena? In order to solve this version of the applicability problem, a number of thinkers from the ancient Pythagoreans over seventeenth century scientists such as Kepler or Galileo to modern-day physicists like Max Tegmark have advocated some version of *mathematical monism*: Although our experiences tell us otherwise, it is argued that reality is ultimately *nothing but* mathematical structure. According to its proponents, the main advantage of this view is that it circumvents the ontological version of the distinctness thesis. And this, of course, also prevents the applicability problem from arising because "our successful theories are not mathematics approximating physics, but mathematics approximating mathematics" (Tegmark 2008, 125).

Given what has been said thus far, one might wonder how the distinctness thesis plays out under different philosophical background assumptions. Consider, for instance, a formalist who denies that mathematics should be viewed as a body of propositions with determinate truth-values, describing an abstract sector of reality.

In order to avoid the metaphysical challenges posed by mathematical realism, formalists think of mathematics as a game-like endeavor in which strings of symbols are manipulated according to freely stipulated rules. Like chess pieces, the symbols with which the game of mathematics is played do not denote anything. The only meaning these symbols have is attributed to them by the mathematician who accepts certain arbitrary rules in order to participate in a game-like activity. Instead of being constrained by a theory-independent reality, the game of mathematics is only driven by inner-mathematical virtues such as rigor, elegance, simplicity, manipulability or formal beauty. For instance, Wigner defines mathematics in precisely this spirit, namely as "the science of skillful operations with concepts and rules invented for this purpose" (Wigner 1960, 2).

It is clear that for formalists, the applicability problem does not arise as the result of the *ontological* version of the distinctness thesis. Since, on their view, there is no abstract realm to begin with, there is also no ontological hiatus that would make the coordination between mathematics and physics appear particularly mysterious. This, however, does not mean that the applicability problem does not arise in a different form. It is easy to see why this is so: From a formalist perspective, mathematics is an arbitrary human creation which is only driven by genuinely inner-mathematical virtues. Physics, on the other hand, is first and foremost constrained by a basic commitment to empirical adequacy[1] and hence by the methodological principle that the ultimate guide in matters of theory-acceptance is adequacy with respect to the segment of reality a theory purports to describe.[2] It is this difference that gives rise to the *praxiological* (or *methodological*) distinctness thesis, and hence to a non-ontological version of the applicability problem: Why is it that a cognitive practice that is guided by virtues like rigor, elegance, simplicity, manipulability or formal beauty proves exceptionally successful in an area where it hard to see why these inner-mathematical virtues are epistemically relevant at all? Seen from this perspective, then, and building on a metaphor used by Wigner, our situation in

[1]To be sure, it could be pointed out that criteria other than empirical adequacy do have their place in physics, especially when physicists invoke super-empirical virtues to break underdetermination on the level of empirically equivalent theories. However, apart from the fact that it is unclear if such super-empirical values should qualify as genuinely epistemic, the role of virtues like rigor, elegance, simplicity, manipulability or formal beauty seem much more fundamental in mathematical research. In mathematics, these virtues are not merely the means to decide between otherwise indistinguishable theories; they are rather the guiding principles for the development and assessment of theories.

[2]Of course, one could question whether such a "empirical paradigm of theory assessment" is adequate in light of more recent developments in contemporary physics. For instance, since the scale of string theory is roughly of the order of the Planck length, the chances of finding direct empirical confirmation of the theory's core claims seem rather remote. Given this lack of empirical backing, it is not surprising that string theorists are guided to a much stronger extent by considerations resembling those used in pure mathematics. However, since the jury is still out on whether string theory is in fact too far detached from the binding norms of experimental science or, alternatively, on how the research practice of string theorists would change if empirical tests were possible, we will ignore this case here (cf., for further discussions, Penrose 2004, Smolin 2006, Dawid 2013 and Hossenfelder 2018).

physics is like that of a person who encounters a bunch of keys on display in an art exhibition. Although the artist who made the keys assures us that she had no practical purposes in mind, we find to our surprise that some of the keys unlock our doors at home. Confronted with this situation, two options seem to be available: either we admit that the situation is indeed highly mysterious; or we abductively infer that there must have been some pre-established harmony between the keys and our doors at home. Mark Steiner argues for the second option, claiming that we should accept that we are living in a "user-friendly universe" i.e. a universe whose deep structure is somehow attuned to the workings of the human mind and hence to the products of mathematical reasoning (Steiner 1998). Steiner's "anthropocentrism" is reminiscent of the old rationalist idea that it is part of God's creation to have made the world and our minds to fit like hand to glove. Hence, if, time and time again, mathematical thinking proves to be the royal road to physical knowledge about the world, then this should be taken to suggest that the deep structure of physical reality is of a kind that makes it amenable to description by mathematical reasoning.[3]

8.3 An Alternative Approach

The aim of the previous section was not to give a complete exposition of the various attempts to find a solution of the applicability problem. The aim was rather to indicate that *the* applicability problem does not exist: Depending on the philosophical background assumptions one accepts, there are different ways of conceiving the relationship between mathematics and physics. And depending on the version of the distinctness thesis one accepts, different versions of the applicability problem arise.

[3]It should be noted for the sake of completeness that there have been skeptical voices as well, effectively denying the distinctness thesis and, consequently, the existence of the applicability problem. The argument, in a nutshell, is this: If mathematics is really just a game-like invention, and if, furthermore, its inventors had genuinely physical purposes in mind, then there is nothing mysterious about the usefulness of mathematics in physical research. This view can easily be substantiated by several well-known examples from science history. Leibniz and Newton invented differential and integral calculus for the explicit purpose to describe systems with trajectories through space and time with forces acting on them. Given this practical background and given the ingenuity of its inventors, the applicability of differential and integral calculus is no more surprising than the fact that hammers are well-suited to drive nails. However, such a deflationary stance toward the applicability problem faces several problems: First, it would be a serious mistake to reduce the role of mathematics to that of a convenient tool for the successful framing of physical descriptions. Quite the opposite: In the face of lacking empirical data, physicists quite often turn to mathematics itself in order to discover novel theories or even previously unknown physical phenomena (cf., e.g., Steiner 1989, 1998 and Colyvan 2001). Second, and closely related, it is also not true that the mathematical tools that proved useful in physics were always developed with genuinely physical purposes in mind. Some of the most productive mathematical innovations such as complex numbers, non-Euclidean geometries or spinors were regarded as purely theoretical first and went on to demonstrate their high practical relevance decades, sometimes even centuries later.

While some may think of the applicability problem as a metaphysical issue that concerns the coordination between two categorically distinct ontological regions, others may focus on the methodological question of why a cognitive practice that is driven by genuinely inner-mathematical considerations proves successful in physics. However, there is one characteristic that is common to all versions of the applicability problem as well as to their attempted solutions: Accepting the success of physics, and building on certain views concerning the nature of mathematics and physics, philosophers consider the applicability of mathematics a phenomenon that requires a philosophical *explanation*. Hence, philosophers pondering over (some version of the) applicability problem are generally in the business of finding an answer to a specific *why-question*—the question as to why mathematics can play the role it does in modern physics.

At first glance, it may seem natural enough to view the applicability of mathematics as a phenomenon that prompts an explanation-seeking why-question. After all, aren't philosophy and the sciences continuous in the sense that both seek to identify sufficiently interesting phenomena for which they then account by means of theoretical explanations? As natural as this view may seem, phenomenologists have traditionally been skeptical of this explanatory paradigm in philosophy. To be sure, the point is not that it is impermissible under any circumstances to construct philosophical theories for particular explanatory purposes. The point is rather that there is a tendency among philosophers to jump to explanations too quickly, thereby ignoring the fact that what is considered to be the explanandum is oftentimes contingent upon presuppositions that are not properly grounded in a faithful and unbiased description of the *things themselves*. Phenomenologically construed, it is only if we abstain from immediately jumping to explanations that we can prevent the risk of engaging in what Husserl has contemptuously called "standpoint philosophy" (*Standpunktphilosophie*): Instead of forcing problems into a particular (and potentially artificial) theoretical mold, phenomenologists are driven by a deep respect for the *phenomena*, i.e., the things exactly as they are given in experience. On a phenomenological view, many philosophical problems could be solved—or even better: made to evaporate—if we resisted the temptation to interfere with pre-established metaphysical, ontological or epistemological schemes and put more effort in a faithful description of the phenomena.

In what follows, we will approach the applicability problem from a phenomenological point of view. This means, first, to shift attention away from the *why-question* and focus on the *how-question* instead. Hence, our aim will not be to explain a problem that, already at the level of its formulation, is contaminated with certain philosophical preconceptions about the nature of mathematics and physics. Rather, utilizing Husserl's idea of "an epoché in regard to all objective theoretical interests" (Husserl, 1970, 135; Wiltsche, 2012, 126–127), ready-made ontologies and epistemologies of mathematics and science will be, to use Husserl's apt term, *bracketed* in order to arrive at a more faithful and unbiased understanding of how mathematics is actually applied in the physical sciences.

Yet, approaching the applicability problem from a phenomenological viewpoint also means, second, to take seriously that mathematics is always applied by a community of historically, culturally and bodily situated subjects. As trivial as this may seem at first sight, it is a fact which, once recognized, opens up two basic avenues for inquiry: Addressing the how-question from a first-person perspective puts us in a position to gain a firmer grasp of the intentional structures that are operative in concrete cases of mathematical-physical theorizing. Drawing from the rich resources of phenomenological investigations of human consciousness, we will seek to identify and describe the sorts of intentional acts by means of which mathematical objects become applied in present-day physical research. However, while unveiling the intentional accomplishments that underlie contemporary physics is without doubt an important task, this *synchronic* view must be supplemented by a *diachronic* investigation into the *genetic origin* of the very idea of an amalgamation between mathematics and physics. It is very natural for us today to take the possibility of mathematized physics simply for granted. However, as the late Husserl of the *Crisis* was at pains to show, the early seventeenth century attempts of a complete mathematization of empirical reality mark a fundamental turning point in intellectual history whose philosophical consequences are not sufficiently understood until the present day. In our view, and for reasons that will become clear, the how-question can only be answered in a satisfactory manner if it is approached from a synchronic as well as from a diachronic perspective, and if ungrounded metaphysical and epistemological preconceptions about mathematics and physics are held to a minimum.

8.4 A Diachronic Investigation Into the Origin of Modern Physics

As the section title suggests, we begin by focusing on the case of modern physics as opposed to other empirical sciences and Aristotelean physics. The goal is to reconstruct or do a genetic investigation of the birth of modern physics: we ask how physics as a mathematized science became possible in the first place.

Aristotelean physics, if you allow the title, was not mathematized. It was Galileo (of course, not single-handedly) who gave birth to physics as a mathematized science. We admit that mathematics can be applied to Aristotelian physics in the same way that logic can be applied to philosophy. But mathematization and application are two distinct issues, as will become clear in the following. Simply put, one can do Aristotelean physics (and that's how it was done for centuries) without ever using mathematics while the same is not possible in the case of modern physics.

How, exactly, is mathematics used in modern physics? To be sure, there is the instrumental use: once we have the mathematical formulation of laws, the observables and the initial conditions, we can use mathematics to make calculations, predictions and inferences. But this role of mathematics is no different from its use in everyday life: If Arezoo puts five apples in the basket, and Harald adds another two, the grand total of apples can be determined arithmetically. But this is not the role of mathematics that is particularly "mysterious": Even though we might be deeply impressed by the effectiveness with which mathematics allows us to draw inferences from known facts or pre-established theories, the ability to do so is not significantly more mysterious than the general human capacity to use the powers of reason to go beyond the immediately given.

In modern physics, however, the relation between mathematics and the objects under consideration is typically much more intimate, especially at the level of fundamental laws. Whereas apples in a basket are accessible independently from the mathematical tools we might apply to them, any attempt to separate Newton's motion, Einstein's curvature of space-time or Schrödinger's wave function from the mathematical formalisms in which they are couched is doomed to failure. We take this as a first indication of what it means to think of modern physics as a mathematized science: Instead of just being applied to objects that could also be ascertained otherwise, mathematics seems to be crucial for the very *constitution* of the objects modern physics purports to describe. Hence, as a first approximation we might say that modern physics is a mathematized science in the sense that, at its core, it deals with idealized, exact objects—objects that are nowhere to be found in our ordinary experience of the world.

Let's begin with Galileo as the initiator of this kind of mathematized physics. The issue here is not Galileo's name, nor his individual contribution or the context of his work. The focus is on a rational reconstruction of a process: the mathematization of physics which is philosophically illuminating to us. Our interest in Galileo is thus similar to that of philosophically-minded historians such as Maurice Clavelin who rightly remarked that "Galilean science was first of all a transition from one conceptual framework to another, the replacement of one explanatory ideal with another and an unprecedented fusion of reason and reality" (Clavelin 1974, xi).[4]

Galileo suggested that nature is written in the language of Euclidean geometry. He regarded nature as composed not of objects of ordinary experience, but of fundamentally different objects such as triangles and circles. Of course he did not discover this as a fact about nature. Galileo mainly relied on theological arguments[5] in order to justify the appropriation of the method of Euclidean and Archimedean proportional geometry for the study of nature. The effect of this methodological innovation was a science whose objects were, at least in part, radically different from the objects we encounter under normal lifeworldly circumstances.

While in his workshop, Galileo worked with surfaces that could be polished in order to increase their smoothness. His remarkable innovation was to proceed to

[4]To be sure, this is also the approach that is taken in the famous section 9 of Husserl's *Crisis* where the name "Galileo [. . .] is the exemplary index of an attitude and a moment, rather than a proper name" (Derrida, 1989, 35; Husserl, 1970, 57).

[5]Galileo's argument, in a nutshell, is that rigorous mathematical proofs allow us to participate in the perfection of God's knowledge. Hence, when an empirical problem can be dealt with mathematically, Galileo feels warranted to regard the geometrical model of the empirical target system as a truthful representation of how God perceives reality (cf., e.g., Galilei 1967, 103; McTighe 1967; Redondi 1998).

the limit of this process in the imagination. As surfaces can be thought of getting smoother and smoother, he declared, there must also be a perfectly smooth surface: a surface with no friction. Of course, such a surface cannot exist in the real world, just as the figures of Euclid—lines, triangles or circles—could not. They too are constructed as the ideal limit of a process: A line with no thickness is an ideality, the limiting pole of a sequence of lines with lesser and lesser thickness. The truly revolutionary aspect of Galilean science is to elevate such constructed idealities to the principle means through which all of reality must be studied. One of the many telling examples in Galileo's oeuvre is his treatment of projectile motion (cf., for the following, Wiltsche 2016, sec. 4): In order to "prove" that projectiles follow a semi-parabolic path, Galileo introduces a scenario which is built up from geometrical objects such as frictionless planes and perfectly spherical projectiles, and in which, consequently, no energy is lost due to friction or perturbation effects, in which objects are not attracted by a common center of gravity and in which the surface of the earth is treated as an Euclidean plane. Although Galileo's "proof" crucially depends on these counterfactual conditions, it is the idealized scenario that becomes prescriptive for our everyday experience: This is because, according to Galileo, the idealized scenario is nothing less than a truthful representation of how projectile motion would *really* look like if all causal impediments and accidents could be put aside. Hence, it is only through idealized objects such as spheres, planes or lines that we can catch a glimpse of the deep structure of reality.

As the late Husserl suggests, Galileo's methodological innovation inaugurated a process in which communities of scientists throughout history replace aspects of the natural surrounding world with mathematical objects. As a result of this process, the world of the modern mathematical sciences is not simply a continuation of ordinary experience and common sense, but rather a radical break. As we will see in more detail later, most of the objects with which we are concerned in physics are constituted very differently from the objects we encounter under normal, lifeworldly conditions. If this is correct, the question naturally arises: What does it mean for our understanding of science and reality that the former seems to proceed through a continuing transformation of the latter?

Galileo himself answered this question in a rather straightforward manner by raising his scientific method (an epistemic achievement) to the status of a fact about nature (an ontological truth). Although, of course, he did not deny that our surrounding world does not appear mathematical at all to us, Galileo argued that the perceived imperfections and irregularities do not belong to reality itself, but rather to our perception of it. It is only because of the limitations of our senses[6] that we

---

[6]It should be noted that there are actually two distinct metaphysical arguments that operate in the background of Galilean physics. First, there is the doctrine of primary and secondary qualities that Galileo introduced in *Il Saggiatore* and that became common currency in philosophical circles through the works of Descartes, Locke, Hume and others. The second argumentative strategy, which seems to play a particularly prominent role in Galileo's scientific practice, is based on the distinction between *natural occurrences* and *phenomena* (cf. Koertge 1977 and McAllister 1996): Natural occurrences are the physical processes, exactly as they occur under normal, lifeworldly

fail to experience reality as it really is: perfect, unchanging and simple. In effect, Galileo thus extended what the ancients had assumed about the heavens to be true of the world in its entirety. While the ancients had restricted the applicability of mathematics to bodies with ethereal composition, Galileo went beyond the dualism of superlunar and sublunar by degrading our surrounding world to a mere veil behind which the real world of mathematical objects is hidden.

Of course, Galileo had no way of knowing that the hidden reality behind the veil of perception is mathematical in nature. And in light of the poor observational-predictive record of early Galilean mechanics,[7] he could not have substantiated this metaphysical assumption in the way it is justified nowadays, i.e. by baptizing it as the best explanation for the success of mathematized physics. What Galileo did, then, was to present as a discovery what was actually a bold methodological conjecture: that the very method Arab/Persian astronomers had used to study the heavenly bodies is applicable to nature in its entirety.

Galileo's systematic replacement of objects of ordinary experience with geometric figures was continued and radicalized in the works of Descartes, Leibniz, Newton and others. Of course, since he simply inherited Euclidean and Archimedean proportional geometry from the tradition before him, Galileo did not have the right tools for his ambitious program of a complete mathematization of nature. It was particularly Newton and Leibniz who took more radical steps by developing their own mathematical machinery which allowed for the idealization of interaction and the replacement of natural motion with differential equations. Utilizing this method, nature turned into a dynamic system radically different from the static order Galileo had envisioned. It is this "dialectical movement" of mathematical innovation and

conditions. Phenomena, on the other hand, are the abstract invariant forms that allegedly underlie natural occurrences. According to Galileo, a natural occurrence is always the result of one or more phenomena and great number of accidents. And although Galileo acknowledges that the accidents are responsible for the huge variety of observable natural occurrences, he claims that they must be systematically excluded from physics through the method of geometrical idealization.

[7] Although it is not an easy task to determine the empirical adequacy of Galilean science from a contemporary perspective, the following episode shows how hard it was to apply Galilean mechanics successfully in the seventeenth century: Four years after Galileo's death the gunner Giovanni Ranieri attempted to apply Galileo's theory of projectile motion to his craft. However, as Ranieri reports in a letter to Evangelista Torricelli—Galileo's successor at the University of Pisa—the experimental results did not even come close to matching the theoretical predictions. Ranieri replicated one of Galileo's geometrical "proofs" by using an elevated gun to perform a number of point-blank shots. While the theory predicted a range of approximately 96 paces, Ranieri achieved ranges of 400 paces and more (cf. Segre 1991, 94–97). Particularly interesting is how Torricelli reacted to Ranieri's complaint: Torricelli explained the empirical inadequacy of Galileo's theory by pointing out "that Galileo [speaks] the language of geometry and [is] not bound by any empirical result" (Segre, 1991, 44). Even more interesting is the fact that Galileo himself was perfectly aware of the practical insufficiencies of his own theory. Shortly after he has presented his "proof" that projectiles describe a semi-parabolic path, he freely admits that the "conclusions proved in the abstract will be different when applied in the concrete and will be fallacious to this extent, that neither will the horizontal motion be uniform, nor the path of the projectile a parabola" (Galilei, 1954, 251).

resulting re-conceptualizations of nature that, in our view, is a defining characteristic of the history of physics: Just as Newton replaced the tools Galileo had used, Einstein paved the way for replacing calculus and differential equations with group theory, thus assigning a more fundamental status to symmetries than to dynamical laws.

Let us conclude this sketch of a diachronic investigation of the genetic origin of modern physics by summarizing the most important results: The objects of modern mathematized physics are not adopted from the world of everyday experience, but are constituted in a fundamentally different way. In physics we do not study motion as it appears to us although the physical notion has its origin in the "intuitive sense of things moving". What we study in physics is rather motion as an idealized entity that is already mathematized—if you are a Newtonian, motion is constituted through differential equations; if viewed from the perspective of relativity theory, motion is constituted through modified equations with Lorentz factor. The possibility of being intentionally directed toward the world in this peculiar manner depends on two fundamental presuppositions: First, it depends on the existence of pure mathematics as the systematic study of idealities such as numbers, lines or polynomials. These idealities are constructed out of intuitively given objects of the life-world whose properties are characterized by a fundamental vagueness and imperfection: While the technologically mediated process of, say, decreasing the thickness or increasing the flatness of life-world objects is essentially open-ended, the true intellectual accomplishment behind pure mathematics lies in the ability of the human mind to jump to the ideal end-point of such empirically interminable processes and to study the resulting ideal "limit-shapes" independently from the concrete particulars that give rise to them. But of course, the mere existence of pure mathematics is not yet sufficient to view reality in a mathematized way. In order for a cognitive agent to immerse herself into the scientific image, as it is displayed in various physical theories, the human mind must, second, also be able to "apply" pure mathematics to the experiential world in a very specific and intimate way. Or, to put it differently: Viewing nature in a mathematized manner is the result of a quite peculiar process of constitution which essentially involves mathematics and which can be further explicated phenomenologically. And to this we will now turn.

8.5 A Synchronic Investigation Into the Constitution of the Objects of Physics

We have argued so far that, first, the objects in physics are constituted differently than the objects of everyday experience and that, second, the physical sciences proceed through a continuing replacement of aspects of our natural surrounding world with mathematical idealities. Given how crucial these two interrelated claims are for our overall argument, it is of utmost importance to be as clear as possible about their implications and philosophical underpinnings. Hence, in this section we will supplement the diachronic investigation from the previous section with a synchronic analysis of the intentional structures that underlie different kinds of cognitive involvement with reality. The primary aim of doing so is to explicate the key notions of *constitution* and *replacement.*

One of the most fundamental insights of phenomenology is that the objects of cognition—in science as well as in everyday life—are not simply given. That objects are given to us is rather a phenomenon that is itself in need of further clarification. What phenomenology seeks to offer, then, is a faithful description of the structures of consciousness that are operative when different kinds of objects are intended through different kinds of intentional acts. Since, phenomenologically construed, these structures are the very condition of the possibility of any directedness towards the world, it is only on the basis of a comprehensive description of intentionality that human cognition, its limits and its potential, can be properly understood.

Let us begin with a simple example from the perceptual realm: Arezoo is supposed to meet Harald at the Double R Diner for coffee and pie. While driving down Main Street, she sees a building some distance away. After pulling into the parking lot she recognizes the building's exterior from a photograph she has once seen: she has reached her destination and Harald is already waving from the inside.

If we adopt a reflective stance toward this perceptual episode, several interesting observations can be made. To begin with, there is a describable difference between what is meant in an act of perception and what is actually sensuously given. During her ride down Main Street, Arezoo is visually attending to a material object that she intends as a three-dimensional thing in space. Yet, if we focus on what is sensuously given, it is clear that the whole object is never visually available to her at once: All that is visually present at any given point in time is a two-dimensional appearance, i.e. a profile of the thing, as it appears from one particular perspective.

Furthermore, the way Arezoo experiences the sequence of two-dimensional appearances is highly structured: As Arezoo drives down Main Street and thus changes the vantage point from which the object is seen, she brings new two-dimensional appearances into view. And although this is hardly ever noticed in the usual course of events, these new two-dimensional appearances fulfill (or frustrate) Arezoo's anticipations of how the object will continue to appear in further acts of perception. For example, in intending the object as a three-dimensional thing in space, Arezoo has the anticipation that there is more to the thing than is revealed in one single glance and that, consequently, new profiles and features will enter her visual field when she changes her position. Anticipations of this kind are not a matter of inferential belief or judgment over and above the experience in which things are perspectivally given; they are rather an essential part of any such experience and hence part of what it means to experience an object as a three-dimensional thing in space.

Finally, there is an intimate relationship between how an object is intended and the structure of the anticipations that are co-given with the sensuous data. For instance, when Arezoo first spots the building from afar, the structure of her anticipations concerning further possible experiences is relatively indeterminate and open-ended: Since, at this stage, she intends the object just as "a building", she wouldn't be too surprised if further experiences revealed a sign that reads "Fire Department". After recognizing the building and perceiving it as what it actually is, however, the structure of her anticipations is much more narrow and specific to what she knows about the Double R Diner.

As these analyses are supposed to show, perception is not merely, or even chiefly, about what is actual. Rather, the sensuously given is always and necessarily embedded into an open, but structured manifold of anticipations concerning further possible experiences. While the structure of these anticipations is what phenomenologists call the *horizon of experience*, the rule that governs the structure of the horizon is called the *sense* or the *noema* of an experience. So, when Arezoo first spots the object from afar, she does so by intending it through a noema that could be linguistically expressed by the term "building". It is this noema that then awakens a structured horizon of possible further experiences against the background of which new sensory data is constantly projected. One can think of the horizon in terms of a *space of possibilities* that plays an instrumental role in the evaluation of perceptual episodes: Whenever Arezoo changes her bodily location and thus receives new sensory input, this input must be harmonized with what is prescribed through the horizon. If the harmonization succeeds, i.e. if the sensuously given is compatible with the possibilities that are laid out in the horizon, then Arezoo's perceptual encounter with the intended thing proves successful. If, on the other hand, the harmonization fails—if, for instance, the building turns out to be an ingeniously designed hologram with no backside at all—, then Arezoo will have to accept that the noema through which the object has been intended must be revised. It is this process of intending objects through specific noemata and then constantly projecting new sensory data against horizons of possible further experiences that phenomenologists call *constitution*. Of particular importance in this context are those aspects of experience that remain invariant across a sequence of changing appearances. When Arezoo perceives the building in front of her from different viewpoints, many aspects of her experience are variable and in a constant state of flux: for instance, the perceived shape of the rectangular building will always be different, depending on the viewpoint from which the building is seen. However, what remains invariant over all varying shape-appearances is, for instance, the lawful angular relations between the perceiving subject on the one hand and each of the sides of the perceived thing on the other. It is invariances of this kind on which the constitution of *perceptual objectivity* is ultimately founded.

It is hard to overestimate the importance of the horizonal structure of intentionality for our understanding of human cognition in general and the notion of constitution in particular. Instead of thinking of intentionality in terms of a static, one-way relation between act and object, intentional directedness turns out to be a dynamic process in which objects are constituted by projecting ever-changing appearances against a horizon of possible further experiences. What makes this insight even more relevant is that the horizonal structure is not just found in perceptual experience. An example will help to illustrate how a phenomenological framework sheds light on the nature of scientific constitution.[8]

Imagine an experimental setup in which an EF-probe is used to measure the strength of an electric field at various points between two charged conductors. Imagine furthermore that two persons, Audrey and Dale, are invited to follow the experiment and describe what they are experiencing. Dale, a complete layman in physics, reports that he is observing a yellow-black piece of electronics whose display shows different digits, depending on where the piece of electronics is put. Audrey, on the other hand, has a PhD in physics and offers a rather different description. She knows that the charged conductors create an electric field that permeates the space between the conductors; she knows that the field exerts a certain force on the EF-probe; and she also knows that the strength of the force acting on the probe depends on two factors, the charge of the probe and the strength of the electric field. Most importantly, however, Audrey is in possession of a mathematical model that allows her to give a quantitative determination of the relationship between the strength of the field, the charge of the probe and the force that acts on the probe. Building on this background knowledge, Audrey is able to describe the situation in front of her as what it actually is: an experimental setup in which two conductors create an electric field whose strength at various points is measured by the EF-probe.

In light of the fixation on propositional knowledge that is still widespread in contemporary philosophy, it might be tempting to explain the differences between Dale's and Audrey's descriptions solely in terms of the background knowledge upon which they draw. However, although there is no point in denying that the available background knowledge does matter in the example at hand, it would be a serious oversimplification to reduce the difference between Audrey and Dale to the differences between what they *know* about physics. Phenomenologically construed, a more complete picture only emerges if we take into consideration how different stocks of background knowledge are used to intend one and the same situation in fundamentally different manners.

As we have seen, Dale intends the experimental setup through a noema with which he is familiar from the context of everyday practice. By intending the EF-probe as a "yellow-black piece of electronics", Dale generates a plethora of more or less determinate anticipations concerning various physical features of the thing. For instance, Dale will have the implicit anticipation that the piece of electronics will reveal a momentarily hidden backside if it is turned around, or that its size will remain the same if it is moved from A to B. Yet, at the same time, other aspects are left unspecified: Since Dale's lack of knowledge about physics only allows him to intend the probe as a normal material thing without special scientific significance, the information it produces lies beyond the scope of his attention. Dale wouldn't be surprised in the least if the numerical values on the probe's display would change erratically, or if the display didn't show any digits at all.

[8]The following example is a modification of an example found in Weyl (1948, 393–397; 1949, 113–114).

Although they are located in close spatiotemporal proximity, and although they seem to visually attend to the same scenario, the *how* of Audrey's intentional directedness is significantly different from the manner in which Dale intends the situation in front of him. This becomes apparent from Audrey's description of what she is experiencing: The fact that she is able to describe the scenario as what it actually is shows that she understands the scientific significance of the experimental setup. However, in order for this kind of scientific understanding to occur, it is not necessary—and, in fact, not even pertinent—to intend the EF-probe as a material thing that will reveal a backside when turned around or whose size will not be affected if it is moved from A to B. In the same sense in which we can "look through" a freehand drawing of a circle and intend an ideal geometrical circle instead, Audrey strips the probe of all its sensible properties such as color or texture, and intends it as a geometrical point with which a scalar factor and a vector quantity are associated. This *shift of attitude* is the result of intending the probe through a very specific noema, namely through the ideal mathematical content "$\mathbf{F}(P) = e \cdot \mathbf{E}(P)$".

What object one experiences is always underdetermined by the experiential data that is available at any given point in time. In the example at hand, the EF-probe could be constituted as a yellow-black piece of electronics, as an aesthetically appealing piece of art, as a paperweight or as a point-like, but otherwise unspecified carrier of certain numerical values. Which object one experiences depends on the noema through which the how of the intentional relation between subject and world is specified. In Audrey's case, it is due to the noema "$\mathbf{F}(P) = e \cdot \mathbf{E}(P)$" that she is able to "look through" the materiality of the probe and to intend it as a geometrical point with which a scalar factor and a vector quantity are associated. At the same time, the noema also awakens a rigidly structured horizon that determines the relations between the probe's charge e, the measured vector force $\mathbf{F}$ and the field strength $\mathbf{E}$ in an unambiguous, precise and mathematically rigorous way. Whenever Audrey receives new data by varying the position of the probe, this new data must be harmonized with what is prescribed through the horizon. If the harmonization succeeds, i.e. if the data is compatible with the space of possibilities that is determined by the noema, then Audrey's encounter with the situation in front of her proves successful. If, on the other hand, the harmonization fails—if, for instance, all properties of the experimental setup are held constant, but the value of the measured vector force $\mathbf{F}$ changes nevertheless—, then Audrey will have to accept that the noema through which the scenario has been intended must be revised.

As the results of our synchronic investigation suggest, there is an important sense in which constitution in science and constitution in everyday contexts are structurally analogous. Being intentionally directed towards reality always means to intend the objects around us through a noema that awakens a more or less structured horizon of possible further experiences. One can think of the horizon as a space of possibilities that is instrumental in the evaluation of any encounter with the world: For whether such an encounter is deemed successful depends on whether the experiencing subject succeeds in harmonizing new incoming experiential data with what is prescribed through the horizon. Note, however, that this harmonization is essentially processual in character and also requires an active role on the part of the observer: The experience of an object as objectively existing is never founded on one isolated perspectival encounter with the object. In order to penetrate the object's full ontological depth, the experiencing subject must "probe" the horizon by constantly gaining new experiential data that can then be projected against the horizon. In simple perceptual cases, new sensuous data is gained through kinesthetic movement, i.e. by varying the location of the observer's body. In the earlier example of the constitution of an electric field, the horizon is explored by varying the location of the EF-probe, the latter serving as a technological extension of Audrey's body.

There is an obvious question that arises at this point: One of the key insights that have emerged from the previous section was that objects in physics and objects of everyday experience are constituted differently. But how can this be the case if, as we have now claimed, there is a structural analogy between constitution in science and constitution in simple perceptual situations? To provide an answer to this question, it is necessary to distinguish clearly between descriptions of the formal (i.e. domain-independent) structures of intentionality and descriptions of the various ways in which intentionality is instantiated in particular domains of cognitive engagement.

Phenomenologically construed, the structural analogy between physical and everyday constitution stems from the fact that the dynamic interplay between noema, horizon and experiential data represents the very core of any intentional relation between subject and any kind of object. However, the differences between physical and everyday objects come into view once we pay closer attention to the noemata through which these objects are constituted in their respective domains. When we are intentionally directed toward individual objects in everyday situations, we always experience these objects as particular instances of more general *empirical types*. For example, in being directed towards the EF-probe through the noema "yellow-black piece of electronics", Dale experiences the intended object as an instantiation of the empirical types "yellow", "black" and "piece of electronics". Empirical types are bundles of anticipations that were formed over the course of previous experiences, and by drawing qualitative analogies between objects that are deemed similar. If empirical types are used to specify the manner in which particular objects are intended, they awake a horizon of further possible experiences. Depending on Dale's previous experiences with other pieces of electronics, he will anticipate, for example, that the intended object will reveal an array of wires and electric components if it is cracked open, or that its perceived shape will change in a specific, but only qualitatively determined manner if it is turned around. Yet, since empirical types are based on association and similarity, such if-then anticipations are characteristically vague and imprecise.

While the manner in which the experimental setup presents itself to Dale is characterized by a fundamental vagueness, Audrey experiences the scenario in a significantly different manner. To begin with, as we have seen, the noema "$\mathbf{F}(P) = e \cdot \mathbf{E}(P)$" functions like a filter, screening off all sensible properties like color or texture. Even though Audrey is still aware of the fact that she is intending a segment of the physical world, the scenario she is experiencing is devoid of the buzzing and blooming confusion of sensible qualities, which is characteristic of normal life-world experiences. The EF-probe is constituted as a geometrical point with only one intrinsic physical property, its having a charge with the numerical value e. Hence, since the probe is constituted as a mathematical ideality, and since mathematical idealities are constituted as remaining self-identically the same, comparisons with other point-like probes are not based on similarity and association, but on an objective ordering of the value of e. However, apart from stripping the intended scenario from its sensible properties, the ideal mathematical content "$\mathbf{F}(P) = e \cdot \mathbf{E}(P)$" also transforms the anticipations that are co-given with the available experiential data. Whereas Dale's anticipations are vague and imprecise, the space of possibilities awakened by the noema "$\mathbf{F}(P) = e \cdot \mathbf{E}(P)$" determines the correlations between all properties of the intended scenario in a mathematically rigorous and quantitatively precise manner. Hence, it is Audrey's ability to intend the experimental scenario through a specific, non-morphological noema that makes reality amenable to a quantitative, mathematically rigorous treatment. Audrey has not just constituted reality: By intending the experimental setup through an ideal, mathematical content, Audrey has engaged in a very peculiar process of constitution which, following Husserl, is called *mathematization*.

In his seminal *Galileo Studies*, Alexandre Koyré notes that "Galileo's [...] mental attitude [...] is not purely mathematical [but] *physico-mathematical*" and that, although "Galileo tells us to start from experience, [...] this 'experience' is not the raw experience of the senses" (Koyré 1978, 108). Not only do we agree; we also suggest that Koyré's remarks can only be fully understood if they are read against a phenomenological background[9]: As far as his work in mechanics and kinematics is concerned, Galileo's primary achievement was neither the invention of new instruments, styles of reasoning or experimental techniques. Nor does the novelty of his approach lie in the mere application of mathematics to empirical phenomena. In our view, what justifies the title "father of modern science" more than anything was Galileo's trailblazing innovation to *mathematize all of reality by intending it through ideal-mathematical noemata*. Paradoxically as it may sound, this revolutionary new way of constituting reality resulted both in an impoverishment and in an enrichment of the empirical world. As our earlier example shows, the segment of reality towards which Audrey is intentionally directed is impoverished in the sense that it is devoid of sensible properties such as color or texture. At the same time, however, Audrey's mathematized world is significantly richer than Dale's:

[9]The suggestion to read Koyré from a phenomenological perspective is by no means far-fetched: Not only was Koyré a student of Husserl in Göttingen; Koyré had plans to write his dissertation on the antinomies of set theory under Husserl's supervision. What is more, as Parker has argued in detail, there are good reasons to believe that Koyré's later interpretation of Galilean physics was heavily influenced by Husserl's take on the issue (cf. Parker 2017). Although the phenomenological traces in Koyré's oeuvre have been overlooked by many, there are, of course, exceptions. For instance, Michel Foucault remarks that "we run across phenomenology in someone like Koyré [...] who [...] developed a historical analysis of the forms of rationality and knowledge in a phenomenological perspective" (Foucault 1998, 438).

Audrey understands the situation in front of her because she not only intends a point-like probe with which a scalar factor and a vector quantity are associated. Audrey also intends additional layers of reality that, while not accessible to Dale, are crucial for understanding the scientific significance of the experimental situation. As long as Audrey immerses herself into the scientific image of reality, the world she experiences is populated by point-like probes, numerical values, vector forces and fields. Following Galileo's footsteps, Audrey has *replaced* the life-world of everyday experience with a scientific image of reality by intending the world through the noema "$\mathbf{F}(P) = e \cdot \mathbf{E}(P)$".

Galileo's achievement of intending reality through mathematical noemata not only initiated the historical process of replacing more and more aspects of our natural surrounding world with increasingly sophisticated mathematical idealities. A crucial by-product of Galileo new scientific vision was also to make these mathematical idealities indispensable for the very definition of objectivity in physics. As we have seen earlier, Galileo did not think of his geometrical models in terms of willful distortions of reality that must later be de-idealized in order to account for the phenomena as they occur under normal life-world conditions. Galileo rather considered his idealized models to be the only way to catch a glimpse of how the deep-structure of reality objectively looks like. A consequence of this interpretation is that the mathematical idealities out of which Galileo's models are constructed become prescriptive for experience: If one accepts, as Galileo does, that the idealized model represents the objective being of reality, then life-world phenomena must be regarded as mere approximations to the ideal case which is nowhere to be found in the realm of everyday experience. Of course, from a contemporary perspective Galileo's identification of simple geometrical models with objective reality must appear somewhat naive. However, the important point is that determining the notion of objectivity solely by mathematical means is still essential to the practice of physical theorizing. Take, for instance, the principle of covariance that lies at the heart of classical and relativistic mechanics: By requiring that the form of the laws of nature must be preserved under transformation from one reference frame to the other, the concept of physical objectivity is solely defined in terms of mathematical transformation rules that specify which properties remain invariant within the allowable group of transformations (Cassirer 1953; Kosso 2003). Hence, the guiding ideal of objectivity turns out to be inseparable from mathematical idealities such as the Lorentz transformations.

8.6 Conclusion

Let us come to a final conclusion. Nature as the subject of modern mathematical science is mathematical because *we have made it so*. That is, the match between mathematics and physics is not a match made in heaven but a match made on earth, through a long and arduous process of mathematization and co-constitution of sciences alongside one another (in this case physics and mathematics). In focusing on the how-question, our aim was to highlight the epistemological aspect of the applicability problem and the role humans play in the ongoing project of the mathematization of nature. Their role is certainly not that of passive spectators of another world that is hidden behind the phenomena. Scientific agents are rather active participants who are constructing and re-constructing objects in order to mathematize different aspects of nature. Seen from this perspective, then, *God* is a mathematician of course, if one recognizes the human as a deity.

Yet, it is understandable how to a physicist like Wigner or Tegmark it might appear that mathematical physics represents the true and actual nature. The process of idealization is hidden from the eye of the scientist for several reasons. In textbooks the views of previous scientists are always cast in modern notation, and reformulated using the current understanding of science. If the history of science is mentioned at all, then its role is that of a confirmation of a cumulative image of science. This history of science for scientists, as Grattan Guinness rightly puts it, aims at portraying *a royal road to us* (cf. Grattan-Guinness 1990, 157).

While for the scientist it is convenient—if not necessary—to forget the origins of her own science, "the original formation of meaning", the philosopher is required to go back and peel away the layers of this already formed "onion" to see what is really inside. While the scientist or the mathematician takes the science of his or her time as a given, the philosopher questions this very science. Such critique and questioning is but the task of the philosophical mind. Otherwise, we too will fall in the trap of miracles and mysteries by forgetting the very origins of physics and the continuous acts of re-conceptualization. Thus, formulating the relationship between mathematics and physics as an *application* is a major source of the problem. Application of one area to another assumes a distinctness which needs to be bridged. In the case of modern theoretical physics, we, unlike other commentators, showed that there is no such distinctness: the objects of physics are constituted mathematically.

Now, this is the beginning of a pluralistic project in which it becomes possible to study the (dynamic) relationship(s) of mathematics with other empirical sciences such as biology, and with other non-empirical sciences such as mathematics itself. We can ask how the previously *unexpected* relationship between different areas of mathematics are possible, for instance, how analytic geometry or algebraic topology are possible. The solution, we conjecture, will arise as a result of the study of the constitution of the objects of these *mixed* fields. In the case of biology, however, particular reasons can be given as to why biological phenomena don't allow mathematization in the same way that objects of physics do (cf. Islami 2017).

Finally, we are perfectly aware that there are many open questions,[10] for example how our phenomenological approach to mathematized physics can account for the

[10] It should also be noted that space constraints will prevent us from commenting on two recent solutions to the applicability problem that are in some ways similar to our own approach, namely da Silva's transcendental-phenomenological account (da Silva 2017) and the account that has recently been developed by Bueno and French (2018).

empirical adequacy of our most successful physical theories. How is it, the critic might ask, that mathematically constituted objects of physics appear in equations that successfully predict the precise value of quantities (with negligible error) in experiments? This is an important and elaborate question which requires the space of its own. Schematically put, the answer involves an account of what an experiment is, how something comes to be a quantity, what we mean by prediction etc. Should it turn out to be impossible to deal with these issues in a constructive way, our position runs the risk of collapsing into an extreme form of idealism. Moreover, and to complete our answer to the *applicability problem*, we need a phenomenological excavation into the origins of mathematics and how it has become the pure abstract mathematics of the twentieth century. Instead of beginning with a readymade ontology and epistemology, we suggest that we study mathematics as used and practiced. Mathematics understood by Wigner was more or less formalist, and only representative of the pure mathematics of the twentieth century. It was then this forgetting of one's own position in history that bred miracles and mysteries. To this ailment, phenomenology has a cure.[11]

References

Bueno, O., & French, S. (2018). *Applying mathematics. Immersion, inference, interpretation*. Oxford: Oxford University Press.

Cassirer, E. (1953). *Substance and function, and Einstein's theory of relativity*. New York: Dover.

Clavelin, M. (1974). *The natural philosophy of Galileo. Essay on the origin and formation of classical mechanics*. Cambridge: MIT Press.

Colyvan, M. (2001). The miracle of applied mathematics. *Synthese, 127*, 265–78.

da Silva, J. J. (2017). *Mathematics and its applications. A transcendental-idealist perspective*. Cham: Springer.

Dawid, R. (2013). *String theory and the scientific method*. Cambridge: Cambridge University Press.

Derrida, J. (1989). *Edmund Husserl's origin of geometry: An introduction*. Lincoln: University of Nebraska Press.

Dirac, P. A. M. (1939). The relation between mathematics and physics. *Proceedings of the Royal Society, 59*, 122–129.

Foucault, M. (1998). *Aesthetics, method, and epistemology*. New York: New Press.

Galilei, G. (1954). *Dialogues concerning two new sciences*. New York: Dover.

Galilei, G. (1967). *Dialogue concerning the two chief world systems*. Berkeley/Los Angeles: University of California Press.

Gödel, K. (1983). Russell's mathematical logic. In P. Benacerraf & H. Putnam (Eds.), *Philosophy of mathematics* (pp. 447–469). Cambridge: Cambridge University Press.

Grattan-Guinness, I. (1990). Does the history of science treat the history of science? The case of mathematics. *History of Science, 28*, 149–173.

Gross, D. (1988). Physics and mathematics at the frontier. *Proceedings of the National Academy of Science of the United States of America, 85*, 8371–8375.

Hilbert, D. (1992). *Natur und mathematisches Erkennen*. Basel: Birkhäuser.

[11]This is the subject of our future work.

Hossenfelder, S. (2018). *Lost in math. How beauty leads physics astray*. New York: Basic Books.

Husserl, E. (1970). *The crisis of European sciences and transcendental phenomenology. An introduction to phenomenological philosophy*. Evanston: Northwestern University Press.

Islami, A. (2017). A match not made in heaven: On the applicability of mathematics in physics. *Synthese, 194*(12), 4839–4861.

Koertge, N. (1977). Galileo and the problem of accidents. *Journal in the History of Ideas, 28*(3), 389–408.

Kosso, P. (2003). Symmetry, objectivity, and design. In K. Brading & E. Castellani (Eds.), *Symmetries in physics. Philosophical reflections* (pp. 413–424). Cambridge University Press: Cambridge.

Koyré, A. (1978). *Galileo studies*. Hassocks: Harvester Press.

Lesk, A. M. (2000). The unreasonable effectiveness of mathematics in molecular biology. *The Mathematical Intelligencer, 22/2*, 28–37.

McAllister, J. (1996). The evidential status of thought experiments in science. *Studies in History and Philosophy of Science, 27*(2), 247–273.

McTighe, T. (1967). Galileo's platonism: A reconsideration. In E. McMullin (Ed.), *Galileo. Man of science* (pp. 365–386). New York: Basic Books.

Parker, R. K. (2017). The history between Koyré and Husserl. In J. A. Raffaele Pisano & D. Drozdova (Eds.), *Hypotheses and perspectives in the history and philosophy of science. Homage to Alexandre Koyré 1892–1964* (pp. 243–275). Dordrecht: Springer.

Penrose, R. (2004). *The road to reality: A complete guide to the laws of the universe*. London: Vintage Books.

Redondi, P. (1998). From Galileo to Augustine. In P. Machamer (Ed.), *The Cambridge companion to Galileo* (pp. 175–210). Cambridge: Cambridge University Press.

Segre, M. (1991). *In the wake of Galileo*. New Brunswick: Rutgers University Press.

Smolin, L. (2006). *The trouble with physics*. Boston: Houghton Mifflin.

Steiner, M. (1989). The application of mathematics to natural science. *Journal of Philosophy, 86*(9), 449–480.

Steiner, M. (1998). *The applicability of mathematics as a philosophical problem*. Cambridge: Harvard University Press.

Tegmark, M. (2008). The mathematical universe. *Foundations of Physics, 38*(2), 101–150.

Velupillai, K. (2005). The unreasonable ineffectiveness of mathematics in economics. *Cambridge Journal of Economics, 29*, 849–872.

Weyl, H. (1948). Wissenschaft als symbolische Konstruktion des Menschen. *Eranos-Jahrbuch, 16*, 375–431.

Weyl, H. (1949). *Philosophy of mathematics and natural science*. Princeton: Princeton University Press.

Wigner, E. (1960). The unreasonable effectiveness of mathematics in the natural sciences. *Communications in Pure and Applied Mathematics, 13*, 1–14.

Wiltsche, H. A. (2012). What is wrong with Husserl's scientific anti-realism? *Inquiry, 55*(2), 105–130.

Wiltsche, H. A. (2016). Mechanics lost: Husserl's Galileo and Ihde's telescope. *Husserl Studies, 33*(2), 149–173.

Chapter 9
The Gauge Principle, Hermann Weyl, and Symbolic Construction from the "Purely Infinitesimal"

Thomas Ryckman

> ***"Only in the infinitely small may we expect to encounter elementary uniform laws;***
> *hence the world must be comprehended through its behavior in the infinitely small."*
>
> *—(Weyl, 1927, 61; 1949, 86)*

Abstract The gauge principle is a broad moniker about invariance properties of fundamental physical laws. It stipulates that every global symmetry of a quantum field theory be replaced by a local one; in effect, that every continuous symmetry of a quantum field (the Lie group under which the field Lagrangian transforms invariantly) become a local symmetry, i.e., an invariance of the Lagrangian under which the smooth Lie group actions are allowed to differ from point to point. It has an unusual "context of discovery": invoked on largely phenomenological grounds by mathematician Hermann Weyl, it emerged in 1918 in the context of classical general relativity and, in Weyl's hands, led to a purely formal unification of Einstein's gravitational theory and electromagnetism. This work prompted Weyl's purely mathematical turn in 1925–6 to Lie theory (on representations of semisimple Lie groups and "Lie algebras", a term later coined by Weyl for the infinitesimal linear algebraic structure of Lie groups). Both Lie groups and Lie algebras play prominent roles in the subsequent development of the gauge principle leading up to the Standard Model (SM), a compilation of quantum field theories that since 1978 is the regnant theory of matter. I suggest that the gauge principle as well as Weyl's predominant interest in Lie theory were motivated by two complementary philosophical demands: (i) phenomenological evidential requirements of "eidetic insight" and "eidetic analysis" imposed on differential geometric construction, and (ii) the metaphysical command of "*Nahewirkungphysik*" Weyl associated with Leibniz, Riemann and Lie: to comprehend the world from its behavior in the infinitely small. The two requirements productively meet in Weyl's notion of

T. Ryckman (✉)
Stanford University, Stanford, CA, USA
e-mail: tryckman@stanford.edu

H. A. Wiltsche, P. Berghofer (eds.), *Phenomenological Approaches to Physics*,
Synthese Library 429, https://doi.org/10.1007/978-3-030-46973-3_9

"symbolic construction": the idea that the sense of a transcendent world portrayed in physical theory can be constitutively understood beginning from a transcendental subjectivity evidentially privileging "radical locality", i.e., the "given to consciousness" epistemic reach ("horizon") of linear relations within the tangent space. Radical locality is the basis of Weyl's constitution of objectivity as an invariance with respect to manifold automorphisms, an intersubjectivity allowing arbitrary coordinates and gauge degrees of freedom of a situated constructing ego. Both concern a necessary redundancy of physical description, a philosophical puzzle that might be elucidated by revisiting the philosophical underpinnings of the gauge principle.

9.1 Introduction

The gauge principle is a broad moniker about invariance properties of fundamental physical laws. It stipulates that every global symmetry of a quantum field theory be replaced by a local one; in effect, that every continuous symmetry of a quantum field (i.e., the field Lagrangian transforms invariantly) be a local symmetry. The gauge principle is not quite *a priori* physics yet remarkably it provides an *a priori* framework for constructing the form of interaction between force and matter fields, a procedure that has been shown to be highly empirically successful. Invoked in 1918 on largely philosophical grounds by mathematician Hermann Weyl, it emerged in the context of classical general relativity and, in Weyl's hands, led to a purely formal unification of Einstein's gravitational theory and electromagnetism. For Weyl, the gauge principle encapsulated two desirable but complementary philosophical demands: (i) the phenomenological evidential requirements of "eidetic insight" and "eidetic analysis" imposed on differential geometric construction, and (ii) the metaphysical command of "*Nahewirkungphysik*", that "the true lawfulness of nature is expressed in laws of nearby action, connecting only values of physical quantities at spacetime points in the immediate vicinity of one another". (Weyl, 1927, 61; 1949, 86) The idea did not work in the context of general relativity but in 1929 Weyl himself carried the gauge principle over to quantum theory, its proper setting. Revived by Yang and Mills (1954), the gauge principle's mandate of *radical locality* plays a central unifying role in the Standard Model (SM) of elementary particles, the quantum field theories describing three of the four known fundamental forces and the regnant theory of matter since 1978. To be sure, the philosophical origin yet astonishing success of the gauge principle presents something of a puzzle. Moreover, a gauge symmetry is widely understood to be a redundancy of description, i.e., an unphysical symmetry merely relating mathematically different representations of the same physical state or history. One can agree with the assessment of a prominent philosopher of physics, that "the elucidation of [the gauge principle] is the most pressing problem in current philosophy of physics". (Redhead, 2002, 299) Our thesis is that revisiting the philosophical underpinnings of the gauge principle can contribute to this elucidation.

9.2 Hermann Weyl and the Philosophical Origins of the Gauge Principle

To the philosopher of science who demurs from scientific realism but finds that the alternatives of antirealism or instrumentalism yield only an anemic understanding of the cognitive role of physical theory, the origin of the gauge principle by Hermann Weyl (1885–1955) presents an instructive case study. A preeminent mathematician of the twentieth century, Weyl also made seminal contributions to the twin pillars of fundamental physical theory, general relativity and quantum mechanics. Nearly a half-century after his death, Fields medalist Sir Michael Atiyah noted the continuing extent of Weyl's influence:

> No other mathematician could claim to have initiated more of the theories that are now being exploited. His vision has stood the test of time. (Atiyah, 2002, 13)

In pure mathematics, Atiyah pointed to Weyl's work on the theory of Lie groups and algebras. In physical theory, Atiyah underscored Weyl's idea of gauge invariance that subsequently became the unifying framework of the SM. In fact, these two currents are thematically and philosophically related. And though not mentioned by Atiyah, Weyl authored a handful of philosophical works giving expression to the reflective musings of a highly innovative scientist. Weyl's philosophical orientation lies far from what logical empiricism regarded as "scientific philosophy"; closely intertwined with his scientific achievements, they are somewhat idiosyncratic. Drawing from figures and traditions largely unknown to contemporary philosophers of science (post-Kantian transcendental idealism including Fichte and Husserl, Nicholas of Cusa, even the medieval mystic Meister Eckert). Weyl's philosophical remarks are broadly stated, purposefully hesitant, and not articulated in any detail. After all,

> an epistemological conscience (*Erkenntnisgewissen*), sharpened by work in the exact sciences, does not make it easy ... to find the courage for philosophical statement. One cannot get by entirely without compromise. (Weyl, 1954b, 648)

Nevertheless, it is not difficult to identify a metaphysics of transcendental subjectivity underlying the two central accomplishments alluded to by Atiyah: the origin of the idea of gauge invariance in 1918 with its restatement in the context of quantum mechanics in 1929, and the 1925–6 purely mathematical work on Lie theory (on representations of semisimple Lie groups and Lie algebras), results that Weyl himself regarded as his greatest mathematical triumph. Both achievements are heuristically motivated by an injunction to comprehend the world from its behavior in the infinitely small, an evidential constraint implicating an intersubjective constitution of physical objectivity related to the intentional analyses of transcendental phenomenological idealism. While Weyl traced the impetus for this "purely infinitesimal" explanatory agenda back to Leibniz, he specifically associated it with Bernard Riemann (in particular, the theory of Riemannian manifolds) and with Sophus Lie (the theory of continuous groups and their infinitesimal generators). Weyl would coin the term "Lie algebra" for the infinitesimal group structure of a Lie group; he also showed that this infinitesimal structure is a real-valued linear space, in fact, a vector space, a concept Weyl was in fact the first to define (1913). It is only fitting that Lie algebras play an important role in the contemporary gauge theories of the Standard Model.

9.2.1 "Idealism in the Infinitesimal"

Following the apt term of Julien Bernard (2015), Weyl's transcendental metaphysics is an "idealism in the infinitesimal". It is a modern descendant of Leibniz's *principle of continuity* ("*natura non facit saltus*") i.e., that all finite changes are to be comprehended as arising through infinitesimal increments acting in sequence.[1] Its modern mathematical setting was provided by two titans of mathematics in the second half of the nineteenth century, Bernard Riemann and Sophus Lie:

> The productivity shown by the differential calculus, by contiguous action [field] physics (*Nahewirkungsphysik*), and by Riemannian geometry certainly rests upon the principle: To understand the world, according to its form and content, by its behavior in the infinitely small, clearly because *all problems can be linearized in passing to the infinitely small.*[2]

In geometry, Weyl observed, Riemann took the step Faraday and Maxwell had taken in physics, according to "the principle: to understand the world from its behavior in the infinitely small."[3] Geometrically, this is the tangent space T_P to each point P of a Riemannian manifold M. Just as in elementary differential calculus, functions differentiable at a point P on the function's graph are locally linear there,

---

[1]On the principle of continuity, see Leibniz's letter to Varignon, 1702: "*Assurément je pense que ce Principe ["mon Principe de Continuité] est general, et qu'il tient bon, non seulement dans la Géometrie, mais encore dans la Physique. La Géometrie n'étant que la science des limites et de la grandeur du Continu, il n'est point étonnant, que cette loi s'y observe par-tout: car d'où viendroit une subite interruption dans un subject, qui n'en admet pas en vertu de sa nature? Aussi savon-nous bien, que tout est parfaitement lié dans cette science, et qu'on ne sauroit alléguer un seul exemple, qu'une propriété quelconque y cesse subitement, ou naisse de même, sans qu'on puisse assigner le passage intermédiaire de l'une à l'autre, les points d'inflexion et de rebroussement, qui rendent le chagement explicable; de manière qu'une Equation Algébraique, qui représente exactement un état, en représente virtuellement tous les autres, qui peuvent convener au même sujet. L'universalité de ce Principe dans la Géometrie m'a bientôt fait connoitre, qu'il ne sauroit manquer d'avoir lieu aussi dans la Physique: puisque je vois, que, pour qu'il y ait de la règle et d'ordre dans la Nature, it est necessaire, que la Physique harmonie constamment avec le Géométrique*; ..." "Brief von Leibniz an Varignon über das Kontinuiätsprinzip", in E. Cassirer (ed.), *G.W. Leibniz: Hauptschriften zur Grundlegung der Philosophie,* Bd. II. Leipzig: Verlag der Dürr'schen Buchhandlung, 1906, 556.

[2]Weyl, 1923b, 9: "*Es beruht ja die Leistungsfähigkeit des in der Differentialrechnung, der Nahewirkungsphysik und der* ***Riemann****sche Geometrie zum Durchbruch kommenden Prinzips: die Welt nach Form und Inhalt aus ihrem Verhalten im Unendlichkleinen zu verstehen, eben darauf, dass* **alle Problem durch den Rückgang aufs Unendlichkleine linearisiert werden**." See also Weyl, 1918, 82; Weyl, 1927, 61; Weyl, 1949, 86.

[3]"Vorwort des Herausgebers", in Weyl (ed.), 1919.

so in a Riemannian manifold the tangent space T_P to each point P is a linear space. These manifolds, according to Riemann, exhibited "planeness in their smallest parts", hence only linear relations are required.[4]The geometric concept of manifold was also the starting point of Sophus Lie's theory of continuous groups. A Lie group is a differential manifold whose points are the group elements, parameterized by continuous real variables. The points are combined by an operation (action) obeying the group axioms; they compose continuously to form a 'space'. Lie's seminal idea was to investigate the group actions on a manifold infinitesimally; indeed, the core idea of Lie theory, as characterized by Weyl, is "descent to the infinitely small".[5]Though generically non-linear, Lie groups can be linearized in passing to a local (infinitesimal) group acting in the tangent space of the group identity. In the simplest case, the local Lie group acts on itself by left (or right) translations forming what Weyl (1934b) termed the "Lie algebra" of the Lie group. The Lie algebra is a much simpler and cognitively more accessible structure than the Lie group yet it contains most of (non-topological) information about the group. This is the crucial fact on which the modern structure theory of Lie groups is based: their respective algebras yield a precise classification of Lie groups (see Sect. 3.1 below).

In the application of both Riemannian geometry and Lie theory to physics, Weyl's "idealism in the infinitesimal" is an epistemological mandate that *comprehensibility* of the physical world, i.e., phenomenological "sense-constitution", is to be gained by bottom-up symbolic construction starting from mathematical relations in the infinitely small. With both Riemannian manifolds and Lie groups in mind, Weyl would regard infinitesimal linear spaces as the legitimate epistemic reach of "eidetic vision" or "insight" or "*Evidenz*" (or whatever one will call it) of the cognizing, constructing subject, an "ego-center" whose secure epistemic range is mathematically understood as the "infinitesimally small" bounded linear region

[4]A crucial feature of Einstein's theory of gravitation is that it allows (pseudo-) Riemannian geometry ("pseudo" since time is treated differently than the three space dimensions) to be the appropriate mathematical framework for the concept of "local inertial frame" and so to uphold the "infinitesimal" validity of special relativity in that theory.

[5]Weyl, 1923b, 34; "*Die Ersetzung der endlichen Gruppe durch die infinitesimal – das ist wieder der 'Rückgang aufs Unendlichkeleine'! – ist einer der Hauptgedanken der Lieschen Theorie.*" Original emphasis. Hawkins (2000, 72) quotes from an 1879 paper of Lie, "In the course of investigations on first-order partial differential equations, I observed that the formulas that occur in this discipline become amenable to a remarkable conceptual interpretation by means of the concept of an infinitesimal transformation. In particular, the so-called Poisson-Jacobi theorem is closely connected with the composition of infinitesimal transformations. By following up on this observation I arrived at the surprising result that all transformation groups of a simply extended manifold can be reduced to the linear form by a suitable choice of variables, and also that *the determination of all groups of an n-fold extended manifold can be achieved by the integration of ordinary differential equations*. This discovery ... became the starting point of my many years of research on transformation groups."

surrounding each point.[6]This is the entry point of transcendental phenomenological sense constitution and a core assumption of "idealism in the infinitesimal",

> Only the spatio-temporally coinciding and the immediate spatial-temporal neighborhood has a directly clear meaning exhibited in intuition. ... The philosophers may have been correct that our space of intuition bears a Euclidean structure, regardless of what physical experience says. I only insist, though, that to this space of intuition belongs the ego-center [*Ich Zentrum*] and that ... the relations of the space of intuition to that of physics, becomes vaguer the further the distance from the ego-center. (Weyl, 1931c, 49, 52)

Phenomenological requirements on evidence, on what is given to consciousness and what can be evidently constituted (in Weyl's mathematically expanded notion of *Wesensschau*) on that basis, are expressly tied to the infinitely small: only in this limited region can a cognizing consciousness impose *evident* elementary and uniform laws. Other mathematical resources may be required for manifolds as a whole; their evidential basis is accordingly less direct. The injunction to comprehend the world from "its behavior in the infinitely small" is a recurrent theme running through Weyl's writings from 1918 to at least 1949. (cf. Weyl, 1918a, 82; 1949, 86).

9.2.2 Transcendental Phenomenological Idealism and "Symbolic Construction"

Weyl's injunction to understand the world from its behavior in the infinitely small is an evidential constraint upon a transcendental idealism according to which objects of knowledge (natural science) are *constituted* via a process Weyl termed "symbolic construction":

> Science concedes to idealism that its objective reality is not given but to be constructed (*nicht gegeben, sondern aufgegeben*), and that it cannot be constructed absolutely but only in relation to an arbitrarily assumed coordinate system, and in mere symbols. (Weyl, 1927, 83; 1949, 117)

Readers of Kant's Transcendental Dialectic (A647/B675) will recognize the passage as an avowal of transcendental idealism. It is also a transcendental *phenomenological* idealism insofar as symbolic construction of the "objective reality" of the purportedly mind-independent objects of physics is, *per* Husserl, a constitution of the *sense* of such objects as having "the sense of 'existing in

---

[6]Husserl, e.g., (1974 [1929], 141), is careful to distinguish the usual (and "countersensical") philosophical notion of evidence as the absolute criterion of truth from evidence as "*that performance on the part of intentionality which consists in the giving of something-itself* [*die intentionale Leistung der Selbstgebung*]. More precisely, it is the universal pre-eminent form of 'intentionality', of 'consciousness of something', in which there is consciousness of the intended-to objective affair in the mode itself-seized-upon, itself-seen – correlatively, in the mode: being with it itself in the manner peculiar to consciousness."

themselves'".[7] Weyl himself expressed this understanding of sense-constitution in a densely exposited account of the phenomenology of perception in the "Introduction" to all five editions of his masterful text on general relativity *Raum-Zeit-Materie* (*RZM*):

> Upon general in-principle grounds: The real world, in each of its components and all their determinations, is and can only be, given as intentional objects of acts of consciousness. Given, purely and simply, are the conscious experiences that I have – as I have them. Certainly, in no way do they consist, as positivists perhaps maintain, only of the mere stuff of sensation. Rather a perception, for example, an object standing bodily there before me, each experience of which is known to everyone but not more exactly describable, is taken up in a completely characteristic manner to be designated, with Brentano, through the expression "intentional object". While I am perceiving, as in seeing this chair, I am thoroughly directed to it. I "have" the perception, but only when I make this perception itself into the intentional object of a new, inner perception (of which I am capable in a free act of reflection), do I "know" something about it (and not merely about the chair) .... In this second act the intentional object is immanent like the act itself; it is an actual component of my stream of experience; but in the primary perceptual act, the object is *transcendent*, i.e., actually given in a conscious experience but not an actual component. The immanent is absolute, that is, exactly what it is as I have it and am able to bring its essence (*Wesen*) to givenness (*Gegebenheit*) before me in acts of reflection. ... The given-to-consciousness (*Bewußtseins-Gegebene*) is the starting point at which we must place ourselves in order to comprehend the sense and the justification of the posit of reality (*Wirklichkeitsetzung*). (Weyl, 1918a, 3–4; 1923a, 3–4)

While the Husserlian resonances are unmistakable; the book's first endnote states that the "precise wording" is "closely modeled" upon Husserl's *Ideen* (1976 [1913]). This passage, utterly remarkable in a monograph establishing much of the modern mathematical machinery of general relativity, is not idle philosophical window dressing. A similar declaration occurs in Weyl's 1930 Rouse Ball lecture at Cambridge, already some years after his most intense period (1918–25) of rather explicit immersion in Husserlian phenomenology,

> "Reality [*Wirklichkeit*] is not a being-in-itself [*Sein an sich*] but rather is constituted *for* a consciousness."[8]

[7] Husserl, ca. (1956 [1908], 382): "My transcendental method is transcendental-phenomenological. It is the ultimate fulfillment of old intentions, especially those of English empiricist philosophy, to investigate the transcendental-phenomenological 'origins' ... the origins of objectivity in transcendental subjectivity, the origin of the relative being of objects in the absolute being of consciousness." Transcendental sense constitution of objective nature is founded on one of *empathy*; e.g., Husserl (1963 [1931], 92): "a *transcendental theory of experience of the other* (*Fremderfahrung*), the so-called *empathy* (*Einfühlung*)" has within its scope "the founding of *a transcendental theory of the objective world* ... in particular, of objective Nature to whose existence sense (*Seinsinn*) belongs *there-for-everyone* (*Für-jedermann-da*) ..."

[8] Weyl, 1931c, 49; compare this passage of Husserl, (1976 [1913], 12), "The existence of a Nature *cannot* be the condition for the existence of consciousness, since Nature itself turns out to be a correlate of consciousness: Nature *is* only as being constituted in regular concatenations of consciousness."

These passages set out a theme crucial to Weyl's *transcendental* philosophy of natural science: the central phenomenological distinction between "objects" beyond (*transcendent* to) experience and those immanent within experience, i.e., "intentional objects" produced in acts of reflection upon experience. The former is the realm of mind-independent objects of the physical world, the target of physical theory; the latter are the idealized mathematical symbolic surrogates of physical theory.

Around 1925 Weyl began to use the expression "symbolic construction" to underscore this distinction between mind-*transcendent* objects, and their symbolic surrogates in physical theory. The term itself originates in Weyl's intervention in the period controversy over foundations of mathematics. His first work in philosophy of mathematics, the predicative analysis of (1918b), drew upon Husserlian phenomenology. By the early 1920s Weyl had become an enthusiastic proponent of Brouwerian intuitionism though largely as interpreted through a phenomenological lens. Still, he was all-too-aware of the severe limitations intuitionistically acceptable methods placed on classical mathematics, writing in 1925 that "full of pain, the mathematician sees the greatest part of his towering theories dissolve into fog". (Weyl, 1926, 534).

Weyl's subsequent divergence from Brouwer was prompted by Hilbert (1922), who had entered the lists against Brouwer and Weyl. Declaring "in the beginning was the sign", Hilbert's idea was to begin with the intuitively given but otherwise meaningless signs of "concrete, intuitive" number theory (elementary arithmetic), a finite part of mathematics, including recursion and intuitive induction for finite existing totalities, grounded in "purely intuitive considerations" (*rein anschauliche Überlegungen*), hence acceptable to intuitionism. This finite formal part is to be supplemented by a strict axiomatic formalization of the rest of the mathematical theory (its infinitary part), including its proofs. Questions of the truth or validity of mathematical statements are to be replaced by the metamathematical demand for a consistency proof of the theory's axioms, to be obtained in a "formal proof theory" in which proofs are rule-governed arrays of concrete and displayable formal signs. A formal consistency proof guarantees the reliability of mathematical theory in yielding the conclusion that the permitted arrays cannot yield a contradiction, such as $0 = 1$.[9] A few years later Hilbert (1926) distinguished between *finitary* and *ideal statements* in an attempt to justify appending to the finitary part of mathematics the contentious infinitary part (Cantorian set theory) by appeal to a Kantian Idea of Reason, i.e., the regulative demand to complete the concretely given in the interest of totality.[10] With this step, Hilbertian metamathematics persuaded Weyl

[9]Famously, Gödel in 1931 showed that any such consistency result could only be relative, since a consistency proof could only be carried out in a stronger theory.

[10]Hilbert (1926, 190): "The role that remains to the infinite is ... merely that of an idea – if, in accordance with Kant's words, we understand by an idea a concept of reason that transcends all experience and through which the concrete is completed so as to form a totality". Weyl (1931a, 28): "Hilbert himself says somewhat obscurely that infinity plays the role of an idea in the Kantian sense, by which the concrete is completed in the sense of totality. I understand this to mean

that the evidentiary demands of symbolic construction could not be reduced to "the demands of open-eyed [*schauenden*] certainty" for each individual statement of a theory. (1931a, 29) The Hilbertian shift from truth of individual statements to consistency, a global requirement on theories, prompted analogy to modern physical theories. As the justification of any particular mathematical statement ultimately requires reference to the entire theory, via a metamathematical proof of that theory's consistency, in theoretical physics evidential justification of particular statements involves an often complex and indirect relation between theoretical terms and their ties to observation and experiment.[11]Following Hilbert, Weyl would take it to be a metatheoretic axiom of "symbolic construction" that *evidence* bears on a theory not statement by statement but only on the theoretical system as a whole. Even so, Weyl dismissed Hilbert's metamathematical "game of formulae" as an adequate philosophical justification of the cognitive importance of mathematics. Instead, he sought to find epistemological warrant for mathematics in its application to the symbolic constructions of physical theory. Of course, unlike mathematics, physics posits a mind-transcendent reality. Still, philosophical reflection (*Besinnung*) on the constructions of physical theory reveals that the physicist must remain content to represent this reality only in symbols (*"das Tranzendente darzustellen ... nur im Symbol*").[12]"Fusing" mathematics with physics thus again brings out transcendental

something like the way in which I complete what is given to me as the actual content of my consciousness, into the totality of the objective world, which certain includes much that is not present to me. The scientific formulation of this objective concept of the world occurs in physics, which avails itself of mathematics as a means of construction. However, the situation we find before us in theoretical physics in no way corresponds to Brouwer's idea of a science. That ideal postulates that every judgment has its own meaning achievable in intuition. The statements and laws of physics, nevertheless, taken one by one, have no content verifiable in experience; only the theoretical system as a whole allows itself to be confronted by experience. What is accomplished here is not the intuitive insight into singular or general contents and a description that truly renders what is given, but instead a theoretical, and ultimately purely symbolic, construction of the world."

[11]Weyl, 1928a, 147–8: "[The] individual assumptions and laws [of theoretical physics] have no separate fulfilling sense [that is] immediately realized in intuition (*in der Anschauung unmittelbar zu erfüllender Sinn eigen*); in principle, it is not the propositions of physics taken in isolation, but only the theoretical system as a whole that can be confronted with experience. What is achieved here is not intuitive insight [*anschauende Einsicht*] into particular or general states of affairs and a faithfully reproduced *description* of the given (*das Gegebene*), but rather theoretical, ultimately a purely symbolic, *construction* of the world." Weyl goes on to state that if Hilbert's view prevails over Brouwer's, as indeed appears to be the case, then this represents "*a decisive defeat of the philosophical attitude of pure phenomenology*, as it proves insufficient to understand creative science in the one domain of knowledge that is most rudimentary and earliest open to evidence, mathematics."

[12]Weyl, 1926, 540; also 1954b, 645. The latter is a lecture entitled "*Erkenntnis und Besinnung (Ein Lebensrückblick)*". *Besinnung*, here translated 'reflection', is a technical term in Husserlian phenomenology, having the meaning of "sense-investigation"; e.g., Husserl, (1974 [1929], 8): "Sense-investigation (*Besinnung*) ... radically understood, is *originary sense-explication* (*ursprüngliche Sinnesauslegung*, orig. emphasis), transforming and above all striving to transform sense in the mode of unclear opinion into sense in the mode of full clarity or essential possibility (*Wesensmöglichkeit*)."

idealism's central theme, that physical objectivity is constituted through symbolic (i.e., mathematical) relations, not "given" through relations of reference and designation.

"Symbolic" signifies more than the truism that mathematics is the necessary instrument of exact natural science. More significantly, the intent of the term is to underscore the conviction that the finite human mind, rooted in "all too human ideas with which we respond to our practical surroundings in the natural attitude of our existence of strife and action" (Weyl, 1932b, 6) *can* attain only a *symbolic* (neither literal nor pictorial) understanding of the infinite (in mathematics[13]) or of the mind-independent real world posited by physical theory.[14] In particular, the twin revolutions of twentieth century physics, relativity and quantum mechanics, have demonstrated that symbolic representation is essential, that "here we are in contact with a sphere which is impervious to intuitive evidence; cognition necessarily becomes symbolic construction."[15] Relativity and quantum mechanics also instruct that physical objectivity is constituted structurally, as "invariance under a group of automorphisms". (Weyl, 1948/9) Transcendental idealist limitations on the scope and character of cognition of nature are thus reformulated through the notions of *symbolic construction* and *invariance*:

> A science can only determine its domain of investigation up to an isomorphic mapping. In particular, it remains quite indifferent as to the 'essence' of its object. .... The idea of isomorphism demarcates the self-evident boundary of cognition.
> (1927, 22; 1949, 25–6)

The term "construction" also echoes Weyl's oft-expressed predilection for constructive vs. axiomatic (i.e., predicative or intuitionist vs. set-theoretic) mathematics. (Weyl, 1985) It reflects evidential preference for theories resting on 'visualizable', iterative or recursive basal structures, for geometry and point-set topology over those of modern abstract algebra even as Weyl insisted that mathematics requires both. (Weyl, 1932a) Above all, it signifies that 'objectivity' in physical theory is constructed as an invariance "for a subject with its continuum of possible positions", and that it arises in step-by-step construction from a basis of what is *aufweisbar* (evident), "something to which we can point to *in concreto*" as demonstrably evident to the constituting consciousness.

> (T)he constructions of physics are only a natural prolongation of operations [the] mind performs in perception, when, e.g., the solid shape of a body constitutes itself as the common source of its various perspective views. These views are conceived as appearances, for a subject with its continuum of possible positions, of an entity on the next higher level of objectivity: the three-dimensional body. Carry on this 'constitutive' process in which

[13] Weyl, 1932b, 7: "*Mathematics is the science of the infinite,* its goal the symbolic comprehension of the infinite with human, that is finite, means."

[14] The influence of Husserl is also apparent in Weyl's use of the term "natural attitude".

[15] Weyl, 1932b, 82. For example, the central underlying theoretical device of quantum mechanics, densities of a complex valued, infinite-dimensional wave function, *can* only be symbolically represented. Dirac, influenced by Weyl (1928b) offers the same philosophical message regarding the necessity of symbolic methods in the first sections of his (1930).

> one rises from level to level, and one will land at the symbolic constructions of physics. Moreover, the whole edifice rests on a foundation which makes it binding for all reasonable thinking: of our complete experience it uses only that which is unmistakably *aufweisbar*. (1954a, 628, 627)

The *fons et origo* of all meaning-constitution, i.e., what is given *in evidence* to ‘pure consciousness’ remains, but “symbolic construction” is the constitutive process wherein concrete symbols go proxy for Husserlian data of ‘pure consciousness’. It is Weyl’s mathematical rendering of Husserlian analysis of essences (*Wesensanalyse*) in terms of a step-by-step symbolic construction beginning from the evidentially privileged standpoint of the “purely infinitesimal”. “Symbolic construction” is then Weyl’s term of art for what is yielded by philosophical reflection upon the enterprise of theoretical natural science. It is his generic term for *sense-constitution* of the transcendent physical world, i.e., a physical theory is a symbolic construct that must not be conflated with that “true real world”.[16]

9.2.3 Transcendental-Phenomenological Origins of Gauge Invariance

We have previously argued that Weyl’s 1918 reformulation of Einstein’s general relativity (GR) within a “purely infinitesimal geometry” was largely spurred by his philosophical orientation to transcendental phenomenological idealism. (Ryckman, 2005) The mandate of *RZM* “to comprehend the sense and the justification of the posit of reality (*Wirklichkeitsetzung*)” beginning from the starting point of the “given-to-consciousness (*Bewußtseins-Gegebene*)” quickly became an explicit recognition that in general relativity this starting point, the primary locus of sense-constitution, lay in the “purely infinitesimal” for the infinitesimally small requires only elementary linear analytic relations, essentially encapsulated in what Weyl termed a linear “connection”. The “purely infinitesimal” thus bounds the immediate evidential reach of a situated ego, the constituting transcendental subject that, through step-by-step construction, invests derived mathematical structures with meaning. In this regard, Weyl’s purely infinitesimal reconstruction of general relativity brings phenomenological reflection upon the levels of sedimented geometrical structures defined on general relativistic spacetime manifolds, a reflection that reveals how there is

[16]Weyl, 1954a, 627: “... the words ‘in reality’ must be put between quotation marks; who could seriously pretend that the symbolic construct is the true real world?” The term ‘symbolic construct’ encompasses not merely the symbolic universe in which physical systems, states, transformations and evolutions are mathematically defined in terms of manifolds, functional spaces, algebras, etc., but also a symbolic specification of idealized procedures and experiments by which the basic physical quantities or observables of the theory are related to observation and measurement. It reflects an insistence, reinforced by quantum mechanics, that physical quantities (beginning with ‘inertial mass’) are not simply given, but “constructed” (1931b, 76; 1934a, b, 109ff).

> a surreptitious substitution of the mathematically substructed world of idealities for the only real world, the one that is actually given through perception, that is every actually experienced, and experienceable ....(Husserl, (1954 [1936-7], 48–9)

The injunction "to comprehend the world" (i.e., structures on the entire manifold) from "the purely infinitesimal" is then just a requirement that sense-constitution of the finite and global mathematical structures defined upon these manifolds ultimately derives from this evidential basis.

General relativity indeed provided the physical and philosophical impetus for the birth of the gauge principle. General relativity preeminently features a local symmetry, the invariance of the form of the Einstein field equations under arbitrary curvilinear coordinate transformations, i.e., the requirement of "general covariance". Einstein embraced this formal requirement as a heuristic corresponding to a supposed generalized principle of relativity, that physical laws appear the same to all observers regardless of their state of motion – accelerating, rotating, inertial – of their reference frames.[17] While the first edition of *RZM* was still in press, Weyl saw how electromagnetism (the only other known interaction in 1918) could also be presented as a manifestation of a kind of local symmetry, analogous to the local symmetry of general relativity.

The additional invariance that Weyl sought to exploit was an invariance with respect to a change of scale ("gauge", hence the name) at each point of spacetime. He did so on grounds that the Riemannian geometry of Einstein's theory was "inconsistent" with the basic thrust of the Riemannian theory of manifolds and the variable curvature of Einsteinian spacetimes. Following Riemann, Einstein allowed the direction of a vector to change as the vector is transported (by an affine connection) "parallel to itself" from point to point around a closed curve in spacetime; the angle between the initial and the returning vector at the same point is the indicator of spacetime curvature. But, of course, vectors have two properties, direction and magnitude, and the Riemann-Einstein geometry required the magnitude of the vector to remain the same while traversing a closed curve. Weyl proposed to rectify this "inconsistency" by requiring a local gauge invariance according to which a "length connection" allows scale changes to vary smoothly from point to point of spacetime. The demand that physical laws remain invariant under these local changes of scale resulted in new degrees of freedom that mathematically, as Weyl showed, brought electromagnetism in addition to gravitation into the metric of spacetime. That is to say, the metric of Weyl's 1918 "purely infinitesimal" geometry of spacetime, unlike Einstein's Riemannian geometry, allowed for variation of unit of scale ("gauge") at each point; this new degree of freedom, a function φ_μ [μ = 1,2,3,4] of the four spacetime coordinates of the point, was shown to be mathematically identical to the vector four potential A_μ of relativistic electromagnetic theory. The end result was that Weyl's geometry yielded fundamental field equations of *both* Einstein's gravitation and electromagnetism.

[17] In fact, general covariance is a purely formal requirement having nothing to do with a generalized principle of relativity.

From Weyl's "purely infinitesimal" starting point, global field laws (the "physical world") are constructed that satisfy not only, as in general relativity, general covariance (freedom to choose spacetime coordinates) but also "gauge invariance" (freedom to choose scale at each point). To be sure, Einstein immediately identified an empirical objection to Weyl's 1918 idea of gauge invariance. So the idea lay dormant until reformulated by Weyl as pertaining to a quantum mechanical factor of phase in 1929 when bringing the relativistic equation of the electron (the Dirac equation) into the four-dimensional spacetime context of general relativity. Indeed, contemporary physical theory pertains to so-called "internal symmetries" of quantum fields rather than to a factor of scale and so from a modern point of view, the gauge principle is misnamed.[18]On the other hand, the basic idea of gauge invariance, that it involves an arbitrary function of the spacetime coordinates, remains. To Weyl, the requirement of gauge invariance, however interpreted, has the "character of a 'more general' ('*allgemeiner*') relativity".[19]

9.2.4 *From the "Raumproblem" to Lie Groups and Lie Algebras*

In a natural development from his 1918 "purely infinitesimal" reformulation of general relativity, Weyl turned to the new "space problem" posed by the variably curved manifolds permitted in Einstein's theory. The old "space problem" concerned the group-theoretical characterization of a geometry: A particular geometry is singled out by its continuous group of motions: figures that transform into one another by the operations of the group are considered equivalent. The Helmholtz-Lie solution to the old "space problem" presupposed a geometric space that permits free mobility of rigid bodies, i.e., a homogeneous space of constant curvature; this is made obsolete by general relativity where the metric is no longer homogeneous (not a positive definite quadratic form but one that is indefinite) and the medium itself is variably curved spacetime.

Weyl's new solution (1921–23) came about by drawing again from the "purely infinitesimal". He noted that the old Helmholtz-Lie solution retained its validity in the infinitely small if posed in terms of a group of rotations defined only in the homogeneous tangent space centered on each point $P \in M$. (Weyl, 1988) At each point, the rotations of a vector at P are assumed to form a continuous group of linear transformations; as the volume of parallelepipeds formed by basis vectors at

---

[18] Internal symmetries refer to the fact that particles occur in multiplets, members of which can be considered as "the same" under the symmetry of the interaction. Mathematically, the multiplets are realizations of an irreducible representation of some internal symmetry group. See further below.

[19] Weyl (1929, 246). Cf. Weyl 1931b, 220: "One can in fact take it as a general rule that an invariance property of the kind met in general relativity, involving an arbitrary function, gives rise to a differential conservation theory. In particular, gauge invariance is only to be understood from this standpoint."

P is assumed preserved by rotations, these linear transformations are a subgroup of the special linear group $SL(n)$. Thus the "nature of space" at each point P is the same, and homogeneous. Metrical relations in neighborhood U of P are then defined on the assumption that rotations at any point $P' \in$ U are obtained from rotations at P by a single linear congruence transformation (affine connection) A taking P to P' by composition with rotations at P. This allows that the rotation groups at different points have distinct "orientations". The affine connection is uniquely compatible with the metric and so passing continuously from P to any other point $Q \in M$ shows that the subgroups at each point have the same metric, i.e., are congruent to the special linear group $SL(n)$. And this means that though the rotation groups at the various points have different "orientations" due to variations of matter and energy, they all share the same infinitesimal Pythagorean (Euclidean) metric group structure. (1923b, 43–61).

Already in the 1880s Sophus Lie had reduced the concept of continuous group to the "germ" of infinitesimal elements that generated it. In abstract form, these "*groupes infinitésimaux*" had been extensively studied and classified some decades before by É. Cartan, building upon earlier results of W. Killing.[20]But Weyl was the first to explicitly recognize that the simple tools of linear algebra could be brought to bear on this infinitesimal structure. Lie's concept of "infinitesimal group" was essentially repurposed in Weyl's solution to the "Space Problem" in the light of the variably curved four-dimensional spacetimes of Einstein's theory. In an appendix, Weyl showed that this infinitesimal group structure could be axiomatically expressed in algebraic terms. (1923b, 82) In particular, the structure of the tangent space surrounding the point that is the group identity (where the Lie group can be considered homogeneous and linear), is a linear vector space, i.e., there is a linear algebra structure in the tangent space to the identity of a Lie group. The axiomatized infinitesimal structure of a Lie group was first termed a "Lie algebra" by Weyl in lectures 10 years later at the Princeton Institute of Advanced Study.[21]In short, Weyl's solution to the new "Space Problem" crucially rested upon the concept of infinitesimal group, recast in language of linear vector spaces. To Weyl this was compelling evidence that"mathematical simplicity and metaphysical primitiveness

[20]See (Hawkins, 2000), and (Eckes, 2013). Cartan's "structure theory" for infinitesimal Lie groups (today, Lie algebras) identifies isomorphic groups through their "structure constants". Cartan worked exclusively at the level of abstract groups; Weyl would translate Cartan's structure theory into the language of matrix groups, group representations by matrices, the language of most interest to physics.

[21]Weyl, *The Structure and Representation of Continuous Groups*. Based on notes by Richard Brauer taken at Weyl's course at The Institute for Advanced Study, 1934–5; reprinted 1955, p.4. In modern terms, a Lie algebra to a Lie group G is usually denoted by the Gothic character g and is defined by three properties: 1) the elements X, Y, etc. of g form a linear vector space; 2) the elements of g close under a commutation relation $[X,Y] = -[Y,X]$, $\forall X,Y \in$ g; 3) the Jacobi identity $[X,[Y,Z]] + [Y,[Z,X]] + [Z,[X,Y]] = 0$ is satisfied. Using Cartan's "structure theory", the structure of a Lie algebra is completely determined by its "structure constants" $c_{ij}{}^k$ that appear in the commutator of any two basis vectors $[X_i, X_j] = c_{ij}{}^k X_k$.

(*Ursprünglichkeit*) are narrowly bound together".[22] The purely infinitesimal solution to the new "Space Problem" in turn led to purely mathematical research on representations of semisimple Lie groups and Lie algebras.[23] And it is in the guise of linear vector spaces that the concept of Lie algebra (and its representations) appears in the contemporary gauge theories of the Standard Model (see Sect. 3.1).

9.3 The Gauge Principle

A symmetry of a (quantum) field theory is a group of transformations that leave the equations of motion of the field (encapsulated in its Lagrangian) unchanged in form. Symmetries may be discrete or continuous, global (applying in the same way everywhere) or local (applying differently at each point of space). The gauge principle refers more specifically to a logic or procedure, or really a *recipe*, for constructing the *form* or *template* of an interaction mechanism into a free (non-interacting) quantum field whose law of motion (Lagrangian density) transforms invariantly under a continuous *global* symmetry (e.g., a rotation in an "internal" dynamical space closed under the action of a given Lie group). A global symmetry is one that applies in the same way at each point of space(time). The form of interaction emerges from "gauging" the global symmetry, i.e., requiring the global symmetry to become a *local* symmetry so that, e.g., the rotations can be different at each point of space(time). In consequence of gauging the global symmetry, a so-called "gauge field" (manifested in particle terms as spin 1 (in units of h), "gauge bosons") appears that is required to transform in a way compensating for the change from global to local symmetry. Supplemented with additional information (appropriate kinetic and coupling terms), the Lagrangian density of the now interacting field is invariant with respect to the new extended group of local transformations.

The so-called "gauge argument" illustrates the canonical way to "gauge" a field theory. The paradigm is set by quantum electrodynamics (QED), the first quantum field theory; in virtue of the logical pattern of the argument in QED, one can view the SM as essentially a generalization of QED. The new gauge degree of freedom here appears not as a factor of scale but as a local phase factor in wave function of electron. One begins with a free electron field $\Psi(x)$ that is determined up to a

[22]Weyl, 1922, 329: "*auf diesem Felde mathmatische Einfachheit und metaphysische Ursprünglichkeit in enger Verbindung miteinander stehen.*"

[23]Weyl, 1925–6. A (non-abelian) Lie group whose Lie algebra cannot be factorized into two commuting subalgebras is called *simple.* A direct product of simple Lie groups is called *semi-simple.* In these papers Weyl supplemented Cartan's infinitesimal abstract group viewpoint with global and topological properties of Lie groups, thus in our view, "comprehending the world beginning from the infinitesimal". In fact, linearizing a Lie group G in the tangent space of the identity to form its Lie algebra g destroys G's global properties, i.e., what happens far from the identity. Hence the need for integral and topological methods.

phase factor θ. The Maxwell-Dirac Lagrangian for the electron field that is the basis of QED transforms invariantly under the *global* phase transformation (applying at each point in the same way),

$$\Psi(x) \Longrightarrow \Psi'(x) = e^{i\theta}\Psi(x)$$

where $e^{i\theta}$ is Euler's formula. $\Psi(x)$ is then invariant under a global U(1) internal symmetry group; the global invariance of the matter system implies, via Noether's theorem, the existence of a conserved quantity, a matter current j^{μ}.

One then "promotes" the global symmetry to a local phase symmetry; this means that an independent U(1) group is associated with *each* space-time point. Then requirement of gauge symmetry demands that the phase parameter θ vary as a function of space-time position *x,* and the phase invariance is *local:*

$$\Psi(x) \Longrightarrow \Psi'(x) = e^{i\theta(x)}\Psi(x) \qquad (9.1)$$

Typically, Lagrangians depend not only on the field magnitudes but also on their (at least first) derivatives. However, by imposing a local symmetry, the derivative of the field $\partial_{\mu}\Psi(x)$ picks up an extraneous term $\partial_{\mu}\theta(x)$ in its transformation; as $\theta(x)$ is a function varying with spacetime position, it is not a covariant object. In order to cancel this unwanted term, the "gauge covariant derivative" is introduced,

$$\partial_{\mu} \Longrightarrow D_{\mu} = \partial_{\mu} - ieA_{\mu}$$

where e is the electric charge of the electron field. The new covariant derivative transforms as

$$D_{\mu} \Longrightarrow D'_{\mu}\Psi = e^{i\theta(x)}D_{\mu}\Psi$$

and $A_{\mu} = A_{\mu}(x)$ is an "invented vector field required to transform as

$$A_{\mu}(x) \Longrightarrow A'_{\mu}(x) = A_{\mu}(x) - \frac{\partial_{\mu}\theta(x)}{e} \qquad (9.2)$$

showing proportionality of the transformation to *e*, the unit of electric charge. The resulting Lagrangian is then invariant under joint local transformation of $\Psi(x)$, given by (9.1), and of $A_{\mu}(x)$, given by (9.2), the added partial derivative in (9.2) *exactly compensating the extraneous position-dependent variation of the phase factor.* Moreover, in imposing the requirement of local symmetry in steps (9.1) and (9.2), the free electron field $\Psi(x)$ is coupled to the electromagnetic field represented by the Faraday tensor, obtained by taking the derivative of the four-potential $A_{\mu}(x)$,

$$F_{\mu\nu} = \frac{\partial A_{\mu}}{\partial x_{\nu}} - \frac{\partial A_{\nu}}{\partial x_{\mu}},$$

and the conserved current now appears in an interaction of the form $ej^{\mu}A_{\mu}$ in conformity with Maxwell theory. For obvious reasons, the new term $A_{\mu}(x)$ is now called a "gauge field". The "gauge argument" then shows how introducing local symmetries dictates the form of the interaction of matter fields. (Yang, 1988, 20).

9.3.1 Lie Algebras in Field Theory: Purely Infinitesimal Operations

As just seen in the case of the gauge group U(1) of QED, local symmetries are introduced by making the Lie group parameters functions of space and time. In more general terms a Lie symmetry group G can be expressed in terms of its infinitesimal generators X^a

$$G = e^{i\theta^a X^a}$$

where θ^a ($a = 1 \ldots N$) are the real parameters of the group, the X^a are linearly independent Hilbert space matrices and there is an implied summation. The set of all linear combinations $\theta^a X^a$ form a vector space; the term generator refers broadly to an arbitrary element of the vector space or specifically to the basis vectors X^a. Letting $t^a = \{X^a\}$ define a set of such matrices, the Cartan structure condition defines the Lie algebra:

$$\left[t^a, t^b\right] = i f^{abc} t^c$$

where $[t^a, t^b]$ is the commutator (Lie bracket) and f^{abc} are the structure constants of the Lie algebra that define the multiplication properties of the Lie group (at least for elements continuously connected to the group identity), constants since the multiplication properties of the group transformations should be independent of any particular representation. As the expression shows, the structure constants themselves generate another representation of the algebra. Local symmetry then requires that the parameters θ^a become functions of x, $\theta^a(x)$. For our purposes, we simply point out that generically one can write the covariant derivative of the field as a linear combination of the ordinary derivative of the field (its infinitesimal displacement in space or spacetime) and a field-dependent infinitesimal gauge transformation

$$D_\mu \Psi \equiv \partial_\mu \Psi - W_\mu \Psi$$

where W_{μ} is a matrix representing the gauge field whose entries are generated by infinitesimal gauge transformations. Therefore the W_{μ} can be decomposed into its generators t_a,

$$W_\mu = W_\mu^a t_a.$$

This shows that the gauge field takes values in the Lie algebra corresponding to the gauge group G. The characteristic of a gauge field, exemplified in its Lie algebra, is that it carries information regarding group structure from one spacetime point to another, a "purely infinitesimal", hence *evident*, operation.

9.3.2 The Gauge Principle Generalized: The SM

The above gauge invariance of QED is only the simplest example of infinite parameter or Lie group symmetry, and it is not typical as its gauge group U(1) is abelian (commutative). The gauge argument was generalized by Yang and Mills (1954) so that Yang-Mills theories of the SM arise in the same logical pattern: Gauging a global symmetry requires (to restore invariance of the field Lagrangian) the introduction of a covariant derivative; the new derivative is required to transform in a manner that introduces a new (gauge) field; the gauge field provides the form of the interaction forces of a matter field. The same mathematical expressions appear with only minor changes, e.g., in place of the phase of the electron field, there are generalized phases associated with the wave functions of multicomponent matter fields. This is the Yang-Mills template for the Standard Model, a spontaneously broken non-abelian gauge theory containing three types of particles: elementary scalars, fermions (spin-1/2) and spin-1 bosons. Spin-1 gauge bosons are the particles of gauge fields. The SM is often represented by its "gauge group", the direct product group SU(3) × SU(2) × U(1) representing the fundamental interactions. Unlike U(1), the special unitary Lie groups SU(2) and SU(3) are non-commutative (non-abelian). SU(2) × U(1) is the symmetry group of the "electro-weak" interactions, where U(1) is the phase symmetry of the weak hypercharge (slightly different from the phase symmetry of the electromagnetic interaction QED) and SU(2) is the isospin symmetry describing weakly interacting particles. The elements of SU(2) are 2 × 2 matrices; in the weak interaction matter particles (*up* and *down* quarks; electrons and electron neutrinos) are sorted into doublets such that the two particles in a doublet are interchangeable, indistinguishable in that interaction. The group SU(3) plays two roles in the SM: as an exact gauge symmetry associated with color for the strong interaction, and as an approximate global flavor symmetry of the strong interactions (the "eightfold way" of Gell-Mann and Ne'eman). SU(3) elements are represented by 3 × 3 matrices, so the symmetry operations pertain to a triplet of particles. Quarks are the matter particles of the strong interaction; each with its own mass and fractional charge comes in one of six *flavors* partitioned into three doublets (*up, down*), (*strange, charm*), (*top, bottom*). *Color*, the strong force analogue to electric charge, itself has three manifestations, *red, green, blue.* Within the same flavor, changing e.g., *red* quarks to *green* quarks leaves the interaction energy of the system unchanged. In general, non-abelian Lie groups yield theories of multiple vector particles, whose interactions are strongly constrained by a gauge symmetry.

In sum, a gauge symmetry is a constraint on the Lagrangian L for any quantum field theory; the first step in constructing a quantum field theory is to ask what gauge symmetry it must obey.[24] A gauge symmetry will dramatically reduce the vast number of theoretically possible Lagrangians. Moreover, both massless and massive gauge theories (provided the masses are generated by spontaneous symmetry breaking via the Higgs mechanism[25]) are renormalizable. In fact, the only way to form a relativistic quantum field theory of spin-1 particles (force-carrying *bosons*) is a gauge theory. (Weinberg, 1995, 340).

9.4 Towards an Elucidation of the Gauge Principle

The gauge principle has been recently described as "the most fundamental cornerstone of modern theoretical physics" (Redhead, 2002, 299). As has been seen, the force fields of the SM are gauge fields, each formally is a variation on the basic template of the gauge principle. Nonetheless, despite its success, few if any theorists believe the SM to be a truly fundamental theory. For one thing, the SM contains approximately 26 free parameters (notably particle masses and Yukawa coupling terms that parameterize the interactions of fermions and the scalar Higgs field). A "free parameter" is one for which there is no theoretical explanation, one whose value has to be put into calculations "by hand", i.e., as determined by experiment rather than predicted by theory. Furthermore, the values of many of these parameters appear suspiciously "fine-tuned" such that if the observed value differed by just a few percent, life as we know it would not be possible. (e.g., Rees, 1999) In addition, the SM features three families of fermions, without any internal account of why this should be so. Moreover, within the SM, there is no clear answer to how many Higgs bosons there are, though only one has been identified (mass 125.1 GeV (1 GeV = 1000 eV), discovery announced by CERN on July 4, 2012); this underdetermination increases the artificiality of the Higgs mechanism for SSB. Finally, the SM says nothing about "dark matter" or "dark energy", i.e., nothing at all about approximately 95% of the energy budget of the universe. For these and other reasons, most quantum field theorists and cosmologists view the Standard Model as merely an "effective theory", a provisional stage in the descent to ever smaller distance scales corresponding to an ascent to ever higher energies. Hence it is widely

[24]The second step is to determine the representations of fermions and scalars under the gauge symmetry; a third step is to postulate the pattern of spontaneous symmetry breaking.

[25]A symmetry of a system is said to be "spontaneously broken" if its lowest energy state is not invariant under the operations of that symmetry. This is an extremely important concept in the weak interaction as the bosons introduced by gauge symmetries are massless, like the photon; their masses arise from the "spontaneous breaking" of the SU(2) × U(1) symmetry through couplings to the scalar Higgs field. See note 26.

assumed that the SM is a "low energy" consequence of more fundamental physics at the higher energies of the early universe, up to Grand Unification (GUT) scales of 10^{15}–10^{16} GeV (at which the gauge coupling strengths of the three interactions theoretically meet;) or even the most fundamental unification Planck scale 10^{19} GeV which necessarily includes gravitation, the remaining known interaction.

To the inquiring philosopher, the SM presents several challenges. For the default scientific realist, some additional work must be done to explain how the known physical laws with their accompanying ontological posits are not truly fundamental but contingent regularities in the sense that their validity is restricted to certain "low energy" epochs. The *pro tem* necessity of these laws, if such there is, may originate only in conditions that are accidental or environmental according to both string theory and most models of inflationary cosmology. More fundamentally, there is the matter of how to understand the empirical success of the gauge principle.

Local symmetry transformations introduce new gauge degrees of freedom that appear to be redundancies or mathematical surplus structure in the physical description of an interaction. The same physical state can be represented by many different solutions of the field equations when these solutions are related by gauge transformations. *Prima facie*, gauge symmetries connect states that are physically the same yet differ in their mathematical description. This prompts an analogy to general relativity, the site of origin of the gauge principle, and a reflection upon Weyl's claim (1929) that the principle of gauge invariance has the "character of a 'more general' ('*allgemeiner*') relativity".

For just as a specific coordinate system must be chosen to extract physical observations in general relativity, so gauge fixing is necessary to retrieve physical predictions from gauge theories. This would seem to indicate that gauge symmetries are not symmetries of nature but of physical description of nature. The redundancy of gauge descriptions itself raises various puzzles. How it is possible for spontaneous symmetry breaking (SSB) of a gauge symmetry to have physical consequences (e.g., gauge bosons acquire a mass)?[26]Further puzzling is the fact that the gauge redundancy of mathematical description appears to be fortuitous since gauge theories of vector particles (photons, and the massive spin 1 bosons of the electroweak theory) as in the SM are *renormalizable*, which means, roughly, that the generic infinities (non-physical divergent integrals) associated with free parameters (such as particle masses) in quantum field theory can be tamed. Thus, local gauge

[26]SSB plays two roles in the SM, giving mass to gauge bosons (other than the photon) and giving mass to fermions (leptons and quarks). In the electroweak theory (unifying the weak and electromagnetic interactions), SSB plays a crucial part, breaking the electroweak $SU(2) \times U(1)$ symmetry into individual electromagnetic and weak forces while enabling mass (the $SU(2)$ multiplets) to emerge spontaneously in theories where initially, there is no mass. A standard story is that SSB invokes the Higgs mechanism to spontaneously break a local gauge symmetry; of course, this a symmetry connecting states that cannot be physically distinguished. In fact, in SSB it is not the local gauge transformations that are spontaneously broken. Rather it is a global symmetry (a unique vacuum state) that is spontaneously broken while the gauge symmetry is explicitly broken by gauge fixing in the Higgs mechanism in order to extract physical predictions. Elitzur (1975) has showed the spontaneous breaking of a local symmetry is logically impossible.

invariance is required for the Lagrangians of the SM to be mathematically tractable and predictive up to higher and higher energies.

Weyl's "purely infinitesimal" generalization of general relativity issued in the demand that fundamental physical theories, in addition to the requirement of coordinate freedom ("general covariance"), should also satisfy the requirement of gauge (more appropriately in its debut, scale) invariance. Both requirements introduce *arbitrary* mathematical degrees of freedom at each point P of the four-dimensional differential manifold representing spacetime; the arbitrariness can be understood phenomenologically, as each point indifferently can be considered the locus of an experiencing, constructing ego.[27] These degrees of freedom arise from two metaphysical prerequisites: (1) a postulate of transcendental phenomenological idealism, that "Reality [*Wirklichkeit*] is not a being- in-itself [*Sein an sich*] but rather is constituted *for* a consciousness", and (2) the *aspiration*, fortified by the successes of differential calculus in physics and indeed of field physics (*Nahewirkungphysik*) itself, that this "reality", constituted as it is by a *situated* consciousness, "can be understood from its behavior in the infinitesimally small", i.e., mathematically comprehended starting from the *evident* simple linear relations within the tangent space. The arbitrary degrees of freedom represent *at each point* particular magnitudes of physical states, *either* as different mathematical functions (scalar, vector, tensor, spinor) of four independent variables (spacetime coordinates) determined by the field laws *or* in terms of an arbitrary vector function of these spacetime coordinates signifying an internal gauge symmetry. In the latter case, these degrees of freedom serve to represent interactions as proceeding through the exchange of particles of spin 1 such as the photon, or the massive gauge bosons of the electro-weak interaction ($W^{\mp}$, Z^0).

At this point we may recall the Weyl-Nozick slogan *objectivity* = *invariance* but signal our departure from the Nozickian realist (Nozick, 2003: Earman, 2004) who maintains that *objective* pertains (by definition) to the mind-independent structure of the world. Our conjecture is rather that of Weyl, that general covariance and gauge invariance are but particular demands of objectivity upon any theoretical construction initiated from the standpoint of radical locality: the constructed physical theory must be independent from any particular starting point from which it is constituted. As a constructive requirement of objectivity, the invariance of laws under arbitrary coordinate and local gauge transformations can be understood in the first instance not as symmetries of nature but of a radically local description of nature. In this way, perhaps, we can see how Weyl's central idea of the gauge principle as a "purely infinitesimal" remnant of sense-constitution is preserved even today in the internal (phase) symmetries of the Standard Model.

[27] Weyl, 1918b, 72: "The coordinate system is the unavoidable residue of the ego's annihilation (*das unvermeidliche Residuum der Ich-Vernichtung*) in that geometrico-physical world that reason sifts from the given under the norm of 'objectivity' – a final faint token in this objective sphere that existence (*Dasein*) is only given, and *can* only be given as the intentional content of the conscious experience of a pure, sense-giving ego."

References

Atiyah, M. (2002). Hermann Weyl 1885–1955. In *National academy of science: Biographical memoirs* (Vol. 82, pp. 1–14). Washington, D.C.: National Academies Press.

Bernard, J. (2015). *L'idéalisme dans l'infinitésimal: Weyl et l'espace à l'époque de la relativité*. Nanterre, France: Presses universitaires de Paris Nanterre.

Earman, J. (2004, December). Laws, symmetry, and symmetry breaking: Invariance, conservation principles, and objectivity. *Philosophy of Science, 71*, 1227–1241.

Eckes, C. (2013). *Les groups de Lie dans l'oeuvre de Hermann Weyl*Nancy: Presses Universitaires de Nancy/Éditions Universitaires de Lorraine.

Elitzur, S. (1975, December 15). Impossibility of spontaneously breaking local symmetries. *Physical Review D, 12*(12), 3978–3982.

Hawkins, T. (2000). *Emergence of the theory of lie groups: An essay in the history of mathematics 1869–1926*. New York: Springer.

Hilbert, D. (1922). *Neubegründung der Mathematik (Erste Mitteilung)*. Abhandlungen aus dem mathematischen Seminar der Hamburgischen Universität 1 (pp. 157–77).

Hilbert, D. (1926). Über das Unendliche. *Mathematische Annalen, 95*, 161–190.

Husserl, E. (1954 [1936-7]). *Die Krisis der europäischen Wissenschaften und die transzendentale Phänomenologie: Eine Einleitung in die phänomenologische Philosophie*. In W. Biemel (Ed.), *Husserliana VI*. The Hague, The Netherlands: Martinus Nijhoff.

Husserl, E. (1956 [1908]). *"Zur Auseinandersetzung meiner transzendentalen Phänomenologie mit Kants Transzendentalphilosophie"*, ca. 1908. Husserliana VII. In R. Boehm (Ed.), *Erste Philosophie (1923–24). Erster Teil* (pp. 381–395). The Hague, The Netherlands: Martinus Nijhoff.

Husserl, E. (1963 [1931]). *Cartesianische Meditationen und Pariser Vorträge* (1931). In S. Strasser (Ed.), *Husserliana I*. The Hague, The Netherlands: Martinus Nijhoff.

Husserl, E. (1974 [1929]). *Formale und Transzendentale Logik* (1929). In P. Janssen (Ed.), *Husserliana XVII*. The Haag, The Netherlands: Martinus Nijhoff.

Husserl, E. (1976 [1913]). *Ideen zu einer reinen Phänomenologie und phänomenologischen Philosophie I* (1913). In K. Schuhmann (Ed.), *Husserliana III, 1–2*. The Hague, The Netherlands: Martinus Nijhoff.

Nozick, R. (2003) *Invariances*. Cambridge, MA: Harvard University Press, revised edition.

Pesic, P. (Ed.). (2009). *Hermann Weyl: Mind and nature. Selected writings on philosophy, mathematics, and physics*. Princeton, NJ: Princeton University Press.

Pesic, P. (Ed.). (2012). *Hermann Weyl: Levels of infinity. Selected writings on mathematics and philosophy*. Mineola, NY: Dover.

Redhead, M. (2002). The interpretation of gauge symmetry. In M. Kuhlmann, H. Lyre, & A. Wayne (Eds.), *Ontological aspects of quantum field theory* (pp. 281–301). London/Singapore, Singapore/Hong Kong, China: World Scientific.

Rees, S. M. (1999). *Just six numbers. The deep forces that shape the universe*. London: Weidenfeld & Nicolson.

Ryckman, T. (2005). *The reign of relativity: Philosophy in physics 1915–1925*. New York: Oxford University Press. (Oxford Studies in Philosophy of Science).

Weinberg, S. (1995). *The quantum theory of fields. Volume 1*. New York: Cambridge University Press.

Weyl, H. (1913). *Die Idee der Riemannschen Fläche*. Leipzig, Germany: B.G. Teubner.

Weyl, H. (1918a). *Raum-Zeit-Materie*. Berlin, Germany: J. Springer.

Weyl, H. (1918b). *Das Kontinuum. Kritische Untersuchungen über die Grundlagen der Analysis*. Leipzig, Germany: Veit.

Weyl, H. (1919). *B. Riemann: Über die Hypothesen, welche der Geometrie zu Grunde liegen*. Berlin/Heidelberg, Germany: Springer Verlag GmbH.

Weyl, H. (1922). Das Raumproblem. *Jahresbericht der Deutschen Mathematikervereinigung, 31*, 205–221. reprinted in GA II, 328–44.

Weyl, H. (1923a). *Raum-Zeit-Materie* (5 Auflage ed.). Berlin, Germany: J. Springer.

Weyl, H. (1923b). *Mathematische Analyse des Raumproblems: Vorlesungen gehalten in Barcelona und Madrid*. Berlin, Germany: J. Springer.

Weyl, H. (1925–6). "*Theorie der Darstellung kontinuierlicher halbeinfacher Gruppen durch lineare Transformationen*" I, I, II, und *Nachtrag*, *Mathematische Zeitschrift* 23–24; reprinted in GA II (pp. 543–647).

Weyl, H. (1926). *Die Heutige Erkenntnislage in der Mathematik. Sonderdrucke des Symposion, Heft, 3*, 1–32. reprinted in GA II, 511–42.

Weyl, H. (1927). *Philosophie der Mathematik und Naturwissenschaft*. München und Berlin, Germany: R. Oldenbourg.

Weyl, H. (1928a). Diskussionsbemerkungen zu dem zweiten Hilbertschen Vortrag über die Grundlagen der Mathematik. *Abhandlungen aus dem mathematischen Seminar der Hamburgischen Universität, 6*, 86–88. reprinted in GA III, 147–9.

Weyl, H. (1928b). *Gruppentheorie und Quantenmechanik*. Leipzig, Germany: S. Hirzel Verlag.

Weyl, H. (1929). *Elektron und Gravitation. Zeitschrift für Physik, 56*, 330–352. reprinted in GA III, 245–67.

Weyl, H. (1931a). *Levels of infinity*. In P. Pesic (Ed.), Translation of *Die Stufen des Unendlichen* (pp. 17–31). Jena, Germany: Gustav Fischer. 2012.

Weyl, H. (1931b). *The theory of groups and quantum mechanics* (2nd ed). Translated by H.P. Robertson. London: Methuen and Co., Ltd. Reprint, New York: Dover, 1950.

Weyl, H. (1931c). *Geometrie und Physik. Die Naturwissenschaften, 19*, 49–58. reprinted in GA III, 336–45.

Weyl, H. (1932a). Topologie und abstrakte Algebra als zwei Wege mathematischen Verständnissess. *Unterrichtsblätter für Mathematik und Naturwissenschaften, 38*, 177–188. reprinted in GA III, 348–58.

Weyl, H. (1932b). *The open world*. New Haven, CT: Yale University Press. Reprinted in P. Pesic (ed.) 2009, 34–82.

Weyl, H. (1934a). *Mind and nature*. Philadelphia: University of Pennsylvania Press. Reprinted in P. Pesic (ed.) 2009, 83–150.

Weyl, H. (1934b–5). *The structure and representation of continuous groups*. Based on notes by Richard Brauer taken at Weyl's course at The Institute for Advanced Study, reprinted 1955.

Weyl, H. (1948/49). Similarity and congruence. *A chapter in the epistemology of science*. lecture at ETH, Zurich. ETH Bibliothek, Hochschularchiv Hs91a:31, 23 pages.

Weyl, H. (1949) *Philosophy of mathematics and natural science*. Princeton, CT: Princeton University Press. Enlarged translation of Weyl, 1927.

Weyl, H. (1954a). *Address on the Unity of Knowledge delivered at the Bicentennial Conference of Columbia University* Reprinted in GA IV, 623–29.

Weyl, H. (1954b). Erkenntnis und Besinnung (Ein Lebensrückblick). In *Studia Philosophica, Jahrbuch der Schweizerischen Philosophischen Gesellschaft* (pp. 153–71); reprinted in GA IV 631-49.

Weyl, H. (1968). *Gesammelte Abhandlungen*. Bd. I-IV (K. Chandrasekharan, Ed). Berlin/Heidelberg, Germany/New York: Springer

Weyl, H. (1985, December). "Constructive versus Axiomatic Procedures in Mathematics", typescript written after 1953. *The Mathematical Intelligencer, 7*(4), 10–17.

Weyl, H. (1988). *Riemanns geometrische Ideen, ihre Auswirkung und ihre Verknüpfung mit der Gruppentheorie*. Berlin/Heidleberg, Germany/New York: Springer. A manuscript written in 1925 for a Russian edition of Lobachevsky's mathematical works; first published in 1988.

Yang, C. N. (1988). "Hermann Weyl's Contributions to Physics", A Hermann Weyl Centenary Lecture, ETH Zürich, October 24, 1985. In K. Chandrasekharan (Ed.), *Hermann Weyl 1885–1985* (pp. 7–21). Berlin/Heidelberg, Germany/New York: Springer.

Yang, C. N., & Mills, R. L. (1954). Conservation of isotopic spin and isotopic gauge invariance. *Physical Review, 90*, 191–195.

Part III
Phenomenological Approaches to the Measurement Problem

Chapter 10
From a Lost History to a New Future: Is a Phenomenological Approach to Quantum Physics Viable?

Steven French

Abstract In 1939 London and Bauer published a short pamphlet on the measurement problem in quantum mechanics (London and Bauer. La Théorie de L'Observation en Mécanique Quantique, Hermann. In: Wheeler JA & Zurek WH (eds) Quantum theory and measurement, Princeton University Press, Princeton, p 252, 1939). For many years, physicists and philosophers took this to be merely a re-statement of von Neumann's view that it is the intervention of consciousness that somehow leads to the wave function collapsing into some definite state. This view was robustly criticised by Putnam and Shimony in the early 1960s and has been generally abandoned ever since. However, before he became a physicist, London studied phenomenology and his work with Bauer is infused with a phenomenological sensibility. In (French, Stud Hist Philos Mod Phys 33:467–491, 2002) I tried to excavate this 'lost history' and articulate the details of London's approach. Here I want to further consider the extent to which this history might be said to have been 'effaced', to use Ryckman's term (Ryckman, 2005) but also indicate how this phenomenological approach might be further articulated in the broader context of recent interpretations of quantum theory and thereby be regarded as a viable alternative.

10.1 Recovering 'Effaced' History

As is well-known, the history of philosophical reflections on physics in the twentieth century has been overshadowed by certain prominent views. Consider, for example, space-time physics and in particular the development of the General Theory of Relativity as appropriated by and presented within the framework of logical empiricism. Alternative and sometimes intertwined strands of thought have been obscured in this history, or, to use Ryckman's term, have been 'effaced'. Thus

---

S. French (✉)
School of PRHS, University of Leeds, Leeds, UK
e-mail: s.r.d.french@leeds.ac.uk

H. A. Wiltsche, P. Berghofer (eds.), *Phenomenological Approaches to Physics*, Synthese Library 429, https://doi.org/10.1007/978-3-030-46973-3_10

he charts the role of neo-Kantian and phenomenological thought in philosophical reflections on Einstein's theory, thereby helping to recover this 'effaced' history (Ryckman, 2005). In particular, he focuses on Weyl's adoption of an explicitly phenomenological stance with respect to both the relevant physics and its philosophical interpretation, where this shift reflect[s] the theory's ambiguous character as lying in the intersection of physics and philosophy' (ibid., p. 159). I shall consider whether a similar ambiguity of character can be ascertained in the case of London and Bauer's analysis of the measurement problem in quantum mechanics.

Here too, when it comes to quantum physics more generally, we have seen the recovery of a history that has effectively been smothered by positivistic construals of the infamous 'Copenhagen Interpretation' for example. Recently, a neo-Kantian perspective has been recovered in the work of Cassirer (1936; see Ryckman, 2018), which has been appropriated by more recent (and broadly realist) philosophical stances (see French, 2014). The natural question arises whether there are also phenomenological strands to this history that can be brought into the light and in (French, 2002) I argued that there are, as manifested most clearly in London and Bauer's pamphlet (1939).[1] In the next section I will sketch the 'usual story' of this problem and the purported role of consciousness in resolving it.

10.2 The Measurement Problem: Usual Story

The usual story that we tell about the measurement problem is often illustrated by Schrödinger's famous 'cat in a box' thought experiment[2]: a cat is placed in a box, together with a portion of radioactive material and a Geiger counter connected to a device that will release poison if triggered. If the material decays, the Geiger counter is fired, the poison is released and the cat dies; if not, the cat continues to live. According to quantum mechanics, the state of the system of radioactive atoms must be described in terms of a superposition of possible states, whose evolution will be governed by Schrödinger's equation. But then, noted Schrödinger, the radioactive material + Geiger counter can also be considered as a system and its state must be described in terms of a superposition, and likewise for the radioactive material + Geiger counter + poison-releasing-device and so on, to include, of course, the unfortunate cat. Thus, Schrödinger remarks, '... an indeterminacy originally restricted to the atomic domain becomes transformed into macroscopic indeterminacy' (Trimmer, 1980, p. 328). But of course, he continues, when we open the box the indeterminacy is resolved – we observe either a live or a dead cat.

[1]Originally published in French and subsequently republished in English translation in Wheeler and Zurek 1983.

[2]This amounts to a kind of appropriation of the thought experiment as Schrödinger's original intention was to undermine Bohr's insistence on the distinction between macroscopic and microscopic systems, with classical physics applying to the former and quantum theory to the latter.

von Neumann, equally famously, enshrined this transformation in terms of his distinction between Processes of the First Kind, which apply to measurement, as represented in Schrödinger's thought experiment by opening the box, and which are indeterministic, discontinuous and thermodynamically irreversible and Processes of the Second Kind, as represented by the evolution of the wave function representing the state of the system, which are deterministic, continuous in time and reversible (1932). However, to formally represent the distinction this way does little, if anything, to resolve the problem, expressed, pithily as ever, by Albert in the following terms: 'The dynamics and the postulate of collapse are flatly in contradiction with one another ... the postulate of collapse seems to be right about what happens when we make measurements, and the dynamics seems to be bizarrely wrong about what happens when we make measurements, and yet the dynamics seems to be right about what happens whenever we aren't making measurements.' (Albert, 1994, p. 79)

von Neumann, of course, would most likely respond that it is not *so* bizarre that the dynamics, as represented by Schrödinger's equation, gets things so wrong when it comes to measurement because measurement culminates in an observation and observation involves a kind of interaction that cannot be captured in quantum mechanical terms, namely one involving a conscious observer.[3] Indeed, he presented an argument – his famous 'Chain Argument' – to that effect: if we assume that quantum mechanics applies to all physical systems,[4] then it will apply to, for example, the radioactive material and the radioactive material + Geiger counter and the radioactive material + Geiger counter + poison-releasing-device and so on up the chain, to include the physical body of the observer opening the box, encompassing her visual system and brain. All the links in this chain will be embraced by the superposition and thus can be taken as subject to Processes of the Second Kind. What then could generate a definite result when the box is opened? Something non-physical, namely the consciousness of the observer, which is not subject to quantum mechanics, cannot be included in the superposition and, in effect, leads to the 'collapse' of the relevant wave function (describing the entire system from radioactive material to the cat and the brain of the observer).

Continuing to follow 'the usual story', this postulation of the role of consciousness, about which von Neumann did not actually say anything (see Bueno, 2019), was (so the story goes) summarised and presented in a little pamphlet by London and Bauer (London & Bauer, 1939/1983)[5] and generated considerable discussion, not all of it either philosophically or physically sophisticated, about the observer-dependence of quantum physics. More significantly it was brought to the attention of philosophers of physics through the advocacy of Wigner (I'll come back to

[3]Albert of course was writing at a time when measurement had come to be regarded as just another interaction.

[4]An assumption that Bohr would reject of course; for a useful account see Freire 2015, p. 147.

[5]This characterisation of the London and Bauer manuscript can be found scattered throughout the relevant literature; see Atmanspacher, 2015.

this below) who invoked it in his own, also famous, argument for the role of consciousness, based on the 'Wigner's Friend' thought experiment: portrayed as an extension of the 'Schrödinger's Cat' case sketched above, we are invited to imagine someone – Wigner's friend – about to open the box containing the cat, the Geiger counter, the radioactive material etc., but in a room that is sealed with a further observer outside the room. The observer outside the room asks her friend whether she saw the cat alive, say, knowing that quantum mechanics predicts a 50% probability of observing such an outcome. The observer now asks her friend what she observed before she was asked that question, and we would expect her to reply, "I already told you, I saw the cat alive", since the question whether she did or did not see the cat alive was already decided in her mind before she was asked (Wigner, 1961). And here Wigner cites in support a specific line from London and Bauer's pamphlet, one that I shall return to later: 'He [the observer in the room] possesses a characteristic and quite familiar faculty which we can call the "faculty of introspection." He can keep track from moment to moment of his own state.' (London & Bauer, 1939/1983, p. 252) Since the issue as to what she saw was already decided in the friend's mind before the question was asked, the state immediately after the interaction between the friend and the whole cat-in-a-box system cannot be a superposition. Thus, Wigner concludes, 'It follows that the being with a consciousness must have a different role in quantum mechanics than the inanimate measuring device ... ' (Wigner op. cit.)

It was primarily by means of Wigner's work that this 'solution' to the measurement problem, and the piece by London and Bauer in particular, was brought to the attention of the likes of Shimony and Putnam (Shimony, 1963; Putnam, 1964) who subjected it to severe criticism, raising the kinds of questions that many of us like to invite our students to consider, such as how, precisely, does consciousness, being non-physical, cause a physical change in the state of the system? And how can the universe as a whole be treated as a quantum system? Wigner, together with Margenau, attempted to respond to these, and other, concerns but Putnam's and Shimony's critiques became entrenched and with its apparent adherence to a philosophically naive form of mind-body dualism, this solution to the measurement problem was subsequently dropped, in favour of the now well-known alternatives. End of story.

However, as I argued in (French, 2002) the 'usual story' is wrong in at least one crucial respect: London and Bauer's pamphlet is not a mere summary of von Neumann's position and interpreted correctly, it offers a much more sophisticated account of measurement which, being grounded in the tradition of Husserlian phenomenology, is capable of responding to Putnam's and Shimony's criticisms. Let me now outline this alternative narrative.

10.3 London and Bauer's Pamphlet: The 'True' Story

Fritz London's scientific biography has been presented in admirable detail by Gavroglu (1995) and can be briefly summarised as follows:[6] he studied with Sommerfeld at Munich and might be characterised as one of the first post-revolutionary quantum physicists, applying the theory to chemical bonding with Heitler and developing quantum models of superconductivity and superfluidity. The development of such models has a direct bearing on the issue of how to approach the measurement problem since they were taken by some as undermining Bohr's approach, grounded in the distinction between the macroscopic and the microscopic, with classical theory applicable to the former and quantum mechanics to the latter. London's work helped to suggest to many that such a hard and fast distinction was simply not viable. Furthermore, the 'Sommerfeld School' was quite distinctive from the Copenhagen group, famously centred around Bohr, not least because of the former's emphasis on puzzle solving rather than broader, foundational issues (Seth, 2010). Indeed, the above characterisation of London as a post-revolutionary is in a sense wide of the mark, since, as Seth nicely sets out, there was no sense of 'crisis' within the Sommerfeld School and hence, he argues,[7] the very attribution of a quantum 'revolution' is inappropriate in this particular academic context.

Significantly, London brought to his work in physics an acute and well-formed philosophical sensitivity that he had begun to develop prior to his scientific studies (for further details see again Gavroglu, 1995). His early essays, written over a period covering his final year of school and the first year of university, reveal Kantian and phenomenological themes (Gavroglu, *ibid*,. esp. pp. 8–23). While at Munich London met Pfänder, the leader of the Munich group of phenomenologists and second only to Husserl within the phenomenological movement (*ibid*., pp. 11–12). Pfänder was so impressed with an essay that London showed him that he urged him to write it up and submit it as a dissertation in philosophy.[8] London's thesis was then published in 1923 in the *Jarbuch für Philosophie und phaenomenlogische Forschung*, which was co-edited by Pfänder with Husserl as editor-in-chief and according to Gavroglu, '[t]he dominant features of Fritz London's thesis place it within the phenomenological movement ...' (*ibid*., p. 15).[9]

[6]As in (French, 2002) I shall not say much about Bauer, although I will add that in 1933 he published an introduction to group theory and its application to quantum mechanics (Bauer & Meijer, 1962) which is significant, of course, because of London's involvement with group theory in the late 1920s (Gavroglu *op. cit*., pp. 53–57).

[7]In a sense this is a warning to those, like Kuhn, who are seduced by what Seth calls 'the romance of revolution' and fail to note or acknowledge the differences in approach and attitude of different 'schools' of physics at the time and, indeed, different physicists.

[8]Shimony suggests that it was London's brother, Heinz, who encouraged him to then go into physics (Shimony in *AIP Oral History Interviews*, 2002).

[9]According to Gavroglu, 'What London was thinking programmatically in 1921 was very close to Husserl's thoughts. In this sense London's problematique was not marginal at all.' (*op. cit.* pp. 13–14).

It is this philosophical sensitivity that London brings to bear on the measurement problem in the pamphlet with Bauer. Before getting into the details, it is worth bearing in mind two points regarding the relationship between his scientific and philosophical work. One might question whether a philosophical view developed in the context of classical physics can offer an appropriate framework for the revolutionary new theory that replaced that context. However, as just noted, the Munich 'ethos' was to treat the development of the new quantum mechanics as another problem solving exercise, using tools adapted from those applicable to classical physics. There was no sense of 'crisis' or even of a revolution taking place[10] and I suggest that just as in his physics, so in his philosophy, London would have felt it entirely appropriate to apply the same philosophical devices as he had used before.[11]

One might also worry that following the publication of Husserl's *Crisis of European Sciences and Transcendental Phenomenology* (1936/1970),[12] London might have had further reason to feel that it may have been inappropriate to apply a phenomenological perspective to this new highly mathematised theory. After all, it is in the *Crisis* that Husserl famously emphasizes the importance of the 'lifeworld', conceived in terms of a 'natural', pre-theoretical understanding that has been overlaid with the 'mathematisation' initiated during the Scientific Revolution. It is ultimately due to the introduction of the infinite manifold via this mathematisation that modern science has been plunged into 'crisis', in the sense of a 'loss of its meaning for life'.[13] Here the suggestion that just as London, following Sommerfeld and his school, saw no crisis in physics, so he would have seen none from a phenomenological perspective, might justifiably be viewed as somewhat facile. What Husserl was concerned with went much deeper and further back in history than the latest developments in quantum physics, back indeed to the arithmetization of geometry which thereby emptied the latter of its meaning. Indeed, Husserl might well have viewed the use of group theory that was so favoured by London as exemplifying this tendency and contributing to the 'crisis'! Of course we could always effectively exclude *Crisis* from consideration in reconstructing the phenomenological basis of London and Bauer's approach, perhaps on the grounds that it appeared long after London's education in phenomenology and at a time when he was fully committed to the quantum project. But perhaps that would be too quick.

[10]Seth suggests that in his interviews with those quantum physicists still alive at the time, one gets a sense of Kuhn posing leading questions in his efforts to elicit a sense that a revolution took place!

[11]Gavroglu (1995) has also emphasised the similarities between London's philosophical and physical concerns, particularly with regard to the treatment of theories as 'wholes'. He cites Mormann's claim that London's 1923 thesis '[. . . I can be considered as a set-theoretic concretization of Husserl's largely programmatic account of a macrological philosophy of science' (Mormann, 1991, p. 70; his emphasis).

[12]Some have argued that this represents a major break with his earlier work; others that it offers a fresh perspective on it motivated by the socio-political context of the time.

[13]Egg draws an interesting parallel between Husserl's concerns and those motivating certain current forms of the 'metaphysics of science' (Egg, 2020).

At the very least we would expect London to be sympathetic to Husserl's insistence on an examination of the 'original meaning-giving achievement' of mathematics as applied to physics. Perhaps, then, we can understand London and Bauer's re-insertion of consciousness into quantum theory as a response to Husserl's call to restore the subjective-relative to physics. Indeed, it is the relative aspect that is absolutely crucial as we shall now see.[14]

Let me begin by noting their reconceptualisation of quantum mechanics as implying a *theory of knowledge*: they write,

> Without intending to set up a theory of knowledge, although they were guided by a rather questionable philosophy, physicists were so to speak trapped in spite of themselves into discovering that the formalism of quantum mechanics already implies a well-defined theory of the relationship between the object and the observer, a relation quite different from that implicit in naïve realism, which had seemed, until then, one of the indispensable foundation stones of every science. (1983, p. 220)

Note the reference to 'a rather questionable philosophy' at the beginning of this passage – it may be that London and Bauer are referring here to the positivistically inclined approach of Heisenberg in his work on matrix mechanics, or the curious admixture of different philosophical strands in Bohr's thought[15] or, more likely perhaps, to a general stance within science of supposing that objectivity meant excising the subjective. Despite such a stance, they write, the formalism itself implies a specific relationship between subject and object. Note also their insistence that this relationship is not that which is supposed by 'naïve realism', underpinned as it is by the firm distinction between the inner (subjective) and outer (objective). And note, in sum and significantly, their core point that quantum mechanics is not to be thought of as merely another theory that can be straightforwardly evaluated in terms of various epistemological approaches; rather, *it itself embodies a particular such approach.*

The nature of that approach is then revealed by consideration of the measurement situation. Here, London and Bauer note 'the essential role played by the consciousness of the observer' in the transition from the superposition, ascribed by the theory to the cat + Geiger counter + etc., to the pure state, in terms of which we characterise a definite result, such as 'cat alive'. Looking at that situation from 'outside', as it were, they write: 'Objectively – that is, *for us* who consider as "object" the combined system [object, apparatus, observer] – the situation seems little changed to what we just met when we were considering only apparatus and object.' (ibid., p. 251). However, they continue,

> The observer has a completely different impression. For him it is only the object x and the apparatus y that belong to the external world, to what he calls "objectivity." By contrast

[14] Føllesdal argues that science and the lifeworld should not be seen as being in opposition, since the latter mediates the reference to reality of concepts of the former and acts as the relevant touchstone through scientific revolutions, say (Føllesdal, 1990); see also Bilban (forthcoming) and Egg (2020).

[15] Although see Bilban (forthcoming) for an interesting and useful analysis of Bohr's thought from a phenomenological perspective.

> he has with 'himself relations of a very special character. He possesses a characteristic and quite familiar faculty which we can call the "faculty of introspection." He can keep track from moment to moment of his own state. By virtue of this "immanent knowledge" he attributes to himself the right to create his own objectivity – that is, to cut the chain of statistical correlations ... ' (ibid., p. 252)

Note here the distinction between the observer's relations with the system and with himself, the latter having a 'very special character'. This is embodied in the 'characteristic and quite familiar' faculty of introspection in terms of which he has immanent knowledge of his own state; that is, knowledge that is indubitable. Here we see London and Bauer's adherence to the phenomenological norm '... to avail ourselves of nothing but what in consciousness we can make essentially evident *in its pure immanence*' (Husserl, 1982, p. 59).

Attention should also be drawn to the emphasis on the free creation of objectivity in this account. In a note added by London we find the following: 'Accordingly, we will label this creative action as "making objective." By it the observer establishes his own framework of objectivity and acquires a new piece of information about the object in question.' (London, added note; *ibid.*) This bears obvious comparison with Husserl's statement that '... we persistently *create for ourselves* new configurations of objects ... which have for us lasting reality. If we engage in radical self-examination – that is, return to our ego ... – then all these forms are seen to be creations of spontaneous "I"-activity ... There we also find all the sciences, which, through my own thinking and perceiving, I bring to reality within myself' (Husserl, 1929/1964, p. 30; my emphasis). Again, I shall come back to this aspect of London and Bauer's account.

It should now be obvious that what is involved in the 'cutting' of the 'chain' of statistical correlations is not as typically characterised on the 'usual story' sketched above, namely consciousness intervening and mysteriously causing the collapse of the wave function. Indeed, London and Bauer themselves are quite explicit on this point:

> ... it is not a mysterious interaction between the apparatus and the object that produces a new y for the system during the measurement. It is only the consciousness of an "I" who can separate himself from the former function $\Psi(x, y, z)$ and, by virtue of his observation, *set up* ['*constituer*'] *a new objectivity* in attributing to the object henceforward a new function $y(x) = u_k(x)$. (1983, p. 252)

In French (2002) I suggested that in the light of this, the transition from a superposition to a definite state might be more suitably characterised in terms of a mutual separation of both the 'ego-pole' and the 'object-pole' through this familiar act of introspection. As a characteristic act of reflection on the observation, this yields a relational act, in which the ego appears as itself related to the object of the act through this act itself. It is of the essence of such an act and of the immanent knowledge that it yields that the ego should appear as one pole but this should not be taken as implying that the ego is to be conceived of as something substantial, over and above or existing prior to this act. Rather it should be thought of as a non-autonomous centre of identity or subject-pole that stands at one end of the relational act, the other relatum of which is the object. The latter is then 'made objective', in the sense of having a definite state attributed to it, by this objectifying act of reflection, thereby cutting the 'chain of statistical correlations'.

Given this, we can now return to the situation of Wigner's 'friend'. Here we need to recall a crucial Husserlian point, namely that between 'living in' the observation, as an experience, and describing it, an essential descriptive change occurs. In making such a description we are no longer 'living in' the observation, but instead we attend to it and pass judgment on it and in doing so we cannot avoid reference to an ego or 'I'. Thus, in such a description, performed after an 'objectifying act of reflection', the ego is 'inescapable' since it *necessarily* appears as related to the object of the act of observation. What the friend set-up illuminates, from this perspective, is precisely that descriptive shift: normally we do not explicitly 'keep track' of our mental states, e.g. in the sense of making a note of them, but what Wigner's argument illustrates is that we do possess this 'characteristic faculty' and can say what our state is, if needs be. Of course, in observing his 'friend', Wigner's consciousness will also separate from the relevant superposition and he will then set up a new objectivity.

We can also see how Putnam's and Shimony's objections are wide of the mark. First of all, the observer *is* included within the remit of the theory – she is not something beyond or outside of it, that mysteriously intervenes to somehow 'cause' the wave function to collapse. Of course there is more to say (see French, forthcoming) but it is also not the case that the separation of the ego places the observer beyond the theory prior to the observation. At the point of observation, there is a separation but only in the above sense that the object and subject poles of the relationship between the knower and the world emerge. It is certainly not the case that the ego or consciousness lies outwith the situation before and after, acting in some way to bring about a definite result. Thus consciousness does not 'affect' nature in a peculiar way and there is no 'mysterious interaction'; rather as sketched above, there is a separation of system and observer. Furthermore, there can be no superposition of mental states of the 'I' since the 'I' can only be said to appear post-separation and relatedly, there can be no (internal) mental process of reduction.

The criticisms are hence side-stepped and Shimony, at least, appears to have acknowledged this, eventually, writing that, 'In view of London's philosophical training as a student of Husserl, however, we now are inclined to believe that the attribution [of the usual story of wave function reduction via consciousness] is incorrect and that the passage quoted [the one above beginning '... it is not a mysterious interaction ...'] should be given a phenomenological interpretation.' (Shimony, 1977, pp. 760–761, fn 7).[16] Likewise, in his interview for the American Institute of Physics Oral History Archives, he says, 'As a student of Husserl,

[16]It is perhaps worth mentioning that this is a bit of an odd paper, especially from today's perspective, concerned as it is with the possibility of using quantum entanglement to demonstrate telepathy. A useful context is Kaiser, 2011.

there were some residues of phenomenology in the little booklet of London and Bauer.' (Shimony in *AIP Oral History Interviews*, 2002; the interview was conducted by Joan Bromberg who unfortunately does not follow up on this remark of Shimony's).[17]

10.4 The 'Effacement'/Co-option of Phenomenology

Given my claim (again, expanded in French, 2002), and the brief discussion above, the question arises: why was the phenomenological underpinning of London's approach to the measurement problem so comprehensively ignored, noted only (so far as I know) by a critical commentator (namely, Shimony) much later?[18] Here I cannot hope to give anything close to a complete answer but can only suggest some relevant strands of thought, of a rather speculative nature.

One feature of the relevant historical period has to do with what might be seen as a move from foundations to pragmatics: with the combination of von Neumann's reconciliation of matrix and wave mechanics and Bohr's apparent victory in his debate with Einstein, attention shifted to the more 'practical' applications of the theory, a shift also powered by the move in centre of gravity of quantum physics from Germany to the USA. With that shift various philosophical nuances may have been lost (see for example, Becker, 2018). There's also the further point that, as Gooday and Mitchell (2013) argue, the distinction between classical and quantum physics itself only emerged over a long period of time, extending into the 1930s, and was dependent on the geographical location considered, a point that meshes with Seth's claim noted above. Thus, although many physics textbooks tend to emphasise the classical/modern distinction as representing a distinctive conceptual break, or revolutionary moment, others, and sometimes the same books, note the

[17]The use of the word 'residue' is interesting here, particularly given Shimony's earlier acknowledgement. He also says that the '... booklet was more explicit about the intervention of mentality in the measurement process than von Neumann is ...' (Shimony in *AIP Oral History Interviews*, 2002) because of London's interest in phenomenology. Shimony goes on to describe how he translated London and Bauer's pamphlet from the original French and used it in his class at MIT in the late 1950s. He also states that Wigner was keen to see the English translation published with an introduction by himself and that Bauer liked the translation (London of course had sadly died by then) but that the original publishers declined, because, Shimony speculates, they wanted to publish it themselves. As he notes, they thereby lost the opportunity to have it published with commentary by Wigner (it was subsequently published in the Wheeler and Zurek collection of course). As Shimiony goes on to admit, it was the London and Bauer pamphlet that led him into the measurement problem and his paper 'On the Role of the Observer in Quantum Theory' was initially presented at a conference on the foundations of quantum mechanics organised by Podolsky in Cincinnati in 1963, with the likes of Wigner, Dirac and Bohm present.

[18]Bueno (2019) suggests that there was no such underpinning in the first place, offering a 'minimalist' interpretation of the London and Bauer text, stripped of any phenomenological reading. I think such a claim not only goes against London's own stance towards his work in physics but renders problematic Shimony's acknowledgement of such a reading.

continuities in theoretical practice. Indeed, the distinction gets applied in different ways to emphasise either continuity or change, depending on the pedagogical or more broadly cultural aims and interests involved, yielding different versions of what was characterised as 'classical' and 'modern' physics.

The conclusion Gooday and Mitchell draw is that classical physics can only be understood to have existed in the limited sense that the label was developed and attributed by theoreticians in the early twentieth century '... who sought to preserve a restricted role for established theory and techniques whilst setting forth a future research programme based on new forms of theorizing' (ibid., p. 751). And of course, this throws further doubt on the reciprocal sense in which 'quantum physics' can be said to have been brought into being by contrast. Thus, rather than the rendering invisible of revolutions by the followers of the new paradigm, as Kuhn would have it, what we observe is physicists constructing a 'classical' identity for their forebears in order to serve their own interests (ibid., p. 722).

Interwoven with this post-hoc establishment of such a contrast are two further strands: first, the construction of the Copenhagen Interpretation itself, as it has come to be understood, via a 'dialogical' process in which different principles and theoretical features were woven together in a manner that was driven by the contingent forces powering the debates at the time (Beller, 1999). Indeed, Beller argues that these principles and features themselves became established as such – that is, as features of the emerging theory – via a process of dialogue between the scientists concerned. Likewise, Camilleri has insisted that the Copenhagen Interpretation understood as a more-or-less unified interpretation 'of' quantum mechanics only came into focus via the opposition of Soviet scientists (Camilleri, 2009; see also Freire, 2015, pp. 79–83). Secondly, the characterization of the measurement problem as a problem, is something that appears quite late in the day as well. de Ronde (personal communication) notes that the phrase 'measurement problem' only begins to appear after the mid-1940s and 'quantum measurement problem' only in the late 1960s. Freire Jr. notes that Wigner was one of the first to use the phrase (ibid., p. 142) and records that '[in] the second half of the 1950s there was a rise of studies on the measurement problem ...' (ibid., p. 86).

This provides some of the background to what might appear to have been an effacement of London's phenomenological approach – instead of the rise of logical positivism, as in the case of Weyl, we have the rise of orthodoxy in the form of the Copenhagen interpretation as quantum theory itself distinguishes itself from its predecessor. However I want to suggest that there was a further factor in play that renders this less of an effacement in the sense that holds for Weyl's case and more like a *co-option* of London's approach, minus its phenomenological core, by no less a person than Wigner.[19] As Freire Jr. nicely sets out (ibid., pp. 149–161) through the 1950s and '60s, Wigner attempted to re-shape the conception

[19] Wigner knew London from their time in Berlin (when Wigner was working on group theory) and described him as 'a very thoughtful, very industrious, thorough, imaginative person.' (Interview with Kuhn, Session II, AIP Oral Histories Archive).

of the orthodox view with von Neumann at its heart and Bohr displaced (just as Kuhn and others were setting up the *Archives for the History of Quantum Physics* which can be viewed as a manifestation of historians' interest in the theory; ibid., p. 153).[20] Thus, in his famous paper with Margenau (who had previously criticised the phenomenological approach to physics; see Margenau, 1978), he wrote 'According to von Neumann and London and Bauer, who gave the most compact and the most explicit formulations of the conceptual structure of quantum mechanics, every measurement is an interaction between an object and an observer.' (Margenau & Wigner, 1962, p. 292). And the following year, he noted 'There is a very nice little book, by London and Bauer, which summarizes quite completely what I shall call the orthodox view' (Wigner, 1963, p. 7).

Here we see quite explicitly the co-option of London and Bauer's approach but in order to 'fit' that conception into the orthodox view the phenomenological element must be quietly shoved off centre stage![21] Subsequently, of course, it is Wigner's 'friend' argument that becomes the focus of attention and also the subject of criticism and debate[22] and over time Wigner came to recant his view of the role of the mind in this context (Freire, 2015, p. 168).[23] My suggestion then is that it was not the case that the London and Bauer pamphlet was itself effaced, as Weyl's work was, but rather that its central point was obscured by Wigner's co-option of it as merely a summary of von Neumann's view as part of his campaign to re-orient the discipline's understanding of its foundations.[24]

10.5 A New Hope?

There is, as there always is, more to say about the history. However, let me now turn to the question: Can we recover, via the London and Bauer manuscript, a phenomenological interpretation of quantum mechanics? I will not pretend to be

[20] Wigner's antipathy to Bohr's philosophy of complementarity is apparent in his own interview with Kuhn from these archives where he notes that, possibly under the sway of von Neumann, the duality inherent in complementarity is not reflected in the formalism where one can easily find three operators that do not commute, such as in the case of spin (Wigner Interview Session III, AIP Oral History Archives). Given what Bilban suggests in (forthcoming), this displacement may be construed as a further effacement of the phenomenological 'strand' of thought.

[21] Here Bueno and I agree on the role of Wigner in this history.

[22] For a recent revival of the argument, that I also think can be handled phenomenologically, see Frauchiger and Renner (2016).

[23] Further evidence of the effect of this co-option can be found in Freire Jr's commentary on the London and Bauer pamphlet, in his chapter on Wigner, which makes no mention of London's phenomenological background.

[24] And as Freire Jr. also notes (2015 p. 150) as part of that re-orientation, Wigner maintained that the measurement problem should not be dismissed as philosophy of physics but should be regarded as a fundamental part of physics itself.

able to offer a complete answer here but I hope I can at least sketch some possible fruitful directions in this regard (see French, forthcoming).[25]

Let me begin by noting that, first, such an interpretation will not fit neatly into the space defined by the axes of the realism-antirealism debate and, secondly, neither will it compare straightforwardly with the most well-known of the current interpretations of quantum mechanics.

With regard to that first point, there have been attempts to render phenomenology (more or less) compatible with realism (see for example Hardy, 2013). It has also been compared to anti-realist lines of thought, such as constructive empiricism (see Wiltsche, 2012). Although there are interesting points of comparison made here, I shall adopt the more widely accepted stance that phenomenology sits askew both (traditional) idealism and current forms of realism and anti-realism, not least insofar as it denies the 'philosophical absolutizing' of the world inherent to metaphysical realism (see Zahavi, 2017). Here I shall take that as amounting to the denial of the 'absolutizing' of the state of the system with the concomitant explication of the constitution of the system as an object of knowledge via the correlative relationship in which consciousness and the system stand (ibid.); that is, in terms of the *mutually dependent context of being* (Beck, 1928).

Regarding the second point, it is commonplace to remark that there is an extensive underdetermination of interpretation when it comes to quantum mechanics (French & Saatsi, 2020). Skipping over a lot of nuances, we can in effect draw another set of axes: along one, we have various forms of 'primitive ontology', based on a consideration of material entities in space-time. For the Bohmian, in her current guise, these will be particles with position as a privileged observable (corresponding to a not-so-hidden variable).[26] For the advocate of the GRW view, these will be rendered either in terms of the matter-field or flashes plus a new physical constant that, in effect, 'clumps' the field and 'sparks' (in a sense) the flashes (at the same space-time points). For the phenomenologist, all such interpretations get off on the wrong foot, of course, not least by assuming an unproblematic reification of the notion of a 'material entity'.

Along another axis we might situate those interpretations that take the theory 'literally' or 'as is', the most prominent being the Everettian or 'many worlds' interpretation. This, perhaps, bears closer comparison with a phenomenological approach than the above interpretations, not least because Everett's core relativisation of the quantum state brings it closer to the correlative framework of a

---

[25]Of course, such considerations should not ignore the prior work of Heelan, for example and in this context see his 2004.

[26]Again in his interview with Kuhn, Wigner asks (AIP session III): 'Why is it that we always see positions macroscopically? Position operator is just an operator like every other operator. What is it that makes our minds principally think in terms of position operators? Why are there macroscopic bodies? Why do they have definite positions rather than having another, arbitrary, wave function, or another, arbitrary, operator measured? I may be completely wrong, but I do feel that there is some mystery here not completely cleared up. Several times I've had ideas on this but nothing really convincing.'

phenomenological view. One might also dwell a little on the fact that in its current revival, the interpretation depends on a decision-theoretic device in order to recover the crucial 'Born rule' of quantum mechanics (which specifies, in its simplest form, that the probability density for finding a particle at a certain position is given by the modulus squared of the wave-function at that position). Here one could speculate that a subjective element creeps into the interpretation, or, at least, a certain view, albeit widely held, of what it is to be rational that underpins this device. I shall come back to this, briefly.

It is also interesting that Everett, in his 'long' thesis of 1956, introduced an 'amusing, but *extremely hypothetical* drama' (Barrett & Byrne, 2012 p. 74) which is, in fact, a version of Wigner's 'friend' argument.[27] However, the upshot of the argument is different: for Wigner it demonstrated the role of consciousness in 'solving' the measurement problem, whereas for Everett it showed what was wrong with the 'orthodox' view as simply stated (and here he followed Wigner in taking von Neumann as representative of that view), thereby clearing the way for his 'relative state' interpretation. There is, again, more to say (not least about the many *minds* variant of this interpretation) but from a phenomenological perspective, the initial move of taking the theory literally also gets off on the wrong foot, albeit a different one!

More fruitful comparisons might perhaps be drawn by focussing on the *correlative* relationship in the context of Dieks' perspectivalism or Rovelli's relationalism. Running throughout such accounts one finds a concern with including consciousness, or not. Thus Dieks writes:

> The appeal to consciousness ... appears to invoke a *deus ex machina*, devised for the express purpose of reconciling unitary evolution with definite measurement results. More generally, the hypothesis that the definiteness of the physical world only arises as the result of the intervention of (human?) consciousness does not sit well with the method of physics. (Dieks, 2018, p. 4)

However, from a phenomenological perspective, of course, consciousness is invoked not as a *deus ex machina* but as that which provides the 'ultimate court of appeal of all knowledge' (Ryckman, 2005, p. 142). If we then take as central the *correlative* relationship by which mind and world are bound constitutively together (Zahavi, 2017, p. 117), understood in the quantum context, we can perhaps retain the advantages of relationist-type interpretations without having to invoke a multiplicity of worlds or of minds. In this regard, as I said, I can only offer a sketch here but the fully-fledged interpretation (if it could be achieved) should at least incorporate the following considerations.

First of all, it goes without saying that the nature of the 'state' in quantum mechanics is problematic (an issue that can perhaps be traced back to Bohr's introduction of the 'stationary' state). From a phenomenological perspective, the

[27] As Barrett and Byrne note (op. cit., p. 29, fn 2), Everett took a class with Wigner on Methods of Mathematical Physics at Princeton in 1954 and presented this version of the 'Wigner's Friend' argument some years before Wigner's appeared in print.

mutually dependent context of being implies that the traditional notion of state (as non-relational or intrinsic or more broadly, mind independent) must be abandoned. From this standpoint, 'systems' do not possess states in and of themselves independently of observers and in this regard, again, there is an obvious point of comparison with perspectival/relational/relative state approaches. However, this should not be understood in terms of some form of 'splitting' of reality; on the contrary, there is but one 'world' in the sense of a reality, comprised of the relevant systems, that is transcendent but there are many contexts of being, in the sense that the states of these systems are dependent on consciousness.

The obvious question, then, is why are certain states preferred (this is, in effect, the so-called basis problem)? One can take a leaf out of the Everettian's book here and appeal to decoherence, whereby the interaction between a system and the environment (where the latter has many more degrees of freedom than the former) leads to the suppression of interference between certain states that are robust in the sense that information about them is stored redundantly in the environment. The observer can then recover that information without further disturbing the system (see Bacciagaluppi, 2016). We can then answer why position is privileged in the way it is (answering Wigner's concern above in fn 27): the interaction potentials are functions of position and thus the states effectively picked out by decoherence tend to be localised in position. Hence, subsequent to the 'separation' of observer and system, position states come to be preferred.

But of course, as is now widely recognised, decoherence in and of itself does not 'solve' the measurement problem, because the combination of system + apparatus + environment will still be in a superposition. It is only through the action of the conscious observer, by engaging in the crucial act of reflection and distinguishing herself as the 'ego-pole', that the relevant separation between system and observer can be achieved.

There remains the further worry, prevalent throughout the discussions of both the von Neumann and London and Bauer approaches, that allowing a role for consciousness in this regard introduces a fatal element of subjectivity and undermines the objectivity of not just quantum mechanics but physics as a whole. Let us return to London and Bauer, who write that understanding this concept of objectivity involves '... the determination of the necessary and sufficient conditions for an object of thought to possess "objectivity" and to be an object of science' (1983, p. 259). They continue, '... Husserl ... has systematically studied such questions and has thus created a new method of investigation called "Phenomenology"' (*ibid.*; here they refer to both the *Logical Investigations* and *Ideas*). The classical concept of objectivity is dismissed as 'useless and even incorrect, [generating] actual obstacles to progress' (*ibid.*). It is the phenomenological concept which is now sufficient for physics' needs, in the sense that '[t]he transcendency belonging to the physical thing as determined by the physics is the transcendency belonging to a being which becomes constituted in, and tied to, consciousness.' (Husserl, 1982, p. 123). Taking objectivity to be cashed out in terms of a transcendency that is independent of or separated from a knowing consciousness is what has generated many of the problems associated with quantum physics (and also, Weyl might say, relativity theory) in the first place. To overcome these problems the phenomenologist insists on objectivity itself being constituted by consciousness.

How, then, is *inter-subjective agreement* to be established? Here we can take a leaf out of the book of relational quantum mechanics (see Rovelli, 1996 and Laudisa & Rovelli, 2013) and note that establishing such agreement itself involved a physical interaction. So, consider a simple arrangement of a system that can be in spin up or down and a measurement device that can indicate 'up' or 'down' (this is adapted from Laudisa & Rovelli, 2013). Assume the interaction between the two is such that when the system is in state 'spin up/down' the measurement device records 'up/down' and observer$_1$ observes a reading of 'up' or 'down' accordingly. The system starts in a superposition of spin up and spin down, interacts with the measurement device, and the observer takes the reading, yielding a particular mental 'state' upon reflection, which of course would be either 'I see a reading of 'up'' or 'I see a reading of 'down''. But we can consider the system + measurement device + observer$_1$ as itself as system, observed by observer$_2$. From this perspective, prior to observer$_2$ taking a reading, the whole composite must be regarded as in a superposition (here we recall the flexibility in the von Neumann 'cut' between measured system and measuring system). Observer$_2$ can then take a reading, there is a reflective act and she too will say either 'I see a reading of 'up'' or 'I see a reading of 'down''. When observer$_1$ and observer$_2$ then compare their results there will be no contradiction because that comparison must itself be considered as a physical interaction describable by quantum mechanics.

As Laudisa and Rovelli remark, 'This internal self-consistency of the quantum formalism is general, and it is perhaps its most remarkable aspect. This self consistency is taken in relational quantum mechanics as a strong indication of the relational nature of the world.' (ibid.) Note that this can be adapted to the phenomenological case precisely because on the London and Bauer picture, consciousness is not set outwith the wave function but, rather, the observer is taken to be embraced by the theory too, so that the latter's internal self-consistency applies in this case also. Of course, for Rovelli the relations that make up the 'nature of the world' are physical relations, understood from a broadly naturalistic viewpoint, but there doesn't seem to be any obstacle in principle to situating them within a correlationist framework.

There remains the issue of accommodating and, more generally, making sense of probabilities within such an interpretation. Here we might recall London and Bauer's emphasis on the *free* creation of objectivity, reminiscent as it is of Husserl's remark that, '... we persistently *create for ourselves* new configurations of objects ... which have for us lasting reality. If we engage in radical self-examination – that is, return to our ego ... – then all these forms are seen to be creations of spontaneous "I"-activity ... There we also find all the sciences, which, through my own thinking and perceiving, I bring to reality within myself' (Husserl, 1964, p. 30; my emphasis). Insofar as we freely create a new objectivity through this regard that separates the ego-pole from the superposition, one can speculate that it is the spontaneous 'I'-activity that generates the relevant quantum probabilities.[28] Note, first of all, that this is not to subscribe to some form of the 'epistemic' approach to probability in quantum mechanics, given, of course, that the distinction indicated by this label is inapplicable in this context. Note, furthermore, that the above freedom does not imply that such creation and the separation (or 'collapse' on a non-phenomenological reading) are somehow subject to our will!

That's all well and good but the question remains how can the probabilities in this interpretation agree with those of textbook quantum mechanics (cf. Greaves, 2007)? Here again we can steal a page from someone else's book, literally. In Wallace's exposition of the Everettian interpretation (Wallace, 2012), as touched on above, the Born rule is recovered via considerations based on decision theory, itself understood as embodying the core features of rational behaviour. It is via our understanding of such behaviour, it is claimed, that probability makes contact with the world. The phenomenologist can appropriate that approach, and the relevant formal proofs, but, of course, would give the underlying understanding of rationality her own interpretation. As Zahavi states, reflection is a pre-condition for the kind of self-critical deliberation involved in such behaviour and, as he says, 'If we are to subject our different beliefs and desires to a critical, normative evaluation, it is not sufficient simply to have immediate first-personal access to the states in question. Rather, we need to deprive our ongoing mental activities from their automatic normative force by stepping back from them.' (2017, p. 23). In other words, we need to effect the core phenomenological move by engaging in a reflective self-distancing through which we enter into a critical relationship with our mental states. Zahavi continues, 'To live in the phenomenological attitude is for Husserl not simply a neutral impersonal occupation, but a praxis of decisive personal and existential significance ...' (ibid., p. 23).

There are also alternatives of course. In their review of Wallace's book, Bacciagaluppi and Ismael note that 'Although the proof of the Born Rule is formulated within the decision-theoretic framework, the mathematical core of the proof does not depend on it: as Wallace remarks, it establishes that if probability basically makes sense, and has the usual qualitative features, in unitary quantum mechanics, then quantitatively it is given by the Born rule (Bacciagaluppi & Ismael, 2015, p. 141). As they go on to note, one could take the Born rule to be merely a phenomenological (not in our sense!) add-on to the theoretical structure of quantum mechanics, but then the worry is that one loses any theoretical underpinning for it. However, '... Gleasons theorem provides another natural way of justifying the Born Rule (perfectly acceptable as part of a pragmatic justification). And to do justice to Everett, he presents such a theoretical argument himself.' (ibid., p. 142; as they note, Everett understands the Born rule in terms of a typicality measure, rather than a credence measure as Wallace does).[29] It seems to me that again, there

[28]The notion of freedom being employed here plays a major role with regard to the phenomenological epoché in general.

[29]Gleason's theorem essentially states that the Born rule follows from the lattice structure of events in Hilbert space. One would of course have to give a phenomenological reading of this structure.

is nothing in principle that prevents the phenomenologist from adapting any of these justifications and indeed, some of the earlier discussions of this issue in the context of the many-minds variant of the Everettian interpretation appear to sail within reach of a phenomenological understanding (see French, forthcoming).

Finally, and briefly, there is the well-known claim that the Everettian or many worlds view is the only interpretation that is relativistically kosher as all other interpretations assume a privileged reference frame (this is a criticism that has been levelled at the Bohmian interpretation in particular given the central role played by the notion of a configuration, in terms of the simultaneous position of all the particles). Here we might bring Weyl back into the picture: the separation of the 'I' from the mutual dependency to yield a definite result should not be understood in terms of establishing such a privileged frame; rather, if we understand any such frame of reference as 'the necessary residue of the ego-extinction', to use Weyl's phrase, we can de-privilege it, as it were, by emphasising its subjective character.

10.6 Conclusion

I noted in the introduction Ryckman's point that Weyl's theory has an 'ambiguous character' in that it lay 'in the intersection of physics and philosophy' (2005 p. 159). The question can be asked, where, then, does the measurement problem lie? Or, relatedly, is the London and Bauer account *physics* or *philosophy*? The answer may seem both contestable and historically contingent. According to many physicists, for many years, the measurement problem was dismissed as a philosophical concern. As indicated above, Wigner disagreed and so, I hazard, would London and Bauer, insofar as they saw physics as implying a theory of the relationship between the object and the observer. In this respect, their approach does not suffer from any ambiguity, since from their perspective, to 'do' physics is to 'do' philosophy!

Sadly, whether it is regarded as effaced or co-opted, London and Bauer's approach to the measurement problem has been lost to the majority of philosophers of physics. As a result, the phenomenological perspective that it embodies has not been properly explored and evaluated. Here I've merely indicated possible avenues down which such an exploration might proceed but even if one is not phenomenologically inclined, as it were, such explorations should be interesting, for the possible alternative understanding of quantum mechanics that they may reveal and for the contrast they thereby offer to current interpretations that have been worn thin through repeated examination.

Alternatively, one might adopt Everett's argument that an observer's relative measurement records in a typical branch would be randomly distributed according to the standard quantum probabilities and establish a phenomenologically appropriate 'typicality measure' (for a useful discussion of such measures see Barrett, 2017.

Acknowledgments I'd like to thank Philipp Berghofer, Tina Bilban, Michel Bitbol, Otávio Bueno, Matthias Egg, Arezoo Islami, Tom Ryckman, Harald Wiltsche and the audience of the conference 'Phenomenological Approaches to Physics', Graz, June 2018, for comments and general support.

References

Albert, D. (1994). *Quantum mechanics and experience*. Cambridge, UK: Harvard University Press.

Atmanspacher, H. (2015). Quantum approaches to consciousness. In E. N. Zalta (Ed.), *The Stanford encyclopedia of philosophy* (Summer 2015 Edition). https://plato.stanford.edu/archives/sum2015/entries/qt-consciousness/

Bacciagaluppi, G. (2016). The role of decoherence in quantum mechanics. In E. N. Zalta (Ed.), *The Stanford encyclopedia of philosophy* (Fall 2016 Edition). https://plato.stanford.edu/archives/fall2016/entries/qm-decoherence/

Bacciagaluppi, G., & Ismael, J. (2015). Review of *The Emergent Multiverse*. *Philosophy of Science, 82*, 129–148.

Barrett, J. A. (2017). Typical worlds. *Studies in History and Philosophy of Science Part B, 58*, 1–80.

Barrett, J. A., & Byrne, P. (Eds.). (2012). *The Everett interpretation of quantum mechanics: Collected works 1955–1980 with commentary*. Princeton, NJ: Princeton University Press.

Bauer, E., & Meijer, P. (1962/2004). *Group theory; the application to quantum mechanics*. Amsterdam: North Holland 1962/New York: Dover 2004

Beck, M. (1928). Die Neue Problemlage der Erkenntnistheorie. *Deutsche Vierteljahrsschrift für Literaturwissenschaft und Geistesgeschichte, 6*, 611–639.

Becker, A. (2018). *What is real? The unfinished quest for the meaning of quantum physics*. New York: Basic Books.

Beller, M. (1999). *Quantum dialogue: The making of a revolution*. London: University of Chicago Press.

Bilban, T. (forthcoming). Phenomenological approach to quantum mechanics.

Bueno, O. (2019). Is there a place for consciousness in quantum mechanics. In J. Acacio de Barros and C. Montemayor (Eds.), *Quanta and Mind* (pp. 129–139). Cham: Springer.

Camilleri, K. (2009). Constructing the myth of the Copenhagen interpretation. *Perspectives on Science, 17*, 26–57.

Cassirer, E. (1936). *Determinism and indeterminism in modern physics*. New Haven, CT: Yale University Press. 1956.

Dieks, D. (2018). Quantum mechanics and perspectivalism. arXiv: 1801.09307v1.

Egg, M. (2020). A revealing parallel between Husserl's philosophy of science and today's scientific metaphysics. In H. A. Wiltsche and P. Berghofer (Eds), *Phenomenological Approaches to Physics* (Synthese Library), Berlin: Springer

Føllesdal, D. (1990). The *Lebenswelt* in Husserl. In: Haaparanta et al. 1990., Haaparanta, Leila, with Martin Kusch and Ilkka Niiniluoto (Eds.), *Language, knowledge and intentionality*, Helsinki (*Acta Philosophica Fennica* 49).

Frauchiger, D., & Renner, R. (2016). Single-world interpretations of quantum theory cannot be self-consistent. arXiv:1604.07422.

Freire Jr., O. (2015). *The quantum dissidents: Rebuilding the foundations of quantum mechanics (1950–1990)*. Heidelberg, Germany: Springer.

French, S. (2002). A phenomenological solution to the measurement problem? Husserl and the foundations of quantum mechanics. *Studies in History and Philosophy of Modern Physics, 33*, 467–491.

French, S. (2014). *The structure of the world*. Oxford: Oxford University Press.

French, S. (forthcoming). *Regarding quantum theory: Phenomenology and the foundations of physics*.

French, S., & Saatsi, J. (Eds.). (2020). *Realism and the quantum*. Oxford: Oxford University Press.

Gavroglu, K. (1995). *Fritz London: A scientific biography*. Camdridge, UK: Cambridge University Press.

Gooday, G., & Mitchell, D. (2013). Rethinking 'classical physics'. In J. Z. Buchwald & R. Fox (Eds.), *The Oxford handbook of the history of physics* (pp. 721–764). Oxford, UK/New York: Oxford University Press.

Greaves, H. (2007). Probability in the Everett interpretation. *Philosophy Compass, 2*, 109–128.

Hardy, L. (2013). *Nature's suit: Husserl's phenomenological philosophy of the physical sciences*. Athens, OH: Ohio University Press.

Heelan, P. (2004). The phenomenological role of consciousness in measurement. *Mind and Matter, 2*(1), 61–68.

Husserl, E. (1964). *The Paris lectures* (P. Koestenbaum, Trans.). The Hague, The Netherlands: Martinus Nijhoff (Original work published 1929).

Husserl, E. (1970). *The crisis of European sciences and transcendental phenomenology* (D. Carr, Trans.). Chicago: Northwestern University Press (Original work published 1954).

Husserl, E. (1982). *Ideas: General introduction to pure phenomenology* (F. Kersten, Trans.). Dordrecht, The Netherlands: Kluwer Academic Publishers (Original work published 1913).

Kaiser, D. (2011). *How the hippies saved physics: Science, counterculture and the quantum revival*. New York: W.W. Norton.

Laudisa, F., & Rovelli, C. (2013). Relational quantum mechanics. In E. N. Zalta (Ed.), *The Stanford encyclopedia of philosophy* (Summer 2013 Edition). https://plato.stanford.edu/archives/sum2013/entries/qm-relational/.

London, F., & Bauer, E. (1939/1983). La Théorie de L'Observation en Mécanique Quantique, Hermann. In J. A. Wheeler & W. H. Zurek (Eds.), *Quantum theory and measurement* (p. 252). Princeton, NJ: Princeton University Press.

Margenau, H. (1978). Phenomenology and physics. In H. Margenau (Ed.), *Physics and philosophy: Selected essays* (pp. 317–328). Dordrecht, The Netherlands: D. Reidel.

Margenau, H., & Wigner, E. (1962). Discussion: Comments on Professor Putnam's Comments. *Philosophy of Science, 29*, 292–293.

Mormann, T. (1991). Husserl's philosophy of science and the semantic approach. *Philosophy of Science, 58*, 61–83.

Putnam, H. (1964). Discussion: Comments on comments on comments, a reply to Margenau and Wigner. *Philosophy of Science, 31*, 1–6.

Rovelli, C. (1996). Relational quantum mechanics. *International Journal of Theoretical Physics, 35*, 1637–1678.

Ryckman, T. (2005). *The Reign of Relativity*. Oxford: Oxford University Press.

Ryckman, T. (2018). Cassirer and Dirac on the symbolic method in quantum mechanics: A confluence of opposites. *Journal for the History of Analytic Philosophy, 6*, 194–224.

Seth, S. (2010). *Crafting the quantum: Arnold Sommerfeld and the practice of theory, 1890–1926*. Cambridge, MA: MIT Press.

Shimony, A. (1963). Role of the observer in quantum theory. *American Journal of Physics, 31*, 755–777.

Shimony, A. (1977). Wave-packet reduction as a medium of communication. *Foundations of Physics, 7*, 759–767.

Trimmer, J. D. (1980). The present situation in quantum mechanics: A translation of Schrödinger's "cat paradox" paper. *Proceedings of the American Philosophical Society, 124*(5), 323–338.

von Neumann, J. (1932). *Mathematical foundations of quantum mechanics* (The English translation, by R. T. Beyer, of the original German edition was first published in 1955). Princeton, NJ: Princeton University Press.

Wallace, D. (2012). *The emergent multiverse: Quantum theory according to the Everett interpretation*. Oxford: Oxford University Press.

Wheeler, J. A., & Zurek, W. H. (Eds.) (1983). *Quantum Theory and Measuremen*. Princeton: Princeton University Press.

Wigner, E. (1961). Remarks on the mind–body question. In I. J. Good (Ed.), *The scientist speculates* (pp. 284–302). Woburn, MA: Heinemann.

Wigner, E. (1963). Problem of measurement. *American Journal of Physics, 31*, 6–15.

Wiltsche, H. A. (2012). What is wrong with Husserl's scientific ant-realism. *Inquiry, 55*(2), 105–130.

Zahavi, D. (2017). *Husserl's legacy*. Oxford: Oxford University Press.

Chapter 11
A Phenomenological Ontology for Physics: Merleau-Ponty and QBism

Michel Bitbol

Abstract Few researchers of the past made sense of the collapse of representations in the quantum domain, and looked for a new process of sense-making below the level of representations: the level of the phenomenology of perception and action; the level of the elaboration of knowledge out of experience. But some recent philosophical readings of quantum physics *all point in this direction.* They all recognize the fact that the quantum revolution is a revolution in our conception of knowledge.

In these recent readings of quantum physics (such as QBism), quantum states are primarily generators of probabilistic valuations. Accordingly, they should not be seen as statements about what is the case, but as statements about what each agent can reasonably expect to be the case.

Three features of such non-interpretational, non-committal approaches to quantum physics strongly evoke the phenomenological epistemology. These are: (1) their deliberately first-person stance; (2) their suspension of judgment about a presumably external domain of objects, and subsequent redirection of attention towards the activity of *constituting* these objects; (3) their perception-like conception of quantum knowledge.

But beyond phenomenological epistemology, these new approaches of quantum physics also make implicit use of a phenomenological ontology. Chris Fuch's participatory realism thus formulates a non-external variety of realism for one who is deeply immersed in reality. But participatory realism strongly resembles Merleau-Ponty's endo-ontology, which is a phenomenological ontology for one who deeply participates in Being.

This remarkable analogy is supported by Merleau-Ponty himself. Indeed, 50 years before QBism, Merleau-Ponty acknowledged the strong kinship between

This work was supported by the Agence Nationale pour la Recherche (ANR-16-CE91-0005-01).

M. Bitbol (✉)
Archives Husserl, CNRS/ENS, Paris, France
e-mail: michel.bitbol@ens.fr

H. A. Wiltsche, P. Berghofer (eds.), *Phenomenological Approaches to Physics*, Synthese Library 429, https://doi.org/10.1007/978-3-030-46973-3_11

the status of quantum mechanics and his phenomenology of embodiment. He did so in two texts that remained unpublished until after his death: *Visible and invisible,* and the *Lectures on Nature.* The final part of this article is then devoted to a study of Merleau-Ponty's conception of quantum physics.

Foreword
Let's imagine that, despite the lack of any all-encompassing picture, an abstract mathematical structure guides our (technological) activities more efficiently than ever, possibly assisted by a set of clumsy, incomplete, ancillary pictures. In this new situation, the usual hierarchy of knowledge would be put upside down. Unlike the standard order of priorities, situation-centered practical knowledge would be given precedence over theoretical knowledge associated with elaborate unified representations; in the same way as, in Husserl's *Crisis of the European Science,* the life-world is given precedence over theoretical "substructions". Here, instead of construing representation as an accomplished phase of knowledge beyond the primitive embodied adaptation to a changing pattern of phenomena, one would see representation as an optional instrument that is sometimes used in highly advanced forms of embodied fitness. As for mathematical formalisms, they would no longer be taken for a structural image of the actual world, but rather understood as a systematic inventory of our most precise possibilities of bodily action in response to a varying phenomenal landscape (along with Jean Piaget's genetic psychology or Andrew Pickering's neo-pragmatism). The Platonician dream of knowing natural forms would be dispelled by the realization that the theories of mathematical physics are variants of a formalized know-how.

11.1 Quantum Physics as Formalized Know-How: What Are the Resistances?

In the eyes of any lucid and unprejudiced thinker, it is obvious that quantum physics is a perfect instantiation of this kind of post-classical conception of knowledge. In quantum physics, it has soon been suspected (especially by Niels Bohr) that a fully coherent all-encompassing representation might well be out of reach. At most, one can provide a mathematical scheme that has the mental function of a "representation" without pretending to be literally a representation *of* the world (Bitbol, 1996, p. 29; Schrödinger, 1951, p. 40). The reason of this apparent limitation of quantum physics is clear to almost anyone. It is the *contextuality* of micro-properties, the fact that micro-properties cannot be disentangled from the material and bodily context of their manifestation. Due to contextuality, no *independent* domain of properties and entities can properly be defined, let alone represented. But physicists soon realized that the lack of a unified representation and the unbreachable contextuality of properties were absolutely no hindrance to the efficiency of the mathematical scheme of the theory. They then used it with amazing success, as a universal chart of microphysical know-how.

It is thus *prima facie* surprising that, for nearly one century, there has been continuous struggle against those unexpected but strong conclusions. Many scientists felt that lacking a unified and consistent narrative about the world is tantamount to falling into non-sense. And they desperately attempted to overcome what they saw as a failure.

Very few researchers tried the opposite strategy that consists in pushing Bohr's approach to its ultimate consequences, thus making sense of the collapse of representations, and looking for a new process of sense-making below the level of representations; say at the level of the phenomenology of perception and action. Very few of them decided to explore the surface of micro-phenomena, and to make sense of their being limited to this so-called surface, rather than trying desperately to tell a tale about the elusive depths hidden behind phenomena. For some time, I had the feeling that this stubborn attitude, this resistance against the radical epistemological lesson of quantum physics, is a constitutive feature of our Western ethos.

Why, if not because of a deeply entrenched cultural obstacle, would a new conception of knowledge that makes sense of our most advanced physical theory not be more widely accepted? Why does it so often find itself opposed by the indignant reactions of certain physicists who reproach it with "betraying the ideal of science" (Stengers, 1997). Why do some mathematicians accuse quantum theory of being unacceptable or even "scandalous" (Thom, 1993)? Why, even when indignation is absent, does the exposition of the minimalist conception of quantum theory give rise to a resigned silence which manifestly expresses a profound disappointment?

This might be due to the fact we are dealing with a breach of several tacit contracts of our civilization. One of these is a fairly recent contract, signed at the end of the sixteenth century, between the desire for a metaphysical breakthrough that motivates scholars, and the craftsmen's need for technological improvement (Scheler, 1993). There is also a much older contract, dating back to ancient Greece, which has made it an obligation to look for a principle of understanding superficial appearances in the inmost depths of things (Schrödinger, 1954), and to seek there the changeless source of any change.

If scientific progress does not help our gaze to penetrate into the very heart of material bodies, and to *definitively guarantee* technological effectiveness by laying bare their secret, what is the point of it? If the progress of knowledge amounts merely to a kaleidoscopic deployment of the phenomenal skin of things, instead of opening up an insight into their very flesh and marrow, does it not seem vain? One may recall that the entities which, in the history of science, were given the title of "realities behind appearances" have turned out to be: (1) *other* appearances (or phenomena) revealed by a new approach or (2) mathematical idealities which express some invariants of the said phenomena. However, this simple reminder is not enough. The "dream of reason" pursues its course; this same dream that Kant upheld at the beginning of his quest, before discarding it in his critical philosophy: the dream of grasping by thought a "representation of things *as they are*" (Kant, [1770] 2004).

11.2 The New Understanding of Quantum Mechanics

But several recent debates and ideas in the philosophy of quantum mechanics suggest, to my delight, that things are now changing at a fast pace. Several new philosophical readings of quantum physics *all point in the same direction,* the direction of full recognition of the fact that the quantum revolution is above all a revolution in our conception of knowledge. Apart from Asher Peres' remarkable "no-interpretation" of quantum theories, which radicalizes the so-called Copenhagen interpretation, the forerunner of this new trend was probably the information-theoretic interpretation of Anton Zeilinger and his school. According to this interpretation, the information made available by experiments exhausts "reality"; and the formalism of quantum theories is derivable (to a certain extent) from a principle of limitation of the information that can be extracted from each system. As for this principle of limitation, it is an operational formulation of the widely acknowledged *contextuality of micro-properties*. From this kind of derivation, or "reconstruction" (Grinbaum, 2007), it becomes clear that quantum theories are not to be understood as an indirect representation of some reality beyond the phenomenal level of experimental information, but as a direct expression of some in-principle bounds of the availability of this information.

Recent developments in the use of quantum theories can be seen as an implicit confirmation of the relevance of an information-theoretic reading. One case is especially striking: it is a recent generalization of quantum theory that applies to several domains of the human sciences (Bruza, Sofge, Lawless, van Rijsbergen, & Klusch, 2009a; Bruza, Kitto, Nelson, & McEvoy, 2009b) such as decision theory, semantics, and the psychology of perception. This application of quantum theory to the human sciences shows that no matter who or what *responds* (human beings or things), the probabilistic structure that is to be used to anticipate the responses is the same. A set of human beings making choices that depend on the options which are presented to them, and on the order of the decisions to be taken, behave exactly like a set of electrons on which one evaluates several incompatible observables (Zwirn, 2009). In particular, a set of speakers who must make a decision concerning the meaning of a polysemic word, according to the propositional contexts, violate the Bell inequalities (Bell, 1987), just as a set of microparticles do (Bruza et al., 2009a, 2009b). There is nothing shocking about the fact that it should be so. For this implies strictly nothing about some alleged similarity between electrons and humans at the level of their *profound being*. There is only a formal isomorphism between the possibilities of epistemological access to electrons and to humans: an isomorphism of their phenomenal reactions to being solicited, and of their informational dispositions. Such universal applicability of quantum theories to any domain whatsoever in which the replies to experimental solicitations depend on their order, strongly suggest that these theories are precisely that, and *only* that: a general procedure for anticipating probabilistically the replies to context-dependent experimental solicitations. They do not even offer a hint in the quest of a faithful representation of some independent reality out there, behind phenomena.

There is also a well-known philosopher of quantum mechanics, who previously held a subtle realist interpretation of this theory, and who now considers that acknowledging the impossibility of a realist interpretation is a condition for the understanding of this theory. This is Richard Healey. His analysis is exceptionally lucid, even though he sometimes balks at its ultimate consequences (possibly a remnant of his former realist self). According to Healey (2017, p. X), "The main barrier to understanding quantum mechanics is not our inability to imagine the world it describes, but the presumption that it be understood as describing the world". In other terms, the feeling that quantum mechanics is still mysterious one century after its first formulation, might well be due to our refusal to accept its non-representational status. Healey thus classifies the many approaches of quantum physics into two (unequal) subsets: those held by the interpreters (the majority by far) and those held by the pragmatists (a minority). Both approaches agree that there is something exceptional in quantum physics, but disagree about the nature of this exception. Interpreters try to figure out a picture of the universe or multiverse that may resemble Alice's wonderland, but that fits more or less squarely with the formalism of quantum mechanics. Interpreters literally look for a proper "interpretation" of quantum mechanics. Pragmatists, instead, examine the use of the formalism, and they conclude that what is truly radical in quantum physics is that there is nothing deeper in it than the rules of its use; nothing like an account of the "universe". Interpreters project the exceptional features of quantum physics onto their objects of thought, whereas pragmatists withdraw from projections and identify the source of this exceptionality in a revelation of the true workings of knowledge that was hidden by classical science. Interpreters are focused onto possible intentional objects, whereas pragmatists come back reflectively towards the condition of possibility of any knowledge of objects. Interpreters adopt a naturalist attitude, whereas pragmatists sketch a phenomenological attitude. To recapitulate, according to pragmatists, quantum physics needs no interpretation, but rather a clear recognition of the reason why there can be none. This is a *motto* that will be developed later in connection with the quest for an unconventional ontology that fits non-representational readings of quantum theories.

11.3 Suspending the Interpretations of Quantum Mechanics: A Phenomenological *epochè* for Physics

What might be a *non*-interpretation of central symbols of quantum mechanics? The standard language of quantum physicists calls "states" certain vectors of a Hilbert space; it then presupposes that these symbols express the intrinsic state of real systems. Instead, in the pragmatist, non-interpretational approach to quantum physics, these vectors have no value as predicates of systems, but only a *function* for those who try to explore these so-called "systems". The function of these vectors is to provide an informational bridge between the preparation and the outcome of experiments, or more generally (to borrow Healey's terms) between the "backing condition and the advice condition" of each physical situation. To sum up, the said vectors of a Hilbert space are no absolute characteristics of systems, but rather predictive tools relative to each agent-situation.

The latter deflationary way of featuring the so-called "state-vectors", that no longer deserve to be called "states", has been pushed even further by another approach to quantum physics (Fuchs, Mermin, & Schack, 2014): QBism, that can be read "Quantum Bayesianism" or "Quantum Bettabilitarianism". QBism in its original form as quantum Bayesianism asserts that "state" vectors are a special variety of Bayesian probabilistic valuations, and that probabilistic valuations are logical constructs rather than physical realities. Here, "state" vectors and probabilistic valuations are not statements about what is the case, but statements about what each agent can reasonably expect to be the case. Ultimately, they are just expressions of what one subjectively expects to be the case; they express each subjective agent's willingness to place *bets* on such and such outcome. Hence the alternative expression for QBism: "Quantum Bettabilitarianism".

Several features of these non-interpretational, non-committal approaches of quantum physics strongly evoke the phenomenological attitude.

The first one is a deliberately first-person approach (be it first-person singular or first-person plural). The project of both phenomenology and non-interpretational approaches to quantum mechanics is to reconstruct a new, self-conscious, type of objective knowledge, starting everything afresh from the first-person standpoint of knowers and agents.

The second common feature is the act of suspending judgment about a presumably external domain of objects, and the subsequent redirection of attention towards the mental or practical activity of *constitution* of these objects. In phenomenology, the suspension of judgment is called the *epochè* (after the Greek skeptics), and the redirection of attention towards the mental acts by which one comes to believe in the existence of perceived objects is called the phenomenological reduction. In QBism, and in Healey's pragmatic approach as well, the suspension of judgment is also clear, since one here considers that *no* symbol of quantum mechanics *refers* to objects or *denotes* predicates of objects. The reflective redirection of attention is just as obvious, since now attention is redirected towards the epistemic function and the practical use of the symbols of quantum mechanics. In QBism, these symbols only represent the probabilistic weights that an agent can use to *bet* on the outcomes of experiments. And in Healey's pragmatic approach, these symbols are relativized to various agent-situations, and their role is to be guides for agents, by giving rigorous prescriptions about how to set their beliefs.

The third common feature, that is specifically developed by Laura de la Tremblaye (2020, in this volume), is the very structure of quantum knowledge as understood by QBism and pragmatism. This structure combines a bundle of expectations, expressed by a state vector and an observable, with some rules to take concrete empirical outcomes into account for redefining the expectations (this is the famous "projection postulate"). Such a combination comes remarkably close to

Husserl's phenomenological theory of perception, with its horizons of expectations that can be fulfilled or disappointed by "intuitive contents" made of sensory *hylè*.

11.4 Reconstituting Ontology After the *epochè*

But the central question I would like to raise goes beyond the level of structure to address the issue of ontology. Even if one adopts Richard Healey's "pragmatist" approach, that is deliberately non-interpretational, there may be a way towards a sort of meta-interpretation of the situation that makes interpretational views so clumsy and so fraught with paradoxes. Even if one considers that quantum physics does not provide us with the smallest hint of a representation of the world, another question can still be asked: what should the world be like in order to display such resistance to being represented as an object of thought? Answering this question would be tantamount to formulating a new kind of ontology, a non-object-based ontology, an ontology of what cannot be represented as an object external to the representation itself. One may think of several candidates to the title of a truly alternative ontology.

One of the most venerable is perhaps that of Schopenhauer. The very title of Schopenhauer's masterpiece evokes such an alternative ontology: *The World as Will and Representation*. Here, the world is not something which must be represented. To a certain extent, it coincides with the lived representation itself; and yet it is more than this representation, since what representation pictorially expresses is the outcome of an obscure impulse that moves everything, and takes in us the form of desire and action: *the Will*. According to this conception, we are not facing the world, as our representation falsely suggests, but we *coincide* with the world *qua* dynamical process and inner impulse; and the representation is just a projection of the oriented power of this dynamical process. In Schopenhauer's philosophy, Kant's "thing in itself" assumes a non-conventional meaning. Schopenhauer's "thing in itself" is not external to the knower but consubstantial with her. Then, the reason why, according to Schopenhauer (and possibly Kant), the "thing in itself" cannot become an object of knowledge, is precisely its lack of distance with respect to us. We tend to be blind to what is too close to our eyes. This is a first sketch of the kind of ontology that would fit a branch of physics such as quantum mechanics, in which the possibilities of separating the known object from the act of knowing are scarce (a feature that is well-known since Kochen & Specker's theorem as "contextuality").

Now, other ontologies that have this special quality are phenomenological ontologies. I will not say much about Husserl's methodologically idealist ontology. Let me just remind you that Husserl took the Cartesian doubt as a starting point. Then, he transformed this doubt into a universal *epochè,* or suspension of judgment about any claim of existence of wordly objects. And he retreated to the only domain of "apodictic certainty" of which any claim of inexistence would be performatively contradictory, namely the domain of "pure conscious life" (Husserl, 1960, p. 21). By contrast, "all positions taken towards the already-given objective world", said Husserl, must be "deprived of acceptance", or "inhibited" (Husserl,

1960, p. 20). The worlds of science and everyday life are then downgraded to the rank of mere phenomena that "claim being" (Husserl, 1960, p. 18), whereas "pure conscious life" is raised to the rank of "the whole of absolute being" (Husserl, 1982, §51). As I have mentioned above, this complete reversal of the ontological hierarchy is usually dismissed as a variety of idealism. Husserl somehow endorsed such a characterization of his philosophy as "idealist", but he gave the latter word a performative rather than doctrinal acceptation. His "... idealism is not a metaphysical substruction ... but the only possible and absolute truth of an *ego* ... recollecting itself on its own doing and its own ability to give meaning" (Husserl, 2007, p. 48). Dogmatic idealism can then be seen as a reification of this performative idealism. It arises when one turns the phenomenological activity that consists of "recollecting" one's own conscious life, and identifying the lived roots of one's "natural" beliefs, into a thing (*res cogitans*).

Would such a conservative ontology be an acceptable ontology for physics? I am not completely averse to it. After all, what else do we have in order to support further ontological claims than this pure conscious experience which, according to Husserl, is "the whole of absolute being"? Positions such as Bohr's, who sometimes declared that the primary task of physics is to introduce some order within human experience, and Chris Fuchs', whose slogan is "experience first", should be considered as reasonable options, rather than being automatically dismissed as "solipsistic".

Yet, there is something specific to physics that is not easily encompassed within a methodological idealist ontology such as Husserl's transcendental phenomenology: it is the role of embodiment and material agency that is so important in laboratories and industries. To a certain extent, Husserl himself was aware of this problem. This is why he dealt with embodiment in his *Ideen II,* and with what he called the lifeworld, the *Lebenswelt*, in his *Crisis of the European Science*. The lifeworld of history, of human work and instruments, includes the historically determined material agency to be found in laboratories. The phenomenological reduction to the lifeworld, after the world of theoretical idealizations has been suspended by the *epoché,* can be seen as an indispensible preliminary act before one performs the transcendental reduction that recollects us within the field of pure consciousness. A reduction to the lifeworld was tacitly performed by Bohr when he insisted that one should suspend any unified representation of the world, and take the formalism of quantum physics as just a "symbolic" anticipation of what can be found in the laboratory; and when he insisted that experimental apparatuses are to be described by using the ordinary language of our everyday life below the level of the quantum symbolism. But such a reduction to the lifeworld tends to be bypassed in the QBist approach to quantum mechanics. Indeed, the main difference between QBism and Bohr is the QBist's explicit refusal of any intermediate domain between the explored environment and the experience of observing and believing.

11.5 Within the Flesh, Within the World

By contrast, embodiment and human agency are natively taken into account by the phenomenological ontologies of Maurice Merleau-Ponty and Michel Henry. Let me start with Michel Henry, before I develop Merleau-Ponty's position at length.

According to Michel Henry (Henry, 1985, 2000, p. 72), *any* experience, including the experience of perceived objects, the experience of abstract forms, or the experience of one's own mental acts, ultimately turns out to be nothing else than an experience of the *self-affection of one's own flesh.* Pleasure and pain are taken as paradigmatic of self-affection, and even intentionality, even the assumption of transcendence of natural objects, must be (and is actually) rooted in the immanent impression that the flesh of a living being is making on itself. In other terms, even the perception of a patch of colour on some "outer" object is considered to be underpinned by a self-perception of the perceiver. One could say, as some contemporary neurobiologists do, that a perception is a guided dream of our flesh (a dream channelled by sensory inputs). Not even the most abstract conception of a mathematical structure can dispense with being rooted in some concrete self-sensitive modality of the living. Intentional consciousness is borne by non-intentional experience, and therefore the deepest layer of consciousness, the purest kind of experience, is nothing else than a naked self-sensitivity. Let me quote Michel Henry on this issue: "*Original* can only mean this: what comes in itself before any intentionality and independently of it, before the space of a gaze, before the 'outside oneself' of which intentionality is only a name; what comes ... before the world, out of the world, foreign to any conceivable 'world', *a-cosmic*" (Henry, 2000, p. 82). In other terms, according to Michel Henry, what comes before intentionality, and before the belief in a world, is the non-directional impression of being there: the awareness of being embodied, without any notion of the separation between one's own body and anything else; the silent voice of the body whose usual name is "cenesthesia". But once again, unlike the naturalistic program, which would try to account for the latter impression in terms of some interaction between natural objects (say the cells and organs) belonging to the human body, the phenomenological program adopts the diametrically opposite stance. It starts from the deepest layer of what we experience, and then tries to justify the belief in a natural world as a consequence of the multifarious differentiations and felt limitations of such a primeval experience.

This phenomenological program was developed and radicalized in Merleau-Ponty's last and posthumous book, *The Visible and the Invisible.* According to Merleau-Ponty, "we can accept a world ... only after having witnessed its arising from our experience of *raw Being,* which is like the umbilical cord of our knowledge, and our source of meaning" (Merleau-Ponty, 1964, p. 209). As a consequence, instead of looking for an ontology of separate observable objects (as in the natural sciences) or an extension of such an ontology that encompasses conceivable atoms of experience (as in panpsychism), Merleau-Ponty's ontology is an ontology of immersion, of connivance, of acquaintance. Merleau-Ponty looks for an "oblique ontology" of intertwining (de Saint-Aubert, 2006), or, in his own terms, an "endo-ontology" (Merleau-Ponty, 1964, p. 279). This is an ontology expressed *from* the inmost recesses of the process of being, rather than an ontology *of* the external contemplation of being*s*. This is an ontology of radical *situatedness*: an ontology in which we are not onlookers of a nature given out there, but rather intimately intermingled with nature, somewhere in the midst of it (Merleau-Ponty, 1964, p. 152). Here, we cannot be construed as point-like spectators of what is manifest; instead, we are a field of experiences that merges with what appears in a certain region of it. This *endo-ontology* is therefore an ontology of the participant *in* Being, rather than an ontology of the observer *of* beings. In endo-ontology, Being is not presented before me as an object of sight, but my vision arises from the middle of Being. "Vision is the tool which allows me to be absent to myself, and to contemplate from within the fission of Being" (Merleau-Ponty, 1985). This is a form of phenomenological embodied non-dualism that was also expressed by Michel Henry (1963, p. 95): "Consciousness is identified with the process of self-tearing (. . .) of being".

In Merleau-Ponty's philosophy, the archetypical element of our world that perfectly fits with an endo-ontology is our *flesh*. We perceive our flesh not as something separate, but as the perceiver that we are: a self-perceived perceiver. The flesh is that strange being endowed with complete reversibility, since it is jointly perceived and self-perceiving. The most obvious case of a two-faced kind of perception is the sense of touch, which, unlike distanced vision, is simultaneously appearance of what is touched and self-revelation of the touching in its carnal thickness. Here, two functions (*toucher* and *touched)* are realized by one and the same body. But this almost trivial remark was considered by Merleau-Ponty as paradigmatic of the true status of the whole world. "My" flesh is witness of the fundamentally fleshy nature of the world. "Where should we locate the boundary between the body and the world, Merleau-Ponty asks, since the world *is* flesh?" (Merleau-Ponty, 1964, p. 182). In an endo-ontological framework, there are no such things as "me", "you", and the world, but a single canvas wherefrom various self-individualizing centers of sensitivity emerge, and which leaves patches of elementarity and half-obscurity between these centers (Barbaras, 1993, p. 304). The role of constituting objectivity, which had been entrusted to the transcendental *ego* by Husserl, and which had been extended to our own-body by Merleau-Ponty at the time of his *Phenomenology of Perception*, was further extended to whatever has the status of a flesh by Merleau-Ponty at the time of *The Visible and the Unvisible.* But since the flesh is boundless, since the flesh is the whole world, any division between the constituter of objectivity and the constituted objects is meaningless. Just as the flesh is self-perceiving, the world *qua* flesh is self-objectifying.

11.6 Endo-Ontology and Participatory Realism

Merleau-Ponty's "endo-ontology" strongly evokes the ontology that was elaborated by Chris Fuchs and other actors of the QBist adventure (despite the fact that their neglect of the intermediate level of bodies and instruments apparently brings them closer to Husserl's transcendental phenomenology than embodied phenomenology).

Fuchs' ontology was inspired by John Wheeler's post-Bohrian idea that quantum mechanics involves "observer-participancy". Just consider this sentence of Wheeler (2016): "The strange necessity of the quantum as we see it everywhere in the scheme of physics comes from the requirement that—via observer-participancy—the universe should have a way to come into being." Observer-participancy was thus ascribed a very strong meaning and a crucial role by Wheeler. For, according to him, any act of observer-participancy is capable of creating the universe, including its past! Some of the naiveté of Wheeler "magical" retrospective creation of the universe by the observer's gaze is avoided by Chris Fuchs' participatory realism, but not all of it. Fuchs (2016) still uses the same word "creation" although in a more moderate form. He indeed accepts that each act of observer-participancy is a present act of creation. Yet he compares this act of creation with an act of reproduction of living beings, in which something new (a child!) arises out of the combination of previous elements (the genetic material of the parents). Instead of "creation", he could thus have used the more modest terms "emergence" or "co-generation".

The idea of participatory realism was born from Fuchs' wish to distance himself somehow from Asher Peres (who was his first teacher and collaborator). Fuchs wished to escape the accusation of "instrumentalism" and to develop his non-conventional version of realism instead. Yet, Asher Peres' endorsement of instrumentalism, and even positivism, is not necessarily to be understood as a philosophical rejection of participatory realism. It might well have been an overcautious statement of this very same doctrine. Consider the following key sentence in a joint paper of Fuchs and Peres (2000): "If the world is such that we can never identify a reality independent of our experimental activity, then we must be prepared for that, too".

From this sentence, one can adopt two strategies:

1. The first strategy, which was adopted by Fuchs, consists in building an overarching metaphysics. It is tantamount to seeking a metaphysical reason for the lack of success of ordinary metaphysics. It is tantamount to saying that the lack of independence of the symbols and statements of quantum theories with respect to our situation and experience reveals something crucial about the nature of reality: that it is highly entangled, thoroughly holistic, and that therefore our knowledge of it can only be participatory rather than representational.
2. The second strategy is anti-metaphysical. It is inspired by Wittgenstein's famous prescription according to which "What we cannot speak about we must pass over in silence". A defensor of this second strategy would express herself as follows. It is true that the characteristics of quantum mechanics irresistibly suggest that there is no such thing as a reality independent of us and our agency. But if

this is so, the only thing we can do is renounce any representation or meta-representation, and devote our efforts to orienting ourselves and surviving in the putative participatory reality. We can hardly speak of what there is as if we were describing it from outside, for, in virtue of the very idea of a participatory reality, this would mean describing ourselves-inside as if we were outside! Then, according to this strategy, instrumentalism is just silent participatory realism.

A mischievous Wittgensteinian philosopher could insinuate that, conversely, participatory realism is noisy instrumentalism. But this might be an exaggeration. Participatory realism is truly useful, because it sketches the only conception of reality that is immediately compatible with quantum mechanics, and by doing so satisfies our want for mental pictures without indulging in wrong representations. Indeed, this mental picture is the only one that fully acknowledges the core reason of Bohr's prohibition of global ontological representations in quantum mechanics. It is a mental picture of the reason of the inadequacy of pictures. We could also say that participatory realism succeeds because it does not ascribe "reality" any positive predicate, but only a negative predicate: the *impossibility* of neatly splitting it into a spectator-like knower and a play-like known. This introduces us to what may be called "negative metaphysics", similar to the famous (or infamous) "negative theology". In the same way as "negative theology" may be taken by some as a good reason to abstain from theology, "negative metaphysics" may be taken by some (the instrumentalists) as a good reason to abstain from metaphysics.

11.7 Merleau-Ponty's Philosophy of Quantum Mechanics (1): Beyond the Classical Paradigm

Let's recapitulate this point. Merleau-Ponty's endo-ontology is an ontology for one who deeply participates in Being. And Chris Fuch's participatory realism is a non-external variety of realism for one who is deeply immersed in reality. This remarkable analogy is supported by Merleau-Ponty himself. Indeed, 50 years before QBism was formulated as a radicalization of the philosophical tendency that was initially adumbrated by Bohr, Merleau-Ponty acknowledged the strong kinship between the epistemological situation of quantum mechanics and his phenomenology of embodiment. He did so in two texts that remained unpublished until after his death: *Visible and invisible,* and the *Lectures on Nature* he delivered during the years 1959–1960 at the Collège de France.

Merleau-Ponty (1995, p. 117) started his reflection by noticing that, even in classical science where it seemed that the whole world could be treated as an object under a neutral and external gaze, there remained a huge blind spot. "Nature resists. It cannot be posited entirely before us. The body is a nature at work within us". The lived body, this double-faced natural reality, this felt locus of feelings, is averse to any conception of nature as a Big Object. Indeed, when we say that we "know" our own body, this is definitely not by taking it as an object of perception or thought, but by being acquainted with it, by coinciding with it, and by letting it self-reveal in our proprioceptive experience. Our own-body stubbornly resists objectification, and therefore the world of which it partakes resists *universal* objectification. Yet, classical science and even science to this day, did not completely renounce the project of universal objectification. As Merleau-Ponty pointed out, science tries to reabsorb its blind spot at any cost. "Science progressively reintroduces what it first excluded as subjective. But science wishes to reintroduce subjectivity as a special case of the relationships and objects that define the world according to it. Then, the world closes on itself, and except by what in us *thinks* and *constructs science*, we become parts of the big object" (Merleau-Ponty, 1964, p. 31). Even the domain of psychology becomes objectified if classical physics is taken as an ideal. (Merleau-Ponty, 1964, p. 36). In other terms, the reabsorption of its own blind spot by scientific research is an ongoing project that takes the form of an attempted naturalization of mind and consciousness. But this project implies (mis)taking mind and consciousness for parts of the "big natural object" called "the world".

However, a momentous episode of the history of science made this project obsolete even before its highly elusive completion. This episode is the advent of quantum mechanics. Indeed, quantum mechanics is simultaneously the most advanced stage of a long-term push towards objectivity, and the place where a fundamental limit of objectification manifests itself, though cryptically. Merleau-Ponty first notices that this manifestation of a fundamental limit of objectification is indeed so cryptic that "science has shown conservatism concerning the theory of knowledge" (Merleau-Ponty, 1964, p. 33). Scientists tried to modify the *contents* of their knowledge without modifying their *conception* of knowledge, and to modify their view of an objective world while maintaining their objectivist naturalism. The reason why scientists have taken this conservative stance is that, without even knowing it, science is "rooted in pre-science"; it is rooted in the pre-science of common sense. However, as Merleau-Ponty explains, no one can truly understand quantum mechanics without accepting a deep transformation of our conception of knowledge. It is in vain that "some physicists frame in an objectivist ontology a physics that is no longer amenable to it" (Merleau-Ponty, 1964, p. 45).

Now, the suitable transformation of our conception of knowledge is deep and devastating. It is one that challenges nothing less than the notions of common sense and the very duality of subject and object. Let me quote Merleau-Ponty once more: "Quantum physics does not put all truths on the side of the 'subjective', which would maintain the idea of an inaccessible objectivity. It rather challenges the very principle of this division and brings the contact between the observer and the observed in its very definition of 'reality'" (Merleau-Ponty, 1964, p. 33). The new physics is then a strong incentive to recognize indirectly a truth that phenomenology knows directly: "that 'objective' and 'subjective' are domains hastily constructed out of a totalizing experience" (Merleau-Ponty, 1964, p. 38).

On the basis of these remarks, Merleau-Ponty sketches the role of philosophy in our understanding of quantum mechanics. Philosophers, he says, can hardly take part in the technical debates about the formalism and its interpretation, let alone about experimental facts. But they become indispensible when "scientific being connects to pre-scientific being" (Merleau-Ponty, 1995, p. 125). Their work becomes crucial when one is concerned with the way scientific knowledge is elaborated out of the pre-scientific layer of the lifeworld, and thereby overrides this pre-scientific layer without being independent of it.

However, when they pursue such an inquiry, philosophers are likely to make a disturbing discovery. By considering attentively the status of quantum mechanics, they discover that this physical theory only "helps us make negative philosophical discoveries"; they discover that "science does not impose an ontology ... It has only the power to remove false evidence from their alleged status of 'evidence'" (Merleau-Ponty, 1995, p. 139–145).

11.8 Merleau-Ponty's Philosophy of Quantum Mechanics (2): Probabilities

This being said, Merleau-Ponty boldly entered into some details of the interpretation of quantum mechanics. He thus noticed that, in quantum physics, "(Probability) does not concern only our ignorance. With (quantum) indeterminism, we are dealing with pure probability ... Probability here enters into the texture of reality" (Merleau-Ponty, 1995, p. 127). But this very statement that "probability enters into the texture of reality" is not to be understood as a claim in favor of "objective probabilities", similar to Popper's propensions or Heisenberg's potentialities. It must rather be understood by due consideration of the exceptional status of the wave function (or state vector) that allows one to calculate quantum probabilities. Notwithstanding their standard interpretation, wave functions are not a description of the state of an object. They rather describe the composite entity indissolubly made of a system, an apparatus and an observer (Merleau-Ponty, 1995, p. 129). As a consequence, quantum probabilities do not reveal the ontological indeterminateness of the known objects out there; they rather express the indivisibility between the act of knowing and what is to be known.

And yet, measurements are still interpreted as if they were measurements *of* something, as if they were measurements that provide us with information about the properties of something. This tension between indivisibility and a project of division is what generates the measurement problem, but also what has the potential to dissolve it.

On the one hand, "The measurement operation, in wave mechanics, is an 'engaged' operation. Every operation of the new mechanics is an operation in the world, which is never foreign to the act of the one who makes the measurement" (Merleau-Ponty, 1995, p. 131). The situation in quantum mechanics is thus reminiscent of our situation of embodiment, of the special status of our own body as a bifacial "flesh", and of the problem of knowing such an own-body: any operation of our own body is an operation within the "flesh of the world". We can then consider that the situation in quantum mechanics is an extension of the archetypal case of the own-body, with its twofold power of touching and being touched, that generated the concept of an endo-ontology. This extension is in perfect agreement with Merleau-Ponty's initial intuition. Indeed, according to him, the case of the flesh is paradigmatic of the true nature of the world, and the whole world should be treated as a big flesh rather than a big object. At the end of the day, quantum physics testifies that the world behaves as a big flesh, of which our flesh is a sample.

But on the other hand, "the role of the observer is to cut the chain of statistical probabilities, to bring out an existence in action. What makes this existence arise is not the intervention of a for-itself, but a thought that annexes a measurement apparatus". Only a thought can cut the chain of possibilities with probabilities and stop the chain at a point that is seen as an actuality; only a thought can cut the indivisible measurement chain, stop mentally the indivisible measurement dynamics, and interpret one aspect of this dynamics, one moment of the subject's experience, as a well-defined actual outcome. After all, something similar arises from many contemporary interpretations of quantum mechanics, such as Rovelli's "relational" interpretation: according to Rovelli, there is no fact of the matter about what is the "real" state of the system (a superposition or a sharp state) after a measurement has been performed. Here, what we still call the "state" is relative to a free option taken by the agent as to which epistemic situation she adopts (as a predictor or as an observer).

This being granted, Merleau-Ponty pointed out that "the problem posed by quantum physics is close to the problem of perception. Its duality is reminiscent of the duality of the perceptual process, in turn global and attentive" (Merleau-Ponty, 1995, p. 135). In particular, the duality between the global expectations expressed by superposed state vectors, and the local actualities brought out by an interpreted measurement, is reminiscent of the Husserlian duality between global perceptive horizons and focused sensory fulfillment of some of the expectations adumbrated by such a horizon.

11.9 Epilogue

Merleau-Ponty concludes his philosophical analysis of quantum physics by siding with a very remarkable (but unfortunately almost forgotten) French philosopher of physics of the mid-twentieth century: Paulette Destouches Février. He quotes her approvingly when she declares: "We are dealing with a human physics, a physics of solidarity. Yet, physics cannot be interpreted in a purely idealistic way. It requires a form of realism that can be called 'participatory'" (Merleau-Ponty, 1995, p. 135). Merleau-Ponty then recognizes explicitly that a good expression of his endo-ontology in quantum physics is the "participatory realism" advocated by Paulette Destouches-Février (1951).

It then turns out that participatory realism, this view entertained by several philosophers of physics, from Paulette Destouches-Février to Christopher Fuchs, who dismiss any attempt at interpreting quantum mechanics as a possible representation of the world, is a perfect match to Merleau-Ponty's endo-ontology of the world-flesh.

References

Barbaras, R. (1993). *De l'être du phénomène*. Grenoble: Jérôme Millon.

Bell, J. S. (1987). *Speakable and unspeakable in quantum mechanics*. Cambridge, UK: Cambridge University Press.

Bitbol, M. (1996). *Schrödinger's philosophy of quantum mechanics*. Dordrecht, The Netherlands: Kluwer.

Bitbol, M. (Ed.). (2009). *Théorie quantique et sciences humaines*. Paris: CNRS Editions.

Brukner, C., & Zeilinger, A. (2009). Information invariance and quantum probabilities. *Foundations of Physics, 39*, 677–689.

Bruza, P., Sofge, D., Lawless, W., van Rijsbergen, C. J., & Klusch, M. (Eds.). (2009a). *Quantum interaction*. Berlin, Germany: Springer.

Bruza, P. D., Kitto, K., Nelson, D., & McEvoy, C. (2009b). Is there something quantum-like about the human mental lexicon? *Journal of Mathematical Psychology, 53*, 362–377.

de la Tremblaye, L. (2020). QBism from a phenomenological point of view. In H. A. Wiltsche & P. Berghofer (Eds.), *Phenomenological approaches to physics*, Synthese Library, 429. Cham, Switzerland: Springer.

de Saint-Aubert, E. (2006). *Vers une ontologie indirecte: sources et enjeux critiques de l'appel à l'ontologie chez Merleau-Ponty*. Paris: Vrin.

Destouches-Février, P. (1951). *La structure des théories physiques*. Paris: P.U.F.

Fuchs, C. (2016). On participatory realism. ArXiv: 1601.04360v3 [quant-ph].

Fuchs, C. A., Mermin, N. D., & Schack, R. (2014). An introduction to QBism with an application to the locality of quantum mechanics. *American Journal of Physics, 82*, 749–754.

Fuchs, C. A., & Peres, A. (2000). Quantum theory needs no interpretation. *Physics Today, 3*, 70–71.

Grinbaum, A. (2007). Reconstructing instead of interpreting quantum theory. *Philosophy of Science, 74*, 761–774.

Healey, R. (2017). *The quantum revolution in philosophy*. Oxford: Oxford University Press.

Henry, M. (1963). *L'essence de la manifestation*. Paris: Presses Universitaires de France.

Henry, M. (1985). *Généalogie de la psychanalyse*. Paris: Presses Universitaires de France.

Henry, M. (2000). *Incarnation*. Paris: Éditions du Seuil.

Husserl, E. (1960). *Cartesian meditations*. The Hague, The Netherlands: Martinus Nijhoff.

Husserl, E. (1982). *Ideas pertaining to a pure phenomenology*. The Hague, The Netherlands: Martinus Nijhoff.

Husserl, E. (2007). *De la réduction phénoménologique*. Grenoble: Jérôme Millon.

Kant, I. ([1770] 2004). *Inaugural dissertation of 1770*. Whitefish, MT: Kessinger Publishing.

Merleau-Ponty, M. (1964). *Le visible et l'invisible*. Paris: Gallimard.

Merleau-Ponty, M. (1985). *L'œil et l'esprit*. Paris: Gallimard.

Merleau-Ponty, M. (1995). *La nature, Notes. Cours du collège de France*. Paris: Editions du Seuil.

Scheler, M. (1993). *Problèmes de sociologie de la connaissance*. Paris: Presses Universitaires de France.

Schrödinger, E. (1951). *Science and humanism*. Cambridge, UK: Cambridge University Press.

Schrödinger, E. (1954). *Nature and the Greeks*. Cambridge, UK: Cambridge University Press.

Stengers, I. (1997). *Cosmopolitiques 4: Mécanique quantique, la fin du rêve*. Paris: La Découverte.

Thom, R. (1993). *Prédire n'est pas expliquer*. Paris: Flammarion.

Wheeler, J. A. (1978). The 'past' and the 'delayed-choice double-slit experiment'. In A. R. Marlow (Ed.), *Mathematical foundations of quantum theory*. New York: Academic.

Wheeler J. A. (2016). Interview, quoted by C. Fuchs, "On participatory realism", ArXiv: 1601.04360v3 [quant-ph].

Wittgenstein, L. ([1953] 1968). *Philosophical investigations*. Oxford: Basil Blackwell.

Zwirn, H. (2009). Formalisme quantique et préférences indéterminées en théorie de la décision. In M. Bitbol (Ed.), *Théorie quantique et sciences humaines*. Paris: CNRS Editions.

Chapter 12
QBism from a Phenomenological Point of View: Husserl and QBism

Laura de La Tremblaye

Abstract The recent introduction of a Bayesian interpretation of the equations of quantum mechanics led us to reconsider the status of physical theories and our relation to knowledge. From the QBist point of view, the quantum state, that generates probabilities by way of the Born rule, is not referring to the microsystem anymore, but rather to the possible results of an observation originating from a certain user. Yet, if the user does not have directly access to the object, how can one be said to acquire knowledge? In order to characterize the QBist position and to explore these issues, we will confront QBism with another approach to knowledge stemming from philosophy, namely: Edmund Husserl's phenomenology. According to Husserl's theory of perception, bodily objects are constituted through the observation of their *profiles*. Varying profiles are connected by the subject into a single unit that is endowed with *identity through time*. In the present moment, the subject anticipates the potential future experiences she will have of this unit, taking into account its noematic structure and her past experiences of earlier profiles. This process is what Husserl calls the "intentional horizon". There is an obvious connection between the intentional horizon and the QBist interpretation of quantum states, that will be explored in details. For both QBism and phenomenology, gaining knowledge consists in making anticipations and actualizing them through observations. QBism, and its phenomenological roots, then reveals a very basic layer of knowledge that can no longer be ignored by theoretical physics.

The introduction of quantum mechanics (hereafter referred to as *QM*) to the field of physics has challenged our classical metaphysical world-view and many of its key concepts such as physical state, particle or measurement. In view of the puzzling

This work was supported by the Agence Nationale pour la Recherche (ANR-16-CE91-0005-01).

L. de La Tremblaye (✉)
Archives Husserl, ENS, Paris, France

Department of History and Philosophy of Science, University of Geneva, Geneva, Switzerland

H. A. Wiltsche, P. Berghofer (eds.), *Phenomenological Approaches to Physics*, Synthese Library 429, https://doi.org/10.1007/978-3-030-46973-3_12

issues surrounding it, quantum theory is a prime entry point to explore more general topics such as the nature of knowledge and objectivity as well as the role of the physicist in the laboratory.

Werner Heisenberg, one of the founders of QM, recognized, inspired by Bohr, its special status already in 1942. In his remarkable *Ordnung der Wirklichkeit* (Heisenberg, 1990), Heisenberg divided the world into three regions, each region being associated with scientific disciplines that differ in the way knowledge is accessed in their respective realms. Quantum theory is associated with the second region, separate from the other fields of physics, and united with, among others, psychology. The following observation was therefore already apparent at the time: there is a fundamental epistemological, and perhaps even ontological difference between QM on the one hand and the rest of so-called classical physics on the other. By locating QM on the same level as psychology, in the category of disciplines whose access to knowledge is not independent of observation, Heisenberg was already asking the question of the relationship of a physical theory to its object of knowledge. What is the connection between quantum theory and the world of the infinitely small it seems to describe? What are the differences between classical and quantum descriptions of the world? These questions are the concern of scientific and philosophical communities since the emergence of these issues almost a century ago. It is therefore a formidable opportunity for philosophy and physics to adopt a reflective attitude, and to again raise several fundamental questions about one of the most successful theories ever devised.

Many interpretations of QM have emerged over the course of the last century. In most cases they were attempts to reconcile a realist position with the structure of QM, while solving some of the enigmas and paradoxes that arise from this realist approach. Other interpretations had a decidedly instrumentalist flavor to them. Yet, there is one particular interpretation that will take center stage in this chapter: *QBism*. Originally called Quantum Bayesianism, QBism adopts a Bayesian point of view regarding the probabilities used in QM. Chris Fuchs, Rüdiger Schack and more recently David Mermin, the founders of this movement, interpret the Born rule as a Bayesian probability rule in its subjective form, as it is usually attributed to Bruno De Finetti and Frank Ramsey (Caves et al. 2002).

From the QBist point of view, the quantum state, which generates probabilities by way of the Born rule, is not referring to the microsystem anymore, but rather to the possible results of an observation made by a certain user, and above all to the expectations this user has about these possible results. Yet, if the user does not have directly access to the object, how can she be said to acquire knowledge? In order to characterize the QBist position and to better reveal its innovative character, we will confront QBism with another approach to knowledge stemming from philosophy, namely: Edmund Husserl's phenomenology.

By comparing QBism with Husserlian phenomenology, and especially Husserl's theory of perception, we will not only be able to discover important points of convergence; we will also be able to take some first steps towards a phenomenological foundation of QBism. Using a phenomenological framework for the analysis of quantum issues, we will shift our attention away from the alleged correlations between the equations and the objects suggested by the theory; instead our focus will be to understand the kind of knowledge that arises from QM.

The ability to anticipate, the updating of our knowledge by observation, and the profound subjectivist description of experience, are three aspects of this truly fruitful analogy. We are then reinforced in our conviction that examining modern physics through the phenomenological prism is thoroughly relevant and useful for epistemology.

12.1 Probabilities, QM and Philosophy: Toward the QBist Approach

12.1.1 QM and Its Interpretations

To give us a general overview of the panel of interpretations still defended today we will use Adàn Cabello's, 2016 classification.[1]Cabello classifies the interpretations of quantum theory "according to whether they view probabilities of measurement outcomes as determined or not by intrinsic properties of the observed system" (Cabello, 2016, p. 2). Here, it is the status of probabilities that gives meaning to the category to which each interpretation belongs. By doing this, Cabello makes a direct connection between the status of probabilities and the relation of each interpretation to the concept of *reality*. The less objective the probabilities, the less realist the interpretation is.

Cabello distinguishes two types of interpretations, those that fall into the *Type I: intrinsic realism* category and those that belong to the *Type II: participatory realism* category. The first category corresponds to the kind of realism that is usually associated with classical physics: the idea that there is an external world made up of objects, and that it is up to physics to study it. Conversely, Type II interpretations refer to conceptions inspired by John Wheeler's participatory universe ontology. These two categories are conceived in opposition to each other: "There are those that view quantum probabilities of measurement outcomes as determined by intrinsic properties of the world and those that do not" (Cabello, 2016, p. 1).

Type I is divided into two sub-sections. These interpretations can be psi-ontic if they consider the quantum state to be an *intrinsic property* of the system. They can also be psi-epistemic if the quantum state refers to our knowledge of an underlying objective reality.

[1]Cabello's classification is recent enough to include the latest version of QBism we will discuss further. Another recent classification presented by Leifer in 2014 puts QBism in the same category (*neo-Copenhagen interpretations)* as the interpretations of Zeilinger and Healey, thus making it a less relevant tool for our study.

Type II interpretations can be *about knowledge* if quantum states refer to an observer's knowledge about the results of future experiments. The standard interpretation of QM, namely Bohr's Copenhagen interpretation, belongs to this category. Alternatively, Type II interpretations can also be *about beliefs*. In this case, the quantum state refers to the beliefs or expectations an observer has about future measurement outcomes. Note, however, that one interpretation stands out as an exception in the Type II/about beliefs category: QBism.[2]

QBism is the strong version of Quantum Bayesianism, adopted by Fuchs, Schack and Mermin. They developed the QBist position as a Bayesian, subjective interpretation of the probability distributions of quantum states. From the QBist point of view "The Born Rule is (. . .) an addition to Bayesian probability (. . .) in the sense of giving extra normative rules to guide the agent's behavior when he interacts with the physical world." (Fuchs, 2010, p. 12). Quantum theory is also considered a "single-user theory" or "single-user manual" that can be used by any individual agent.[3]

Let's take a closer look at the Bayesian interpretation behind QBism.

12.1.2 Bayesian, Personalist, Probabilities

The ideas defended by Fuchs and Schack are based on the Bayesian interpretation of probabilities defended by Bruno de Finetti (1935) and Frank P. Ramsey (1926). It is the strong, so-called subjectivist or *personalist* form of Bayesianism, which is based on decision theory. "Probability (. . .) is constituted by a degree of doubt, of uncertainty, of conviction, which our instinct makes us feel in thinking of a future event, or, anyway, of an event whose outcome we don't know." (de Finetti, 1989, p. 175). On this view, probabilities are rational estimates that an agent assumes in order to make the best possible decisions. Another way of understanding would be to consider an estimate made by an agent as reflecting the amount she would be willing to bet on the occurrence of a future event, and this estimate is based on her beliefs.

[2]QBism was developed in 2001 by Carlton Caves, Chris Fuchs and Rüdiger Schack (Caves et al. 2002). It was first called Quantum Bayesianism but became QBism in 2010. Fuchs and Schack however decided not to relate the B in QBism to Bayesianism anymore because of the number of different Bayesian interpretations of probabilities. As suggested by David Mermin, QBism can be understood as Quantum Brunoism for Bruno de Finetti (Mermin, 2013) or by Fuchs as Quantum Betabilitarianism (Fuchs, 2015, 2016). These are indications on what type of interpretation of probabilities QBism supports.

[3]However, QBism should not be interpreted as an instrumentalist approach of QM; it rather draws a new ontology called *Participatory Realism* (Fuchs, 2016).

Horse	Odds offered	Implied prob.	Amount bet	Payout if horse wins	Net loss
1	Even	1/2	$120	$240	$20
2	1: 2	1/3	$80	$240	$20
3	1: 3	1/4	$60	$240	$20
	Total	13/12	$260		

This construal is based on the famous *Dutch Book argument*. It was formulated by Ramsey and de Finetti to justify the postulate that if a rational agent bets according to her degrees of belief about an event and if these degrees of belief violate the axioms of probability, then it is possible to propose a Dutch Book, namely a bet in which the monetary loss for the gambler is certain. Let us consider an example developed by Rüdiger Schack[4].

Imagine a gambler attending a horse race and a bookmaker offering the odds listed in the "odds offered" column of the table above. The gambler decides to bet money on the horses by making a compound bet. This is to say that he spends 260 dollars in total by betting 120 dollars on the first horse, 80 dollars on the second horse and 60 dollars on the third. If horse one wins, the payout is 240 dollars and he loses twenty dollars. The same conclusion is reached if the second or the third horse wins. In any case, he is sure to lose 20 dollars.

The problem here is that the probabilities do not add up to one, but to 13/12. In this case, the rules of probability, or the Kolmogorov axioms, are violated: we are dealing with a Dutch Book scenario in which the bettor is certain to lose.

By contrast, we define the notion of *Dutch Book coherence* (*coherent* according to the Dutch Book argument) as a bet where the possibility of a creating a Dutch Book is ruled out. We can prove that an agent's betting odds are Dutch Book coherent if and only if they conform to the standard probability rules. Isabelle Drouet specifies: "The arguments of the Dutch Book establish that there is a lack of rationality in maintaining non-probabilistic degrees of belief, as well as not revising one's degrees of belief in a Bayesian way." (Drouet, 2016, p. 10) This theory is also based on the ability to revise an agent's degrees of belief based on the availability of new information. Therefore, on Ramsey's and de Finetti's account, probabilities are a tool for assessing the consistency of an agent's beliefs, but not for producing any judgment on the outside world. The probability that horse one wins does not refer to its intrinsic ability to win but rather to what the bettor estimates to be the probability of its victory.

[4]Rüdiger Schack presented this example in 2016 in the *Information-Theoretic Interpretations of Quantum Mechanics* conference at the University of Western Ontario in London, Canada.

12.1.3 Applying Bayesian Rules to Quantum Theory

Fuchs and Schack propose to interpret Born's probabilistic rule (stated in what is often referred as the 4th postulate of QM) in terms of a Bayesian rule. The coefficients associated to each eigenstate composing the state vector here correspond to the probability of measuring each eigenstate. These measurement probabilities are estimates that the agent, as a QBist physicist, makes on the possible outcomes of an experiment; the outcomes being represented by the eigenstates. The Born rule coupled to the state vector, is therefore a coherent betting structure that each physicist uses independently to bet on the results of her own experiments.

Applied to Quantum theory, this Bayesian interpretation transforms QM into a generalization of the theory of probabilities applied to physics. This, by the way, is not a recent idea. Two remarkable French philosophers of physics, Paulette Destouches-Février and Jean-Louis Destouches, already understood quantum theory as a generalization of the probability theory since 1951 (P. Destouches-Février, 1951, p. 275). Consider the following passage which brings out the basic idea quite nicely: "(. . .) According to the so-called subjectivist approach, probabilities derived from wavefunctions should be considered as predictions assessments, i.e. the chances that the observer associates with obtaining a result *if he performs the measurement*; the probability assessments are assumed to be appropriate, i.e. consistent (it is impossible to bet against these assessments in any case) and to be appropriate regarding the experiment (the probability of the difference between the frequency and the probability tends towards zero with the number N of cases observed) (. . .)" (P. Février, 1956, p. 82) This approach leads her to conclude that quantum theory is "essentially indeterministic", or "irreducibly subjectivist" (P. Destouches-Février, 1951, p.276). Similar ideas are also found in Bitbol (1998, 2014).

The subjective approach developed by Paulette Février, Jean-Louis Destouches or Michel Bitbol is the basis of the philosophy behind QBism and it fits quite well with the basic gesture of phenomenology to start out from the first-person standpoint. QBism is more than a generalization of probability theory: it constitutes a true ontological shift leading to the construction of Participatory Realism. While (Bitbol, 2020) develops the phenomenological roots of the new ontology outlined by the QBist approach, I will examine the compatibility between this new approach and the epistemology of Husserl's phenomenology.

12.1.4 QBist's Epistemology

If the Bayesian, or subjective, interpretation of the Born rule is accepted, it not longer makes sense to draw conclusions about the state of the world from the quantum state. Traditionally, the quantum state vector is supposed to offer a description of the system at a given time. Its continuous change according to the

Schrödinger equation is then considered to describe the evolution of the system over time, until the moment of measurement. Yet, when a measurement occurs, a sudden change in the state of the system is imposed (cf. the discussion of the fifth postulate of QM in Cohen-Tannoudji et al. 1977, pp. 220–221), taking the form of what is often called "the reduction of the wave packet". This strange feature no longer exists in QBism because the state vector does not describe the real state of a perfectly determined system in the outside world or a state of (objective) knowledge, but only the agent's beliefs about her future measurement outcomes. There is no "real reduction", but rather an update of the agent's beliefs according to the information available to her. The state vector does not refer to any information or knowledge about the studied system but only to the agent's degrees of belief about future measurement outcomes. Therefore, the so-called "independent reality" and quantum states are no longer related.

How, then, does QBist interpretation maintain the connection between the quantum states on the one hand and *reality* on the other? What kind of information do QBists deal with?

QBism is established on the basis of several assumptions, but the main argument is that the Born rule is a *normative* rule. It is the most accurate decision-making tool we have for betting on measurement outcomes; a norm for betting. It describes the decision-making behavior to which every agent must aspire, but does not correspond, as Type-I-interpretations in Cabello's table would assert, to any "law of nature". Born's norm for betting is shared by physicists, but the assignment of coefficients in a state vector is personal.

When a physicist plans to measure some properties of a microsystem, such as the polarization of a photon, she prepares it for the subsequent measurement of some observable. She thus establishes a state vector, which is usually considered as the description of the real state of the system before any measurement (what is often referred to as the 1st postulate of QM), but that can be reduced to an equivalence class of preparations (Peres, 1995). The Born rule then gives the probabilities of measuring the particle in each eigenstate of an observable (or, more neutrally, the probability of measuring a certain eigenvalue of that observable). The physicist will then go on to carry out the measurement. While making this measurement, she is not expected to observe any *reduction of the wave packet*, since she does not consider that the quantum state established just before the measurement correlates to a real state. But she does notice the following: since her measurement result corresponds to one of the eigenstates of the quantum state vector, this information must be considered as relevant. The QBist physicist then modifies the state of her system on paper, re-writing the quantum state taking into account the outcome of the measurement she has just made for any future estimates.

To summarize, the physicist bets on the results of a measurement using the state vector and the Born Rule, then modifies her degrees of beliefs *a posteriori* depending on the result of her experiment, in the same way as the Bayesian revision of probability estimates. This cycle – prediction / experiment / revised degrees of belief – is the basis for any prediction about future experiments.

In other terms, the kind of knowledge QBists deal with is created through an interaction between an agent's estimations and her experimental results. It is perfectly clear now that QBism no longer deals with the properties of the world, but rather with the relationship between estimations and responses to experiments. Fuchs specifies: "I now think it is much better to reserve the word KNOWLEDGE solely for the outcomes of quantum measurements once they become part of the mental makeup of an agent interested in them" (Fuchs, 2002, p. 49). In that sense, knowledge is what is given by experiments; it is what I can agree on with other agents. But "what either of us may do with that knowledge is a different story" (Fuchs, 2002, p. 49). Any outcome (or knowledge) can affect one's degrees of belief differently.

12.1.5 Philosophical Consequences

By associating the use of probabilities with decision theory, Fuchs, Schack and Mermin conceive of quantum theory as a powerful tool for orienting oneself in the world. Yet, at the same time, they are denying that we should think of QM in terms of a representation of this world, and they provide instead a new definition of the concept of knowledge.

By calling a probability assignment a "tool an agent uses to make gambles and decisions" or a "tool he uses for navigating life and responding to his environment." (Fuchs, 2010, p. 4) Fuchs has often been told that QBism is a form of instrumentalism. Yet, some important features distinguish QBism from instrumentalism. The most important of them is that QBism is associated with a non-standard conception of the world and of our position within it. As mentioned earlier, this non-standard conception is *participatory realism.* And participatory realism arises from a clear understanding of what a measurement is: not a revelation of a property that was already there, but a moment of creation. It is due to the insistence on the role of the agent and her participation in the universe of experience around her that QBism prioritizes measurements as a means for an agent to take action on the world around her. As QBists we can no longer uphold the metaphysical assumption that there is a definite reality independent of us. We rather sketch the idea of a reality whose form is continuously created as we interact with a non-specified, possibly formless, environment.

So, this strongly subjective view of probability distributions and the Born rule is hard to reconcile with a classical physicalist ontology according to which reality is external and independent of the observer. But Fuchs has a reply in store to the objection that this point of view is too subjective. He points out that it is necessary to distinguish two aspects of probability ascriptions. Each ascription of probability coefficients is subjective in the sense that it depends on the agent who uses them, and especially on her situation and on the information available to her. What is not subjective, however, is the general rule for ascribing probabilities in any well-specified situation. This rule is shared by each and every physicist who utilizes QM to make predictions.

It is no trivial task to draw a clear line between the subjective and the objective aspects of the Born rule. In order to account for this problem, Fuchs and Schack invoke a completely new form of intersubjectivity. It is through the use of Bayesian probabilities that the multiplicity of subjectivities elaborates a reasoning that can be shared by everyone, and that, consequently, can be called "objective" in precisely this limited sense. Again, this leads to the new conception of knowledge: knowledge is no longer understood in terms of an objectively true description of the intrinsic properties of the world; it is rather understood as the kind of knowledge that is needed to guide the future research of any agent, thus implying a weaker form of objectivity.

Why, then, should we embrace such an unconventional interpretation? The strength of QBism lies in its ability to dissolve the paradoxes and enigmas that are usually thought to haunt QM. The principles underlying QBism make it possible to develop a simple conceptual apparatus that provides simple answers to the problems associated with Schrödinger's famous cat thought experiment, Wigner's friend paradox or the Einstein-Podolsky-Rosen enigma (Einstein et al. 1935).

Three fundamental, non-mathematical principles govern the understanding of quantum theory through QBism on a philosophical level. If the first two tenets have already been discussed (namely the normative reading of the Born Rule, and the agent-related probability assignments), the third tenet takes one more step towards a first-person interpretation of QM.

In this third principle, Fuchs goes so far as to claim that the measuring device is the analogue to a sensory organ of the agent (Fuchs, 2018, p. 20). Therefore, for a QBist, "(. . .) the outcome of a quantum measurement is a *personal experience* for the agent gambling upon it." (Fuchs, 2018, p. 21). Here, we see a two-step reduction. According to a QBist, knowledge reduces to the outcomes of experiments, and these outcomes reduce to personal experiences. The most basic knowledge for a QBist is precisely the agent's personal experience; it is her primal stuff. "This is because QBism's understanding of quantum theory is purely in first-person terms" (Fuchs, 2018, p. 20). The personal experience thus defined is very similar to the Husserlian concept of *lived experience* and to William James' concept of pure experience.[5]

Husserl emphasizes in his famous *Principle of all Principles* "*that every originary presentive intuition is a legitimizing source of cognition,* that *everything originarily* (so to speak, in its 'personal' actuality) *offered* to us *in 'intuition' is to be accepted simply as what it is presented as being*, but also *only within the limits in which it is presented there.* We see indeed that each theory can only again draw its

[5]Fuchs quoted the following sentences of W. James : "My thesis is that if we start with the supposition that there is only one primal stuff or material in the world, a stuff of which everything is composed, and if we call that stuff 'pure experience,' then knowing can easily be explained as a particular sort of relation towards one another into which portions of pure experience may enter. The relation itself is a part of pure experience; one of its 'terms' becomes the subject or bearer of the knowledge, the knower, the other becomes the object known" (Fuchs 2018, p. 32).

truth itself from orginiary (sic) data." (Husserl, 1983, p. 44). Similarly for Husserl, our immediate and situated experience is what is taken for granted.

This parallel leads us to take a look at the Husserlian phenomenology, and more specifically to Husserl's theory of perception.

12.2 Quantum Observation or Sensory Perception ?

12.2.1 Husserlian Theory of Perception

According to Husserl, sensory perception is composed of a set of *profiles* (or *adumbrations*, *"Abschattungen"* in German). Roughly speaking, a profile is what we call a "facet" of a thing. However, it is important to bear in mind that, genetically speaking, the facets come first: things, as we experience them under normal circumstances, are "constituted" on the basis of a series of facets. Indeed, it is from a set of profiles that the subject will be able to extract an invariant that remains identical throughout a series of changing facets. A set of these profiles is connected by the subject into a single unit that is deemed to have *identity through time* beyond these intrinsically different perceptions. "It is unity in multiplicity, i.e., identity in continuity; it is what presents itself in the continuity of appearance as a continuity of adumbrations of the object, and only in this presentation does it come to givenness and demonstrate itself for what it is." (Husserl, 1997, p. 85). It is this identity on which the constitution of things is ultimately founded. The constant identity, which arises from the connection between past profiles and expectations about the future, establishes a meaning: the meaning of a certain object, or a "thing", thus "constituted".

However, aiming at a constant identity crucially implies a horizon of expectations associated with the profiles: the (confirmed or disappointed) anticipation of future profiles. Indeed, when we observe an object, we accumulate several sensations (visual, auditory, olfactory, etc.) and several profiles that we associate with this object. At a given time and in the light of this identity constituted by the plurality of experiences and his past experiences, the subject tends to anticipate in a certain way the potential future experiences of the object.

Let us consider a concrete example. When I observe a cup on a table, I associate with this cup all the perceptions and sensations it evokes over time. Each visual perception is different from one moment to the next, but I keep looking at the same cup (or, rather, I associate the variety of percepts to what I consider to be a single, permanent, self-identical, cup). I project a future for this object thus defined. The cup is placed on a table and I look at it. Even though I only see a facet of it, I have the distinct feeling that I have the ability to *anticipate* my future perceptions of the cup. I can imagine what I could see if I looked at the back of the cup, or underneath. If, all of a sudden, my cup is about to fall, I anticipate the fall. The cup might fall down and break, but there is a chance it hits the floor without breaking. An infinity of anticipations belong to my experience of an object. This "open manifold" of anticipations constitutes what Husserl calls the *intentional horizon*. This horizon represents all future potential worlds that are compatible with the subject's present experience.

As the perception flows on, the subject is directed in her anticipations. The movement in the flux of experience gives the possibility to the subject to constitute the intentional horizon. "If, in a perception, the series of appearances runs its course in continuous unity, then the first determination of the change, the so-called differential of movement, already defines the "direction" of the course, and thereby is given a system of intentions that are continuously setting out and continuously getting fulfilled." (Husserl, 1997, p. 86). This horizon being constituted, the subject tends now to *fulfill* these anticipations.

This horizon of an object is constantly modified by present observation. In the continuum of experiences, through ever-developing sensory observation, the subject updates such horizon of possibilities. By filling some of his anticipations with a present sensory perception, the subject reduces his horizon of possibilities so that only one of these possible worlds is actually perceived. Now my cup is actually falling down, and it breaks. Of all the past anticipations, only one is actually perceived. The anticipations are being fulfilled by a present perception.

There is more that can be said about the structure of the Husserlian intentional horizon. First, there is an *internal horizon* that is part of the intended object. This horizon corresponds to the possibilities of future perceptions that I attribute to the object itself. However, this horizon cannot be separated from, second, an *external horizon*. The external horizon corresponds to the possibilities of future experiences that I attribute to the spatio-temporal context of the intended object. According to Husserl, the ability to anticipate future perceptions is an integral part of the constitution of an object. I cannot intend an object without co-intending a horizon that I attribute to it: "Thus every experience of a particular thing has its *internal horizon*, and by "horizon" is meant here the *induction* which belongs essentially to every experience and is inseparable from it, being in the experience itself" (Husserl, 1973, p. 34).

This perceptual horizon is then confronted with sensory observations that continuously modify it, constituting the "perceptual process". To recapitulate, perception is a matter of updating the horizon of possibilities through present observation. With these remarks as a backdrop, we are now in a position to make the similarities between Husserl's theory of perception and the QBist understanding of knowledge explicit.

12.2.2 Is the Husserlian Perceptual Process the Equivalent of the QBist Experimental Process?

According to Fuchs and Schack, the quantum state is presented, through the Born rule, as a rational tool for expressing the beliefs of an experimenter about future measurement outcomes of an experiment she performs. In this context, the bundle of expectations expressed by a quantum state can be understood as the quantum equivalent of what Husserl calls the intentional horizon. To be more precise, we can say that the state vector is the analogue of the *internal horizon* of an object.

Following a QBist interpretation of quantum theory, the anticipated possible profiles correspond to the eigenstates that compose my quantum state before the measurement.

When my cup is about to fall, I can either anticipate that it will break or that it will remain undamaged, but never both. Similarly, when I am about to measure the spin of a particle in my laboratory, I can either anticipate my system in the eigenstate *Up* or in the eigenstate *Down*, but never both. My Husserlian anticipations are based on my past experiences with falling cups and on a tacit recognition of the environment. It is from this information, and from my own impressions, that I determine a horizon of possibilities for my cup.

As a QBist I would say that the possibilities for the physical system, expressed by the quantum state, are determined by my beliefs. In establishing the state vector, I express in a formal way my beliefs about the future of my measurements; and these beliefs arise by due consideration of my own past experience (including the experience of preparation). The (probabilistic) estimates of subsequent measurements are thus analogous to estimates of future perceptions, namely the internal perceptual horizon of an object.

From this perceptual horizon, only one of the possible scenarios is eventually perceived. My cup finally broke. The possibility that it could also have survived the fall remains virtual. It then appears that the role of the present perception is of paramount importance and interacts with my previous perceptual horizon. If the perceptual horizon is an integral part of my experience of the object, then the present perception has a direct effect on the constitution of the object itself, through the determination it imposes on the perceptual horizon.

In the same way, only one eigenstate, *Up* for example, is perceived, this outcome being understood as my personal experience in a QBist framework. Observing a trace on a screen or hearing a click from an experimental apparatus is an experience that is analogous to the sensory nucleus of perception, as it is understood within Husserl's phenomenology. Prioritizing experienced perception in this way is in perfect agreement with the main idea behind QBism.

Proponents of QBism have the following slogan: "experience first". This can straightforwardly be compared with Husserl's epistemology, which puts lived experience (and especially perception) at the centre of the knowledge process, instead of centering knowledge around its objects. Husserl's conception of sensory perception is then echoed by the QBist understanding of measurements made by physicists in their laboratory. The parallel we are alluding to lies in the similarities between sensory perception on the one hand and quantum observation (as seen by QBists) on the other.

When the physicist finally observes some result with her measuring device, she confronts the outcome with the state vector that she has previously established. The quantum state that consisted of *expectations* of one or the other eigenstates, is now determined as *one* of the corresponding eigenstates. The physicist then modifies on paper the "state of her system", which was a composition of several incompatible states, into a single state, namely the one that has just been measured.

There is no longer any mystery in the measurement act for a QBist physicist. As we mentioned previously, there is no "disturbance" of the system, no sudden collapse of its state, but only an *update* of the information allowing to bet in an optimal and coherent way about the future. Relying on a distinction made by Cora Diamond (1995), one can say that QBism still retains a peculiar form of *realism*: Of course, QBism is not "realist" in a quasi-platonic sense, because it does not give the formal symbols of state vectors any significance as descriptions of the reality of physical systems. But it is "realistic" in the most concrete sense of the term, because it is limited to the immediate empirical reality of laboratory observations and the anticipatory activity of physicists.

To recapitulate, the QBist will simply modify her state vector according to the results just obtained, in order to optimize his subsequent bets. The same method is found in Husserlian phenomenology. Several futures were open to me at the time of the fall of my cup. A horizon of still undetermined possibilities was present before the cup hit the ground. Among the endless possibilities prefigured by the horizon, there were two scenarios: the scenario in which the cups breaks, and the scenario in which the cup survives the fall. After the cup burst into pieces, the anticipation of the cup remaining undamaged is frustrated, and other anticipations come to the fore. A new horizon is opening up, just as undetermined as the previous one, but redefined. Each new observation allows me to correct, and hence to re-direct my future expectations. In this sense, perception is a guide to evolve in the ever-developing world of experience.

The cup is the analogue of the microsystem, the perceptual horizon parallels the QBist quantum state, the perceptual act corresponds to the physicist's measurement and the modification of my possible horizon corresponds to the modification of the state vector after the measurement.

This leads to the following proposal: QBism can be understood as a phenomenological reading of QM. By prioritizing the experimenter, and by giving experimentation a prominent place, this interpretation attracts criticism from those who seek a "realist" reading (in the ordinary sense) of QM. QBism is not part of a realist/anti-realist debate since in this view QM is no longer related to any external reality but rather to the agent's personal experience. What used to be called "external reality" or "subject-independent world" is replaced by the evidence of the Jamesian (or Husserlian) *pure experience*. *Objective knowledge* is also redefined as the degree to which a discourse on an experience can be shared (Fuchs, 2018, p. 20).

QBism is an attempt to re-found QM on the basis of the only ultimate certainty we have: the agent's personal experience. QBism here agrees with Husserl who aimed at re-founding science (and the natural attitude) on the evidence of pure experience. This particular case also suggests that an important historical step could be taken to reconnect science and philosophy: QBism represents an innovative attempt to reconcile the most basic philosophy of knowledge with fundamental physics. QBism succeeds in putting oneself *in* the philosophical attitude, which is represented, from our perspective, by the phenomenological attitude.

12.2.3 Can We Take the Comparison Any Further?

Husserl's phenomenology and QBism are in perfect agreement with the fact that processes of knowledge acquisition are not only characterised by their dynamic nature, but also require an active role on the part of the agents who strive for knowledge. Consider, for example, the following passage from *Experience and Judgment*: "Thus, a horizon of validity is continuously present, a world posited in the validity of being, an anticipation which, in the continuous movement of realization that specifies and confirms, goes beyond what is at any given time grasped in singularity and relative determinateness and accepted as such" (Husserl 1973, p. 34) It is essential to emphasize the use of the expression "continuous movement of realization". It is this continuous movement, guided by our expectations and fulfilled by our present perceptions, that is the central characteristic of lived experience. And it is the permanent excess of expectations with respect to any possible fulfillment that generates the feeling of the "transcendence" of things. Husserl insists: "One the one hand, this transcendence is relative to the continuously anticipated potentiality of possible new individual realities and of groups of such realities which are to be experienced in the realization of the process of their entering into consciousness *from* the world; on the other hand, this transcendence is the internal horizon, the complex of characteristics not yet perceived, associated with every real thing offering itself to experience" (Husserl, 1973, p. 34).

A similar dynamic interpretation can also be found in QBism. When Fuchs and Schack claim that the state vector is a tool for "orienting oneself in life", they state that each observation is analysed in relation to past experience in order to make the most appropriate decisions in the future. And this chain never breaks. A present observation instantly becomes a past knowledge that then is used for the next observation. The decision-making system is constantly guiding the agent.

In phenomenology as in QBism there is no value judgment on what the best decisions are, or on what the best path of future anticipations might be. The subsequent modification of expectations is only motivated by the former fulfilment. In phenomenology, we cannot dissociate this continuous process from its own time-development, nor can we overcome the ambiguity of a present moment overlapping the future moment. The same observation can be made on the side of QBism: the passage from a future moment to a present moment allows the present observation to influence our expectations about the future of experiments.

The final point we would like to insist upon is that the subjective interpretation of QM is fully assumed by QBists, and echoes Husserl's first-person starting point. Since, for a QBist physicist, the quantum state corresponds to the formal expression of her estimates for future measurement outcomes, this state is completely personal to her. "When I contemplate a measurement, I contemplate its results, outcomes, consequences *for me*, no one else—it is *my* experience" (Fuchs, 2018, p. 20). In other words, the place of the agent is not interchangeable with another agent. It makes no sense to have a third-person discourse about an agent's personal experience. As a QBist agent, from my point of view, I only have access to *my* personal experience, and potentially, always being part of my experience, to an agent's speech, this time telling me about *her own* experience.

As Fuchs put it:

> There's no transformation that takes the one personal experience to the other personal experience. William James was just wrong when he tried to argue that two minds can know one thing. (Fuchs, 2015).

In this sense QBism differs from the Copenhagen interpretations. The Copenhagen school had already pointed out the importance of the observer in the practice of QM, but the observer was considered as some kind of impersonal agent practicing quantum experiments, using classical concepts to describe them, and manipulating the quantum "symbolism". The need for objectivity as defined by classical physics was transferred to QM, at the level of the classical concepts it uses to formulate well-defined results of experiments. Then, whereas, for the defenders of the Copenhagen school, the measurement had a crucial part to play, the agent who performs the measurement was entirely irrelevant. According to Mermin, "[. . .] every version of Copenhagen takes a view of the world that makes no reference to the particular user who is trying to make sense of that world" (Mermin, 2017, p. 8). From this point of view, the Copenhagen interpretations are third-person discourses on a macroscopic representative of agents' experiences. From a QBist perspective, subjectivity then seems to be only partially integrated into the Copenhagen interpretations, which then elaborate discourses on the subjective but not a subjective scientific discourse.

12.2.4 Then, Why Should QBism Be Linked to Husserl Rather than William James?

To investigate QBism's philosophical context, we highlighted a structural similarity between Husserl's theory of perception and the QBist understanding of knowledge acquisition in QM. We also insisted on the dynamic structure of both QBist's participatory process of knowledge acquisition, and Husserl's perceptual process. And we then pointed out some differences with Bohr's interpretation to emphasize the consistently and thoroughly subjective character of QBism. However, we must not forget Chris Fuchs' repeated references to William James' philosophy.[6]

It is now known and accepted that James had an influence on Husserl despite the later separation between Husserl's phenomenology on the one hand and James' psychology on the other. Nevertheless, they are close enough to share some fundamental ideas. Husserl's student, Dorion Cairns, relates: "In 1894 Stumpf called Husserl's attention to James' *Psychology*, and Husserl felt on reading it that James was on the same track as he. The notion of horizon and many others he found there" (Cairns 1976, p. 36). Husserl and James being in agreement on fundamental notions such as *pure experience* or *horizon*, it seems reasonable to seek in QBism the signs of a phenomenological way of thinking.

Yet, it is important to bear in mind that Fuchs does not endorse all of James' thinking: "William James's 'pure experience' and Alfred North Whitehead's 'actual occasions' have something of this character, but QBism has more: It has the guidance of the quantum formalism for shedding light on the notion. When a 'pure experience' is polarized into an agent-object division quantum theory rears its head by telling us the normative relations the agent should strive to satisfy with his beliefs. These normative constraints cannot be independent of the character of the underlying primal stuff" (Fuchs, 2018, pp. 32–33).[7]

However, while the normative character is not present in James' thinking, it is in Husserl's. On this particular point, Jocelyn Benoist puts it quite nicely: *"Husserl's and James' thoughts seem to share the same descriptive purpose originally. They intend to return to the purity of experience as it is given. The ways in which they interpret this experience, however, are quite different: characteristically, phenomenology deals with the so-called experience as necessarily being 'intentional', to the effect that experience always complies with a formal structure: intentionality. On the contrary, James would evolve toward a more and more contentual conception of experience*" (Benoist, 2006, p. 441). Intentionality, in its particular form of perceptual intentionality that allows the perceptual process, is what structures our relationship to experience from a Husserlian point of view. In the same way, for a QBist, quantum theory gives the normative rules that are profoundly linked to the very nature of the so-called "underlying primal stuff".

12.3 Conclusion

It is now clear that phenomenology could be a philosophical assistance for a better understanding of QM. Exploring the structures of Husserlian notion of

---

[6]Almost all of Fuch's papers contain references to the philosophers who influenced him, and in particular the American Pragmatists. A good presentation of these influences is found in Fuchs 2012 and 2018.

[7]Similar can be found in (Fuchs, 2015).

experience as well as studying the differences between Husserl's and James' thinking are suggestions to promote a phenomenological reading of QBism. Of course, understanding QM according to the QBist point of view gives the impression that quantum theory loses part of its explanatory power since in this view QM is no more a science relating to microsystems but rather to the agent's ability to predict experimental outcomes. But, actually, if interpreted phenomenologically, QBism reveals the special relations that unite the physicist and her experience, as well as the nature of the knowledge of which theoretical physics is a special case. From a QBist point of view, QM ought not to be intepreted independently from concrete subjects, and it is this point that is also emphasized by phenomenology.

References

Benoist, J. (2006). *Phénoménologie ou pragmatisme? Archive de Philosophie, 69*(3), 415–441.

Bitbol, M. (1998). Some steps towards a transcendental deduction of quantum mechanics. *Philosophia Naturalis, 35*, 253–280.

Bitbol, M. (2014). Quantum mechanics as generalized theory of probability. *Collapse, 8*, 87–121.

Bitbol, M. (2020). *A phenomenological ontology for physics: Merleau-Ponty and QBism*. In this volume.

Cabello, A. (2016). *Interpretation of quantum theory: A map of madness*. ArXiv: 1509.04711v1 [quant-ph].

Cairns, D. (1976). *Conversations with Husserl and Fink*. The Hague, Netherlands: Nijhoff.

Caves, C., Fuchs, C., & Schack, R. (2002). Quantum probabilities as Bayesian probabilities. *Physical Review A, 65*(2), 022305.

Cohen-Tannoudji, C., Diu, B., & Laloë, F. (1977). *Quantum mechanics* (Vol. 1). New York: Wiley.

De Finetti, B. (1989). *Probabilism. Erkenntnis, 31*(2-3), 169–223.

Destouches-Février, P. (1951). *La structure des théories physiques*. Paris: Presses Universitaires de France.

Diamond, C. (1995). *The realistic spirit: Wittgenstein, philosophy, and the mind*. Cambridge: MIT Press

Drouet, I. (2016). *Le bayésianisme aujourd'hui*. Paris: Éditions matériologiques.

Einstein, A., Podolsky, B., & Rosen, N. (1935). Can quantum-mechanical description of physical reality be considered complete? *Physical Review, 47*(77), 777–780.

Février, P. (1956). *L'interprétation de la mécanique ondulatoire et des théories quantiques*. Paris: Gauthier-Villars.

Fuchs, C. (2002). *Quantum states: What the hell are they?*http://www.physics.umb.edu/Research/QBism/WHAT.pdf

Fuchs, C. (2010). *QBism, the perimeter of quantum Bayesianism*. ArXiv:1003.5209 [quant- ph].

Fuchs, C. (2012). *Interview with a quantum Bayesian*. ArXiv:1207.2141 [quant-ph].

Fuchs, C. (2015). A private view of quantum reality. *Quanta Magazine*. https://www.quantamagazine.org/quantum-bayesianism-explained-by-its-founder-20150604/#

Fuchs, C. (2016). *On participatory realism*. ArXiv:1601.04360v3 [quant-ph].

Fuchs, C. (2018). *Notwithstanding Bohr, the reasons for QBism*. Arxiv:1705.03483v2 [quant-ph].

Heisenberg, W. (1990). *Ordnung der Wirklichkeit*. München u.a: Piper.

Husserl, E. (1973). *Experience and judgment*. London: Routledge.

Husserl, E. (1983). *Ideas pertaining to a pure phenomenology and to a phenomenological philosophy*. The Hague, Netherlands: Nijhoff.

Husserl, E. (1997). *Thing and Space: Lectures of 1907*. Dordrecht, The Netherlands: Springer.

Leifer, M. S. (2014). *Is the Wavefunction real*? In 12th Biennial IQSA meeting quantum structures. Olomouc, Czech Republic.
Mermin, D. (2013). *QBism as CBism: Solving the Problem of "the Now"*. Arxiv:1312.7825v1 [quant-ph].
Mermin, N. D. (2017). *Why QBism is not the copenhagen interpretation and what john bell might have thought of It*. ArXiv:1409.2454v1 [quant-ph].
Peres, A. (1995). *Quantum theory, concepts and methods*. Dordrecht, The Netherlands: Kluwer.

Index

H. A. Wiltsche, P. Berghofer (eds.), *Phenomenological Approaches to Physics*, Synthese Library 429, https://doi.org/10.1007/978-3-030-46973-3

The manufacturer's authorised representative in the EU is Springer Nature Customer Service Centre GmbH, Europaplatz 3, 69115 Heidelberg, Germany. If you have any concerns regarding our products, please contact ProductSafety@springernature.com

Printed and bound by CPI Group (UK) Ltd, Croydon, CR0 4YY
05/10/2026
02228121-0014